Titles in the Series

C. S. Bertuglia, M. M. Fischer
and *G. Preto* (Eds.)
Technological Change,
Economic Development and Space
XVI, 354 pages. 1995. ISBN 3-540-59288-1
(out of print)

H. Coccossis and *P. Nijkamp* (Eds.)
Overcoming Isolation
VIII, 272 pages. 1995. ISBN 3-540-59423-X

L. Anselin and *R. J. G. M. Florax* (Eds.)
New Directions in Spatial Econometrics
XIX, 420 pages. 1995. ISBN 3-540-60020-5
(out of print)

H. Eskelinen and *F. Snickars* (Eds.)
Competitive European Peripheries
VIII, 271 pages. 1995. ISBN 3-540-60211-9

J. C. J. M. van den Bergh, P. Nijkamp
and *P. Rietveld* (Eds.)
Recent Advances in
Spatial Equilibrium Modelling
VIII, 392 pages. 1996. ISBN 3-540-60708-0

P. Nijkamp, G. Pepping and *D. Banister*
Telematics and Transport Behaviour
XII, 227 pages. 1996. ISBN 3-540-60919-9

D. F. Batten and *C. Karlsson* (Eds.)
Infrastructure and the Complexity
of Economic Development
VIII, 298 pages. 1996. ISBN 3-540-61333-1

T. Puu
Mathematical Location and
Land Use Theory
IX, 294 pages. 1997. ISBN 3-540-61819-8

Y. Leung
Intelligent Spatial Decision Support Systems
XV, 470 pages. 1997. ISBN 3-540-62518-6

C. S. Bertuglia, S. Lombardo
and *P. Nijkamp* (Eds.)
Innovative Behaviour in Space and Time
X, 437 pages. 1997. ISBN 3-540-62542-9

A. Nagurney and *S. Siokos*
Financial Networks
XVI, 492 pages. 1997. ISBN 3-540-63116-X

M. M. Fischer and *A. Getis* (Eds.)
Recent Developments in Spatial Analysis
X, 434 pages. 1997. ISBN 3-540-63180-1

R. H. M. Emmerink
Information and Pricing
in Road Transportation
XVI, 294 pages. 1998. ISBN 3-540-64088-6

F. Rietveld and *F. Bruinsma*
Is Transport Infrastructure Effective?
XIV, 384 pages. 1998. ISBN 3-540-64542-X

P. McCann
The Economics of Industrial Location
XII, 228 pages. 1998. ISBN 3-540-64586-1

L. Lundqvist, L.-G. Mattsson
and *T. J. Kim* (Eds.)
Network Infrastructure
and the Urban Environment
IX, 414 pages. 1998. ISBN 3-540-64585-3

R. Capello, P. Nijkamp and *G. Pepping*
Sustainable Cities and Energy Policies
XI, 282 pages. 1999. ISBN 3-540-64805-4

M. M. Fischer and *P. Nijkamp* (Eds.)
Spatial Dynamics of European Integration
XII, 367 pages. 1999. ISBN 3-540-65817-3

J. Stillwell, S. Geertman
and *S. Openshaw* (Eds.)
Geographical Information and Planning
X, 454 pages. 1999. ISBN 3-540-65902-1

G. J. D. Hewings, M. Sonis, M. Madden
and *Y. Kimura* (Eds.)
Understanding and Interpreting Economic
Structure
X, 365 pages. 1999. ISBN 3-540-66045-3

A. Reggiani (Ed.)
Spatial Economic Science
XII, 457 pages. 2000. ISBN 3-540-67493-4

P. W. J. Batey, P. Friedrich (Eds.)
Regional Competition
VIII, 290 pages. 2000. ISBN 3-540-67548-5

Donald G. Janelle · David C. Hodge (Eds.)

Information, Place, and Cyberspace
Issues in Accessibility

With 77 Figures
and 27 Tables

Springer

Prof. Dr. DONALD G. JANELLE
University of California, Santa Barbara
Center for Spatially Integrated Social Science
Santa Barbara, CA 93106-4060
USA

Prof. Dr. DAVID C. HODGE
University of Washington
College of Arts and Sciences
Seattle, WA 98195-3765
USA

ISBN 3-540-67492-6 Springer-Verlag Berlin Heidelberg New York

Library of Congress Cataloging-in-Publication Data applied for
Die Deutsche Bibliothek – CIP-Einheitsaufnahme
Information, place, and cyberspace: issues in accessibility; with 27 tables/ed.: Donald G. Janelle; David C. Hodge. – Berlin; Heidelberg; New York; Barcelona; Hong Kong; London; Milan; Paris; Singapore; Tokyo: Springer, 2000
(Advances in spatial science)
ISBN 3-540-67492-6

Springer-Verlag Berlin Heidelberg New York
a member of BertelsmannSpringer Science+Business Media GmbH

Cover design: Erich Kirchner, Heidelberg
Typesetting: Camera ready by editors
SPIN 10733972 42/2202/xz Printed on acid-free paper 5 4 3 2 1 0

Foreword

The use of the term *the information age* to describe the period that we now find ourselves living in is open to misinterpretation. Society has always been based on exchanging information, and our libraries have long been rich sources of vast quantities of readily available information; it is information technologies that have changed rapidly since the invention of the digital computer. These technologies are themselves products of long-term societal processes: The economic desire to shorten the time that lapses between production and consumption of commodities, annihilating space with time; the political desire to control such large-scale systems as commodity chains, nations, and the military; and the human desire to liberate ourselves from the constraints of our local daily lives. They also have had profound effects on societal processes. One of the most widely discussed effects, and a consistent theme of this volume, is that the information age is bringing about the end of geographical distance as a significant barrier of human interaction.

This claim underlies prognostications about the information age: That this will be the age of globalization; of the global village; of the liberation of human interaction from the tyranny of space; of the dissolution of cities and workplaces; of the plugged-in society; and of the surveillance society. If these prognostications were true, then the topic of accessibility would indeed be a disappearing research program and this book a marker of its disappearance. Yet, things are much more complicated than this; the demise of distance has been greatly exaggerated. While there is a germ of truth to these prognostications, as there must be for them to resonate as they do, they often disguise more than they reveal. Flows of information are possible almost immediately over distances of arbitrary length, but this does not mean that everyone is equally accessible to everyone else. Rather, the geometry of the information age approximates the hypothesized wormholes of quantum physics – instantaneous connections between those who are plugged in to the right equipment, while neighbors remain off-line and inaccessible. Geographic and non-geographic information are available in unprecedented quantities, but they accumulate in the hands of certain social actors whereas others are excluded – creating black holes where information seems to disappear from social view. Even the ultimate distance-less society, cyberspace, becomes un-navigable without using spatial metaphors to make sense of it, and is connected in complex but predictable ways to the differentiated material spaces of society.

This collection of essays takes up the challenge of rethinking what accessibility means and how to measure it in the information age, with particular attention to geographic information. It addresses not only accessibility between those who are plugged in, and the geography of cyberspace, but also differences in accessibility to information technologies and the relationship between cyber-accessibility and accessibility on the ground. In doing so, the authors revive what has been an important but theoretically moribund concept; breathing new life into the concept of accessibility, and challenging preconceptions about its demise. They also move beyond attempts to equate accessibility with an exogenous Newtonian metric of Euclidean distance to unpack how accessibility is a construct of social practices.

The conversations that lie behind this book were catalyzed and made possible by a conference organized under the auspices of the National Center for Geographic Information and Analysis. This conference, *Measuring and Representing Accessibility in the Information Age*, was held in November 1998, at the Asilomar Conference Center in Pacific Grove, California. It was one of a series of nine meetings organized by NCGIA between 1997 and 1999 to advance the research agenda of geographic information science, under the Varenius Project (funded by the National Science Foundation, NSF Grant SBR-9600465). These nine meetings were equally divided among three areas of focus: Geographies of the Information Society; Cognitive Models of Geographic Space; and Computational Implementations of Geographic Concepts (http://www.ncgia.ucsb.edu:80/varenius). The accessibility meeting was held within this first area. The Geography of the Information Society Panel, chaired by Eric Sheppard and including Helen Couclelis, University of California at Santa Barbara, Stephen Graham, Newcastle University, UK, J.W. Harrington, Jr., University of Washington, and Harlan Onsrud, University of Maine at Orono, also organized meetings on *Place and Identity in an Age of Technologically Regulated Movement*, and *Empowerment, Marginalization and Public Participation GIS*. The Panel conceived the topic of measuring and representing accessibility, but the success of this meeting was due to the efforts of David Hodge and Donald Janelle in bringing the idea to fruition. Under their exceptional organizational skills, together with those of NCGIA staff LaNell Lucius and Abby Caschetta, a stimulating three-day meeting occurred at which preliminary versions of the chapters that follow were presented. This book is exemplary of how the Varenius Project is catalyzing and making available new research within areas central to geographic information science.

Michael F. Goodchild[1] and Eric Sheppard[2]

[1] Director of Project Varenius and Chair of the Executive Committee of the National Center for Geographic Information and Analysis. Department of Geography, University of California, Santa Barbara CA 93106-4060, USA. Email: good@ncgia.ucsb.edu

[2] Chair of the Varenius Panel on Geographies of the Information Society. Department of Geography, University of Minnesota, Minneapolis MN 55455, USA.
Email: shepp001@maroon.tc.umn.edu

Preface

The objectives of this book are to broaden understanding of conceptual and analytical approaches for accessibility research appropriate to the information age, and to demonstrate possible contributions for geographic information science in representing the geographies of the information society. In seeking to meet these objectives, the editors and authors highlight significant linkages among information resources, traditional places, and cyberspace, and focus on expanding models of space (and time) that encompass both the physical and virtual worlds.

The origins of this book stem from two multi-disciplinary conferences sponsored by the National Center for Geographic Information and Analysis (NCGIA). The first was the September 1996 conference in Baltimore on *Spatial Technologies, Geographic Information and the City.* The second, from which the chapters of this book originate, was the November 1998 conference at the Asilomar Conference Center in Pacific Grove, California on *Measuring and Representing Accessibility in the Information Age.* This book is structured around the primary themes of that meeting. Part I explores the conceptualization and measurement of accessibility; Part II focuses on the visualization and representation of information space within Geographic Information Systems (GIS) and other computerized display systems, and Part III considers the social issues that should inform the measurement and representation of accessibility. Each of these parts is preceded with an integrative essay that links the individual chapters to the broader literature on accessibility – primarily from geography, regional science, and planning. In Chapter 1, the Editors offer an explanation for the book's title, casting a wide perspective that focuses on the resource role of information, the importance of accessibility in the everyday life of places, and the co-adaptation of societal structures and cyberspace.

Special recognition is given to Helen Couclelis, who inspired the proposal for a Varenius initiative on accessibility in the information age. She organized the conference in Baltimore and was instrumental in placing accessibility on the agenda of the Varenius project. In Part IV, the Conclusion, she broadens the scope of this collection, raising issues regarding the sustainability of current societal accessibility practices in the interrelated realms of transportation and communication.

We thank those who made this book possible, beginning with Michael Goodchild (Director of the NCGIA's Varenius Project); he orchestrated the preconditions for sponsoring a broad range of research and conference initiatives. Members of the Varenius Panel on Geographies of the Information Society accepted a proposal to foster research on issues relating to accessibility, and we owe special thanks to the Chair of the Panel, Eric Sheppard, for support and advice at all stages of this project. The Steering Committee for organizing the conference at Asilomar – Michael Batty, Helen Couclelis, Arthur Getis, Harvey Miller, and

Mark Wilson – made substantial contributions in conceptualizing the principal issues for discussion and research; they are the authors of this book's principal integrating chapters – the introductions to its first three principal Parts.

LaNell Lucius and Abby Caschetta (NCGIA staff) provided essential logistical support for organizing the Asilomar conference, and Karen McFarland (Department of Geography, University of Washington) structured and maintained the conference Web site. Cartographic support from Patricia Chalk and Susan Muleme of The University of Western Ontario's Geography Department, and editorial advice from Marianne Bopp of Springer-Verlag, are also acknowledged. We are grateful for your interest and help.

Finally, we are indebted to the authors for sharing their research papers, for engaging freely with ideas and good humor through three days of spirited and productive discussion, and for revising their manuscripts for this book. We hope that this collection of original work will extend the discussion to a broader audience and encourage additional research.

Donald G. Janelle[1] **and David C. Hodge**[2]

[1] Center for Spatially Integrated Social Science, University of California, Santa Barbara CA 93106-4060, USA. Email: janelle@ncgia.ucsb.edu

[2] College of Arts and Sciences, University of Washington, Seattle WA 98195-3765, USA. Email: hodge@u.washington.edu

Contents

Introduction

1 Information, Place, Cyberspace, and Accessibility

Donald G. Janelle[1] and David C. Hodge[2]

[1] Center for Spatially Integrated Social Science, University of California, Santa Barbara CA 93106-4060, USA. Email: janelle@ncgia.ucsb.edu

[2] College of Arts and Sciences, University of Washington, Seattle WA 98195-3765, USA. Email: hodge@u.washington.edu

1.1 Introduction

This chapter provides a broad overview of alternate ways for seeing the operative linkages between human experiences on the ground (in place) and user experiences in cyberspace. Information is treated as the resource that binds these realms into functional human systems, while computer, telecommunication, and transportation technologies are viewed as tools of accessibility that are allocated differentially among people, institutions, and regions. Two general propositions guide the discussion. First, there are significant structural linkages among information resources, traditional places, and cyberspace; and second, grasping these linkages requires expanded models of space (and time) that encompass both the physical and virtual worlds.

The operational definitions of 'information', 'place', 'cyberspace', and 'accessibility' are seen as open-ended, subject to new interpretation in different social situations and technical domains. *Information* lies on a continuum from raw facts to knowledge and can be see as the outcome of creative manipulation of data and previous insight to assist our arrival at new thresholds of understanding. In the context of this book, information is one of the critical resources of the new economy. Information figures centrally in the processes for producing and allocating goods and services.

Place is seen as an extended locale of human activity imbued with the heritage, identity, and commitment of people and institutions. It is often linked with notions of family, neighborhood, and community (e.g., my hometown). However, the distinctive qualities of some places extend their recognition far beyond the local domain of immediate and long-term experience. Thus, Manhattan is a place in the minds of those who have never been there and for those who share its ambiance in a more transitory state (e.g., visitors). As with information, the meaning of place is subject to transformation through social and technological innovation, and through various levels and means of association and experience.

Cyberspace is an innovation that impacts significantly on our notions of place. Cyberspace may be seen as an electronic linkage of computers and their users that facilitates interaction through shared hardware, software, and protocols for communication. These interactions use facilities for e-mail, chat-rooms, bulletin boards, data and information exchange, and e-commerce. Cyberspace is also home to virtual worlds (e.g., virtual cities and virtual landscapes) that parallel the behavioral settings and rules of places and social networks in physical space, and some that don't. Dodge explores some of these worlds in Chapter 11. The interfacing of the physical and virtual worlds is seen in the possibility to link creatively a vast quantity and diversity of Web sites, and in the use of the Internet to deliver services (cyber-medicine, distance education, electronic commerce, entertainment, and so forth) and to search out information. However, since tools for the exploitation of cyberspace are not available universally, serious questions arise about accessibility to this important new medium of exchange. As President Clinton remarked in his State of the Union Message (27 Jan 2000), 'access today depends on having a computer. We must close the digital divide between those who have the tools and those who don't.'

Accessibility shares some of the ambiguity associated with 'information', 'place' and 'cyberspace'. Ideal or acceptable levels of accessibility are moving targets, changing continually in response to new technical possibilities and to socially defined standards. In the most general sense, accessibility relates to the ease by which people and institutions gain use of what they need to survive and to share in the opportunities of a society. As such, accessibility is a function of several factors. These include physical proximity to opportunities, the technical capability to overcome distance (e.g., automobile ownership), and the ability to surmount barriers to entry (e.g., income to pay for theatre tickets, laws to prohibit discrimination, or knowledge to use effectively a computer).

This book addresses the multi-dimensional character of accessibility. The chapters that follow invoke an extensive social science literature on accessibility, derived largely from the disciplines of economic geography, regional science, and urban and regional planning. It is the objective of the editors and authors to identify changes in accessibility brought about through spatial technologies. *Spatial technologies* are defined 'as the complex of transportation, communication, and information technologies that together modify spatial relations'.[1] Priorities include the need to differentiate accessibility levels among a broad range of social groupings, and the need to study disparities in electronic accessibility. In doing so, the authors investigate new measures and means of representing accessibility to capture the effects of information technologies, along with those of more traditional means of spatial interaction.

Concepts of potential and realized interaction, and of accessibility, are central to geographic theory and models. Current models are based, however, on physical

---

[1] Helen Couclelis, ed. 1996. *Spatial Technologies, Geographic Information, and the City, Technical Report 96-10* (December). (Santa Barbara: National Center for Geographic Information and Analysis), p. 6.

notions of distance and connectivity that are insufficient for understanding new forms of structures and behaviors characterizing an information age. Accessibility and spatial interaction in the traditional physical sense remain important, but information technologies are dramatically modifying and expanding the scope of these core geographical concepts. Through technological, structural, and social developments, an increasing range of transactions takes place in virtual space, or in some new hybrid space combining the physical with the virtual. Of importance also is the influence of new forms of communication on the use of and investment in traditional transportation infrastructure. Moreover, just as space can be fragmented so too can time, as activity rhythms in one place become increasingly synchronized with those in distant places. Geographic information science and technology, themselves products of this new information age, potentially have a major role to play in helping to reconceptualize, measure, represent, monitor, and plan for the new emergent geographies of accessibility.

1.2 Information and Accessibility

Goodchild and Sheppard (see Foreword) recognize correctly that society has always depended on information and has gone to great lengths in promoting significant information resources – from the notable centralized repositories at Alexandria to the Library of Congress. However, the nineteenth-century movement for public libraries in every city and town and the recent emergence of a globally interconnected Internet are models of information resources that transcend many of the barriers to building informed, knowledge-empowered societies. Hence, it is not information alone that distinguishes the early twenty-first century Information Age; rather, the Information Age is distinguished by the technologies for disseminating information and by the growing dominance of information industries over the economy at large. Batty and Miller (Chapter 8) observe that information is replacing energy as the basis for organizing economies and societies. The power and freedom that information offers motivate the quest for technologies that provide nearly instant connectivity among institutions and individuals.

The reality of technology's promise to annihilate the effects of distance on human commerce and discourse falls short of universal application. Societal processes that allocate the tools and knowledge to make this happen are embedded in legacies of differential access to infrastructure that are difficult to overcome. Market forces allocate information differentially over space and selectively among people and institutions, and the commodification of information poses a price barrier to access. Constrained by limits on financial resources and knowledge, many people and institutions are bounded by systems of social and political discrimination.

Information technology provides the potential to make all places equally accessible to opportunity. However, society's allocation of the tools for overcoming the intrinsic barriers of distance may feed growth in the relative inequality among places, thus increasing rather than diminishing the underlying complexity to the human geography of accessibility. Assessment of such a proposition makes the measurement and representation of accessibility important scientifically and significant socially.

It is important to assess critically the assumption that equal and unlimited accessibility is necessarily in the long-term best interest of society. For instance, for access to information, quantity may be less important than quality and timeliness. Sui (Chapter 7) sees the inundation of information posing threatening prospects for information-overloaded and dysfunctional societies. Separating out the creative use of information in ways that enhance individual accomplishment and societal benefit may not be easy. Thus, ways to filter information may be as, or more, important to some than accessibility. In the realm of cyberspace, as Dodge (Chapter 11) observes, the ratio of noise to information may be very high.

Other attributes of information are also important and need to be factored into our discussion. Hanson (Chapter 16) speaks of 'collective information assets' and maintains that diversity of information sources enhance the choices and range of opportunities within communities. The social context of information access also derives from constitutional rights and legislative initiatives that can shift the control of information among public and private sector agents (see Onsrud, Chapter 18). These many facets of information resources suggest that there is no single, easy way to measure and represent accessibility.

1.3 Place and Accessibility

The roles of places as repositories and conveyors of social capital are central to any consideration of accessibility. In Chapter 16, Hanson sees place as a fundamental construct that needs to be strengthened, possibly through selective use of communication technologies that enhance community building and human networking capabilities. She sees jobs as important stabilizing linkages to place for most people. Scott (Chapter 3) and Shen (Chapter 4) make explicit attempts to differentiate job accessibility at metropolitan levels. While transportation is the operative tool for achieving job access, they recognize that telecommunications and Internet access represent a growing factor in the ability to match employees with job opportunities.

Adam's (Chapter 13) offers a highly explicit representation of human travel and communication behavior. His use of activity diaries and CAD-based visualizations illustrate how individual daily activities are embedded in both virtual and place-based networks. Extensibility linkages (the ability of people to engage with distant locations) yield virtual presence or participation beyond the local realm, but the bounded nature of everyday life at the local level remains a paramount factor in the lives of all subjects in his investigation. Kwan reinforces this observation in

Chapter 14, illustrating through innovating mapping approaches the ability of people to operate simultaneously at different spatial scales, from local to global levels.

Arguably, geographical location, per se, is of less importance today than in previous decades for determining spatial interaction patterns. For example, in Chapter 2, Couclelis and Getis note the potential for a growing dissociation between places and functions, as activities become more person-based rather than place based. Thus, ordering a book over the Web might lessen the likelihood of place-based interaction. Another aspect of potential dissociation with place relates to variations in the ability of people to engage in advanced forms of telecommunications and data transfer. Will societal cleavages based on race, ethnicity, or income intensify as those with advanced telecommunication capability distance themselves physically through household relocation from social groups and activities that they wish to avoid?

It remains to be seen if networks in cyberspace alter our identities with grounded locations in physical space. Will the emergence of cyber-networks supersede our heritage of place-based memories and traditions? Certainly, more complex forms of organization are emerging in ways that extend the functional-physical continuum, bypassing customary spatial relations and embedding traditional places in broader networks of linkage beyond the physical reach of daily transport systems. At issue is whether or not Internet usage results in a bonding of people and institutions within the local region that is as strong (or stronger) as any bonding that may occur with distant opportunities through cyberspace.

1.4 Cyberspace and Accessibility

Although they clearly have the capability of complementing one another, place and virtual space share an uneasy alliance. In part, this is because of the recent, rapid, and potentially destabilizing consequences of cyberspace. However, this uneasy relationship may also reflect insufficient attention to analyzing the structure of cyberspace and its patterns of use. Uncovering the structure of cyberspace poses problems in the use of traditional concepts of morphology and distance. The physical hardware has the character of traditional infrastructure networks – paths, switching centers, relays, and capacity limits. But the process of use is invisible and does not conform to any strict Newtonian metric. The lack of a centralized system to monitor flows of activity over the network makes geographical interpretation difficult. Thus, Batty and Miller (Chapter 8) express concern for vagueness, fluidity, and low accountability for either the content or the quality of interactions on the Internet.

The scarcity of reliable quantitative indicators for analysis may be a major inhibitor to research on the nature of cyberspace. However, this book offers a few important examples of different approaches. Moss and Townsend (Chapter 10) compare metropolitan centers in the United States by the number of computers connected to the Internet, Internet domain name registrations, and capacities of backbone networks. By focusing on tangible evidence of infrastructure, their analysis reveals the intrinsic linkage of cyberspace with the national, urban economic landscape. Dodge's survey of literature in Chapter 11 explores some of the attempts to map cyberspace, including his own mapping of relative accessibility (virtual distance) based on the analysis of hyperlinks among academic Web sites in the United Kingdom. Murnion illustrates another innovative attempt in Chapter 12. He focuses on a distance measure based on the speed of Internet connections – delays in the response times (latency) among a set of globally dispersed computers.

In contrast, Forer and Huisman use a simulation approach to assess accessibility patterns associated with the virtual delivery of courses to university students in Auckland. Comparisons of accessibility patterns for students in different parts of the city are simulated for different times of the day for standard classroom courses offered at the university and for virtual courses accessed from home. They illustrate how patterns of behavioral response and changes in individual accessibility are easily mapped in conventional formats that could assist the evaluation of options for the delivery of education and other services.

There is significant potential for widespread service delivery and for public participation in making decisions through cyberspace. However, the prerequisite accessibility to make this possible may be constrained by inadequacies of current Internet navigation and communication tools. There is a need for tools and protocols for efficient use of the Web. Idiosyncrasies of web-page designers can thwart the establishment of clear standards. Designs for the users of advanced systems frequently preclude access by those with average equipment and skill. And, the on-line persistence of Web sites, or lack thereof, raises issues about the dependability of Internet information and contact sources. Thus, there is a need to go beyond basic connectivity via a computer – browsing, searching, and communicating are not easily represented through traditional social science surveys or geographical mapping techniques; yet, they constitute an important aspect of accessibility in the information age. Adams (Chapter 13) and Kwan (Chapter 14) illustrate how people operate simultaneously at multiple temporal and spatial scales, and how they may be interactive with several cyber locations at the same time. Their visualization techniques capture some of this complexity, but not all of it.

Not withstanding the possibilities that new cyber navigation tools will stretch even further the ability of people to operate at multiple scales, there is need to remind ourselves that the structure of cyberspace and its use are embedded in the realities of physical constraints. These include people bounded by biological needs for sleep and food, and by scheduling constraints of jobs, and schools. While society may be able to transfer many functions to the cyber realm, most activities still operate in a world of material flows constrained by transportation and land use patterns, which must be accounted for in any measurement schema.

1.5 Conceptualization, Measurement, and Representation

Separating out the lack of interaction independently from the lack of access is problematic in any effort to measure and represent accessibility. Inherited urban structure, for example, poses a significant constraint and should be considered in any attempt to assess the real impact of virtual technologies on future structures. With the right hardware and software tools at their disposal, individuals now have the capacity to selectively turn *on* and *off* their engagement with the world of information, but there is significantly less flexibility in their ability to escape the structural confines of their immediate physical environments. This book illustrates theoretical approaches for dealing with such issues, seen for example in Heikkila's development of a fuzzy-logic framework for understanding the complexity of accessibility (Chapter 6). In addition, Adams (Chapter 13) and Kwan (Chapter 14) illustrate the uses of 3-D CAD and GIS representations to depict how space-time diaries help capture the dynamics of behavior at the individual level. These authors do not claim final solutions to the problems of representation, but they do demonstrate that new space-time typologies, and new topologies, are needed to accommodate the interdependence of both the physical and virtual worlds of everyday life.

A hybrid blend of physical and virtual space may now constitute the new geography of the information age; and, as Batty and Miller note (Chapter 8), it may reinforce patterns of physical infrastructure that impinge on the comparative levels of physical accessibility for different regions. For example, on-line book purchases still require transshipment facilities to accommodate the physical movement from the producer to the consumer. Thus, it is interesting to consider how virtual spaces map onto traditional conceptions of geographic space, and vice versa, and to address the issue of how we handle such complexities analytically. For example, how can traditional spatial interaction and spatial gradients within hybrid spaces (and space-times) be visualized with GIS?

Are there information counterparts to the accessibility and potential surfaces developed by regional scientists for interaction in physical space? Scott (Chapter 3) considers the issue of scale and of potential versus realized accessibility in looking at job access in the Los Angeles region, and Shen (Chapter 4) has devised indexes and composite measures of access to consider both commuting and telecommuting options within the Boston metropolitan area.

Data availability is a constraint on advances in the area of accessibility research. The traditional national census fails to capture information on telecommunication and Internet activity, and correspondingly omits possibilities to account for the dynamics of cross-border commerce or for individual lifestyle changes in response to the potentials of information-age technologies. Survey designs illustrated here by Adams (Chapter 13), and Kwan, (Chapter 14), and simulation methodologies suggested by Forer and Huisman (Chapter 5), extend the scope of time-geography framework for depicting human space-time behavior. And, as illustrated by Harvey and Macnab (Chapter 9), real-time accessibility for face-to-face or interactive voice communication is constrained by the proportion of the world population that is awake across the different time zones at any given time of the day. Their innovative study poses yet additional need to extend the empirical base for understanding current issues of global interdependence and accessibility.

1.6 Societal Issues

Shen (Chapter 4) is explicit in seeing a new geography of opportunity arising from the transportation and telecommunication options available in metropolitan settings. This parallels closely the opening remark by Couclelis in the final chapter of this book – 'accessibility is the geographic definition of opportunity'. Models of how institutional and other contingencies influence who has access to whom, what, when, and where, via physical and especially via virtual contact, are required for assessment of policy approaches to reduce inequalities in opportunities for social and economic interaction. Analytical measures and computerized visualizations of accessibility are needed to reflect hardware and software availability, inadequacies of education and training, cultural factors, and differential relevance of the Internet to everyday life. Such measures and representations of accessibility will contribute insights and reference points for judging efforts to mitigate the perpetuation of *information poverty* for certain places and social groups.

Discussions by Occelli (Chapter 17), Onsrud (Chapter 18), and Couclelis (Chapter 20) reveal that it is just as important to understand how societies shape technologies as technologies shape societies. Information access is seen as a resource and as a contributor to differential levels of social power. The control of information technology and of information itself can be a means of social control that undermines cultural diversity and that limits the range of the common mindset. Questions regarding continued free access to the Internet, charges for use, and the potential for government taxation impact on levels of accessibility in ways that could alter the very nature of cyberspace and its relationships to social structures and processes.

1.7 Conclusion

Analytical and theoretical approaches to measuring and modeling accessibility are certainly at a threshold for change. The chapters that follow point to innovative possibilities to enhance understanding of how cyberspace and telecommunications might be embedded in more socially informed assessments of changing accessibility patterns and processes. However, Sui (Chapter 7) and the authors of chapters in

Part III present critiques of methodologies that fail to account for broad social policy and theoretical perspectives.

Sui raises the possibility that society has changed beyond the point of relevance for worrying too much about accessibility measurement. He sees greater validity in a more evolutionary pattern of change that is best characterized as ‘adaptation’ instead of accessibility. Hanson (Chapter 16) argues that there are 'silences' in our existing measures of accessibility, omission of which could foster narrowness of interpretations regarding the fundamental importance of place-based networks of people and institutions. Indeed, Wilson (Chapter 15) reminds us that the opportunity costs of accessibility should not be neglected. Mugerauer (Chapter 19) extends this argument, presenting a forceful case for empowerment in the use of measurement and descriptive tools. He raises concerns about the culturally homogenizing influences of common information pools and analytic methodologies. He questions whether or not current spatial analytic tools, with built-in assumptions about the nature of space, such as GIS, can be adapted to account for local interests and cultural perspectives.

In the concluding section, Part IV, Couclelis adopts a policy perspective that focuses on the potential importance of accessibility analysis to inform the transformation of spatial structures in ways that are more sensitive to environmental and societal cost constraints. She argues that automobile dependence might be lowered through new ways of uniting transportation and communication technologies. This view places the accessibility theme in a central position for the transportation and communication planning of cities and regions. She observes how transport uses of information technologies offer a new order of flexibility for automobile users and a new order of uncertainty for planners. Nonetheless, she does not foresee an inevitable outcome of deteriorating, uncontrolled environmental change. She reminds us that societies can shape these technologies and their uses to achieve sustainable environments. Similarly, Occelli (Chapter 17) sees the policy uses of informed measures and models as meriting a high priority in the research agenda of regional science and planning. Enhanced measurement tools and modeling concepts may open the way and the willingness to use information technologies to address issues of social equity, imbalances in regional economic development, and threats to the sustainability of local and global environments.

Part I

Conceptualization and Measurement

2 Conceptualizing and Measuring Accessibility within Physical and Virtual Spaces

Helen Couclelis[1] and Arthur Getis[2]

[1] Department of Geography, University of California, Santa Barbara CA 93106-4060, USA. Email: cook@geog.ucsb.edu

[2] Department of Geography, San Diego State University, San Diego CA 92182-4493, USA. Email: arthur.getis@sdsu.edu

2.1 Introduction

The study of accessibility in geography and related disciplines has a distinguished history dating back to Ravenstein's work over a century ago. In the late 1940s to the 1960s, scholars such as Zipf, Stewart, Warntz, and Wilson theorized about the way individuals and aggregates of individuals respond to the constraints of cost, time, and effort to access work, shopping, recreation, and other spatially distributed opportunities. Since that time accessibility has been closely related to but also distinguished from such key geographic concepts as mobility, nearness, and the friction of distance. The models developed for its study belong for the most part in a large class of constructs known as spatial interaction models because they represent the patterns and intensity of interactions among locations in geographic space. Different forms of spatial interaction models have been successfully used to study accessibility at the aggregate level, while the study of individual movements in space-time has provided insights into the significance of accessibility in people's daily lives. One of the most robust findings of modern quantitative geography has been that interactions decline sharply with increasing distance, which is another way of saying that there is less and less contact between or among people or places as these become less and less accessible from one another. These kinds of models have proved extremely useful not only for understanding how people are spatially related to their economic and social activities, but also for the help they gave planners in designing transportation systems and land use structures that satisfy general accessibility needs.

Recent technological and societal developments require us to rethink the concept of accessibility at all scales. In technologically advanced societies, there is mounting evidence that urban areas are being restructured and that many kinds of social relations are being re-shaped as the new communication and information technologies increasingly permeate society, culture, and the economy at all scales. Moreover, the meaning of accessibility itself appears to be changing, as individuals or groups increasingly are able to access far-away people, goods and services without recourse to physical movement. It would appear that mobility, nearness, and the friction of distance no longer are a necessary part of the definition of accessibility; or rather, that accessibility in physical space is now being complemented by accessibility in virtual space, which seems to defy basic principles of spatial interaction. Researchers and planners thus have to deal with something of a paradox: just as the data and technical tools required for more thorough explorations of urban and regional phenomena become widely available, there is serious concern that the trusted concepts and models of yesterday may be letting them down. Most authors in this section seem to agree that given appropriately updated and operational definitions of accessibility, geographic information science can play a leading role in helping us understand, explain, and perhaps even predict some of the actual and potential implications of the new technologies for the spatial organization of our cities and regions.

Three recurring major themes emerge from the chapters in this section. First, issues surrounding the diverse conceptualizations, definitions, and measures of accessibility; second, the distinction between individual and aggregate accessibility; and third, the changing spatial relations within the information society itself, and their reciprocal relationships with changing accessibility conditions. Such issues for the most part have been raised in traditional accessibility research. It is well known that problems of definition have limited the effectiveness of accessibility measures and inhibited their successful application in planning. Equally troublesome have been the issues of potential versus revealed accessibility, of structure versus agency, of the demand versus the supply view of accessibility, and several more. However, the societal and technological developments of the information age tend to add new dimensions to the old dilemmas.

2.2 Conceptualizations, Definitions, and Measures of Accessibility

Accessibility has always been an elusive concept. As Scott notes, accessibility is not a distinct physical entity easily counted or measured, ' . . . it is a concept, a perception, something each one of us will experience, evaluate, or judge differently.' (Scott, Chapter 3, p. 28).

Accessibility is indeed a relative and contextual notion, and the 'correct' definition largely depends on the scope and context of the investigation. It is unlikely that any particular definition will satisfy all research and policy needs. Thus a first task would be to survey the main types of situations within which the question of accessibility arises and clarify the meanings and role of the term within these contexts. Several new problem contexts arise in the information age: differential access by economic, educational and cultural background, as well as by geographic location and socioeconomic group, to the technologies allowing virtual access;

access within cyberspace; the possibility to substitute virtual for physical access; the comparative quality of the access experience in case of substitution; and so on.

Another problem is that accessibility has often been treated as a purely spatial issue. Yet individual scheduling of activities is not only a spatially constrained process but one that is also strongly time dependent. More than ever before, accessibility should be approached as a time-space phenomenon. While the idea is not new (it goes back to Hägerstrand and the 1950s), it has new implications at an age when virtually instant access to some opportunities frees up considerable time and thus enables access to others. The scheduling of activities may no longer be constrained by the spatial logic of multi-purpose trips, or the temporal constraints of business hours. New temporal constraints come into play as we access people and places in very different time zones across the globe - and so on. It is thus with relief that we see in the paper by Forer and Huisman that the familiar time-geography framework developed by Hägerstrand can be extended to take into account some of these novel situations. It is unclear whether it can be equally well adapted to all of them.

Among more recently formulated concepts, a particularly useful one appears to be that of *proximal space*. This defines a place as part of its (physical or functional) vicinity and thus allows us to see places in the context of the other places with which they interact (or to/from which they are accessible). In traditional geographic terminology, proximal space embodies both the site and situational characteristics of locations. For example, a site may provide good bus service to a set of other locations that together constitute its proximal space from the point of view of physical accessibility; or it may provide the means to connect to the Internet, allowing users to access specific other locations that together help form the proximal space for these users at that particular location. How can we compare these two instances of proximity or accessibility? In her paper (Chapter 3), Scott makes creative use of the proximal space notion to compare accessibility in physical versus functional space in the Los Angeles area. Proximal space (characterized by various degrees of proximity) is also implicit in Heikkila's geographic *fuzzy clubs* (Chapter 6), where greater accessibility (however defined) to a place is represented as a higher degree of membership in a corresponding 'club'.

Scott (Chapter 3) and Shen (Chapter 4) make some of these abstract ideas more concrete by focusing on the role of new technologies in affecting accessibility for employment. With still small but increasing numbers of people turning to telecommuting, models must be able to indicate what employment opportunities are reachable (physically, virtually, or both ways) by a person at a given location. A composite measure of accessibility devised by Shen takes into account the varying mix of physical and virtual accessibility to jobs for Boston residents. The model presented by Scott combines a traditional spatial interaction model with the G_i^* local statistic devised by Getis and Ord (1992) to shed new light on the notorious spatial mismatch phenomenon (a classic case of deficient accessibility) in the Los Angeles area. Unlike its predecessors, Scott's model can be extended to the study of accessibility to employment in virtual and hybrid (physical and virtual combined) spaces.

As they should, the chapters in this section raise a number of questions, some very theoretical, others very practical and technical. To what extent do we need new concepts and measures of accessibility, rather than adaptations of existing concepts and measures? What theoretical criteria might guide prediction or explanation in accessibility research, and what technical criteria might help us choose among different measurement approaches? Given the obvious relevance of geographic information science and technology, how can GIS functions and operations be used to represent and measure the expanded notions of accessibility discussed here? And how can other promising formalisms (e.g., local statistics, fuzzy set theory) enhance the usefulness of GIS for accessibility research?

2.3 The Distinction Between Individual and Aggregate Accessibility

A recurrent theme in the papers is the problem of scale. This is expressed as either the micro-macro problem in a spatial context or the individual-based versus aggregate measure problem more generally. Disaggregate and aggregate-level measures serve different functions; each allows for different types of questions. The behavioral, decision-making strengths of the micro approach are often undermined by the unavailability of individual data. The aggregate approach, while meeting many analytical requirements (e.g., more manageable sample sizes) is by its nature limited to the study of problems dealing with group behavior and averages. The age-old question of the appropriate spatial resolution continues to be debated. The debate is complicated by the apparent space-lessness (and scale-lessness) of many of the newer telecommunications and information technologies. Multi-scale or scale-free approaches to analysis do not necessarily do justice to the complex effects of the new technologies on accessibility and human interaction.

The papers presented in this section provide some promising clues. Several of them successfully accomplish the transition from individual to aggregate and from local to metropolitan levels of analysis. The approach presented by Forer and Huisman (Chapter 4) models individual access to educational opportunities as operational trajectories within time-space prisms, and then generalizes these into aggregate patterns of accessibility for the student population. Heikkila's idea of fuzzy clubs also appears to bridge the gap between individual circumstances and the characteristics of clubs with respect to accessibility for their membership. Still, the micro/macro problem in complex systems, as urban and regional systems invariably are, reaches beyond the issues of aggregation and scale. A very important related aspect is that of emergent properties and behavior. As an example, it is well known that the individual increase in accessibility afforded by widespread automobile ownership in North American cities resulted in urban structures characterized by low overall accessibility scores. What the corresponding emergent phenomena for the information-age city may be, is anybody's guess. Further, by focusing too hard on how best to aggregate individual choices and paths in space-time, one may lose sight of the fact that these choices and paths themselves are to a large extent socially constructed (more simply put, enabled and constrained by aggregate behavior and its spatial consequences). Finally, we should keep in mind that a large part of urban structure may be explained not by the accessibility needs of individuals but by their distancing needs, as they strive to avoid the vicinity of less desirable groups, land uses or environments. While these centrifugal tendencies have always been present, the technological and organizational possibilities of the information age may greatly amplify these phenomena.

2.4 Changing Spatial Relations

Inevitably, we will need to move beyond physical proximity in our representations of accessibility in order to understand how old and new kinds of spatial relations come together in post-industrial landscapes. Some key questions suggest themselves: How are the traditional spatial relations studied in socioeconomic geography responding to new conditions of access to goods, services and information? What new kinds of relations are replacing, complementing, or otherwise affecting traditional spatial relations? What other spatial and non-spatial relations become especially important in the information age? The papers in this section hint at these kinds of questions, which are more thoroughly explored in Part III of this book.

At the root of many of the spatial changes taking place in the post-industrial city is the increasing dissociation between places and functions: activities are becoming more and more person-based rather than place-based, so that where you are is less and less a reliable indication of what you may be doing. The assumption of a strong structural correspondence between spatial and functional relations, on which much of traditional socio-economic geography is based, can no longer be taken for granted. At the Asilomar workshop several novel (and at times quite exotic) concepts were proposed by participants grappling with these new types of spatial relations: places as networks; the extensible individual; 'wormholes' in functional space, like tunnels in space-time allowing instant access to activities at physically distant places; 'real virtualities' as well as virtual realities, recreating at a place the feel or function of processes taking place elsewhere. Papers in this section directly or indirectly raise more down-to-earth questions regarding spatial relations. What interactions must remain spatial and thus continue affecting spatial structure the old-fashioned way? What are the social costs and benefits of vastly increased opportunities of interaction within the functional/physical continuum, and how does that structure of opportunities vary over space? How is the quality of interaction affected for different kinds of functions when the spatial relation is by-passed? Last but not least: What is the practical relevance of defining and measuring accessibility when spatial relations no longer necessarily correspond to the most important functional relations? Here Daniel Sui (Chapter 7) gives a particularly provocative answer. He argues that we may be barking up the wrong tree by focusing on a mechanistic notion such as accessibility, when the evolutionary notion of adaptability may be much better suited to the fluid, flexible, immaterial and strongly cognition-oriented types of relations fostered by the information age.

To the extent that further research on accessibility is worth pursuing, there is consensus that the focus should be on a generalized notion of accessibility as *process* taking place in physical, virtual, or hybrid (physical/virtual) spaces. As Occelli (see Chapter 17) notes, accessibility as a concept has always occupied the intersection of physical and functional (including virtual) spaces, the place where the socio-economic relations woven in functional space touch ground, and where the spatial relations still constrain the functional. Studying that intersection has always been a considerable challenge, because functional space potentially has indefinitely more dimensions than physical (geographic) space. The difficulty increases manifold in the information age with the appearance of virtual space, which shares properties of both the physical and the functional. Thus the challenge will be to find consistent conceptualizations that can handle the dynamic interconnectedness of physical, functional, and virtual relations over space, as well as the infinite capacity of individuals and societies to both adapt to and modify the ever-changing contexts of their interactions.

References

Getis, A. and Ord, J.K. 1992. The analysis of spatial autocorrelation by use of distance statistics. *Geographical Analysis* 24(3): 189-206.

3 Evaluating Intra-metropolitan Accessibility in the Information Age: Operational Issues, Objectives, and Implementation

Lauren M. Scott

Environmental Systems Research Institute, 380 New York Street, Redlands CA 92373, USA. Email: LScott@ESRI.com

3.1 Introduction

Suburbanization, economic restructuring, globalization, and rapid developments in transportation and telecommunications technologies have had dramatic impacts on the urban landscape, fundamentally altering the spatial and organizational composition of where we work and where we live. How have these broad spatial processes impacted intra-metropolitan accessibility? How are these impacts expressed physically on the urban landscape? Who, in terms of both geographic location and socioeconomic groups, is affected? How? What are the implications for urban development and planning policy?

The emergence of the multicentric metropolis – within the context of globalization, economic restructuring, suburbanization, and rapid technological developments – has prompted these and other questions relating to intra-metropolitan accessibility. Some researchers argue that since transportation and communications are already so highly developed in most U.S. cities, physical accessibility has become somewhat ubiquitous (Giuliano 1995, Chintz 1991). These technologies have developed unevenly over time and space, however, creating a complex patchwork of different spaces associated with disparate, sometimes contradictory, patterns of intra-metropolitan accessibility (Graham and Marvin 1996, 322). While on the one hand advances in telecommunications and transportation technologies allow individuals to extend their geographic reach, on the other they facilitate consolidation and dispersion of urban activities. For individuals lacking access to computers, to automobiles, or even to effective public transit, the ever-expanding spatial separation facilitated by technological developments may serve to diminish, rather than to enhance, intra-metropolitan accessibility.

Notions about intra-metropolitan accessibility provide the basis for a variety of urban policy and transportation planning decisions; they represent key components in urban economic theory relating to land use and urban development; they serve as a common focus for geographic research concerned with economic growth, transportation patterns and infrastructure, metropolitan form, urban efficiency, and social equity. Despite the centrality of these ideas to urban research agendas, accessibility remains a difficult construct to both operationalize and define (Pirie 1979, Helling 1996). The challenges become more pronounced in the *information age* where access to urban activities and spatial opportunities is no longer necessarily constrained by physical space, but increasingly takes place via electronic telecommunications networks.

This chapter suggests an analytical framework for evaluating and monitoring changes in intra-metropolitan accessibility associated with broad urban restructuring processes. The chapter is structured into three sections. The first section reviews operational issues associated with measuring and modeling intra-metropolitan accessibility. Section two outlines the proposed analytical framework. The final section applies this analytical framework to employment data in the Greater Los Angeles region. Three different analyses are presented, each highlighting a specific component of intra-metropolitan accessibility: (1) urban spatial structure; (2) transportation infrastructure; and (3) the complex attributes (the resources, individual constraints, and preferences) associated with different groups of individuals. The objective of this final section is not to fully analyze the Los Angeles data set, but to demonstrate the effectiveness, flexibility, and potential of the proposed analytical framework for addressing a wide variety of research questions, for contributing to urban theory, and for evaluating urban planning strategies.

3.2 Operational Issues

This section of the chapter reviews the accessibility literature in order to identify definitional and representational issues related to operationalizing an effective analytical framework for evaluating intra-metropolitan accessibility.

Defining Accessibility

> Accessibility… is a slippery notion… one of those common terms that everyone uses until faced with the problem of defining and measuring it (Gould 1969, 64).

A large body of formal urban theory contends that accessibility is at the very core of processes shaping urban spatial structure: people choose residential locations that satisfy both housing needs and workplace access, and employers choose work locations that are accessible to employees, urban infrastructure, and consumer markets (Giuliano 1995, 3). Consequently, many researchers indicate that the concept of accessibility is fundamental to definitions and explanations of urban form, function, and efficiency, arguing that a location's access to economic and social opportunities largely determines its value, development intensity, and economic, social, and political uses (Wachs and Kumagai 1973, 438; Knox 1980, 368; Koenig 1980, 169).

It is not surprising, therefore, to find the term *accessibility* appearing frequently in local, regional, and national documents; accessibility is commonly cited as a fundamental objective for urban and transportation planning. Nonetheless, the concept of accessibility is seldom given an operational definition in these documents, and accessibility measures are rarely used to monitor urban system performance, to construct regional profiles, to compile social inventories, or to evaluate proposed planning strategies (Knox 1980, 367).

On the other hand, the concept of accessibility, and a variety of different measures, has been applied liberally in research studies. Within these studies accessibility takes on a large number of different definitions. The concept of accessibility, for example, has been used to describe the physical proximity between two or more locations[1], to represent the freedom of individuals to participate in urban activities[2], and to reflect the set of activity sites or opportunities available at particular locations in space[3] (Burns 1979). A review of the literature on accessibility measurement reveals three primary issues relating to definition:

(1) Potential vs. Outcome. At a very broad level, measures of accessibility may be grouped into one of two definitional categories (Breheny 1978): *potential* measures and *outcome* measures. Potential measures consider accessibility to be a property of specific locations or individuals, and may involve counting spatial opportunities and/or measuring distances between origins and destinations, but they do not incorporate actual travel behavior, or use observed travel flows to calibrate or to simulate measure components. Isochronic accessibility measures, for example, define accessibility in terms of the total number of spatial opportunities within a specified distance or time cost of a particular location *i*, regardless of whether or not individuals at *i* actually use these spatial opportunities. These potential measures define accessibility in terms of the potential for spatial interaction.

Outcome measures, on the other hand, define accessibility in terms of behavior, as expressed through observed travel patterns. These outcome approaches consider proof of accessibility to be a function of *realized accessibility* – the actual use of services or actual participation in activities surrounding specific origins (Morris, Dumble, and Wigan 1979, 92). Network models of accessibility relying on actual travel flows to identify highly accessible nodes, for example, reflect this outcome definition of accessibility.

Spatial interaction models – the most commonly used accessibility measures – incorporate elements of both the potential and the outcome definitional categories. While these accessibility measures generally define accessibility in terms of the

---

[1] See Ingram (1971), Haynes and Fotheringham (1984), and Jones (1981).

[2] For example, Hanson and Schwab (1987) examine the link between accessibility and individual travel behavior using data from Uppsala, Sweden.

[3] Knox (1980), for example, investigates accessibility to primary medical care in Edinburgh, Scotland, ultimately creating a map showing both under- and over-serviced regions.

potential for spatial interaction, they calibrate model parameters using actual travel behavior (the outcome-measure strategy).

Whenever accessibility models rely on actual travel behavior, either directly or for model calibration, it becomes troublesome to disentangle structure from agency. Suppose, for example, we find that journey-to-work distances have increased. Using an outcome definition of accessibility (i.e., using actual travel flows), it is difficult to determine whether the longer commutes are the result of improved accessibility (an improved transportation system, for example, may provide access to better jobs or to better homes at a farther distance away) or whether the longer commutes are the result of diminished accessibility (workers may be required to travel farther because suitable employment or housing is just not available nearby) (Knox 1980, 369).

(2) Mobility vs. Accessibility. Handy (1994) notes that accessibility has only recently become a focus in transportation planning. Traditionally, transportation planners have emphasized mobility and infrastructure performance over concerns about accessibility. Mobility refers to the ease of movement or the physical ability to transcend space (facilitated travel), and encompasses monitoring the infrastructure for travel (road capacities, speed limits, and congestion, for example). Accessibility, on the other hand, extends this concept of mobility, to include examination of the *context* for travel (Helling 1998). Travel is rarely undertaken for the sake of movement alone, but instead takes place within specific contexts (Hanson 1998), motivated by the desire or need to satisfy a variety of economic, social, or psychological objectives (Wachs and Kumagai 1973, 439). The concept of accessibility fully encompasses this notion of *context*, extending the scope of concern associated with mobility to include the spatial/temporal opportunities provided at destinations and the social, economic, political, and psychological capability to reach destinations (Handy 1994). As a planning goal, then, a focus on accessibility reflects a broader, more inclusive concept that has advantages over an exclusive focus on mobility.

(3) Definitional Components. Whenever we model the concept of accessibility, we are implicitly asking three questions: *Accessibility to what?*, *By whom?*, and *How?* Destination choices, access costs for different individuals, and mode of access, each represent important components of accessibility (Handy 1994, 5). A review of the accessibility measurement literature, however, suggests it is rather difficult to integrate this full range of accessibility elements; rarely are more than one or two of these components incorporated into accessibility scores. Consider the example of measuring accessibility to employment opportunities. With many accessibility models, accessibility scores will increase as job counts increase, regardless of variations in the number of workers competing for jobs.[4] Formulations attempting to integrate worker supply into the accessibility index often muddy

[4] Consider 3 isolated grid cells, each with 100 jobs. Even if the first cell has no workers, the second has 100 workers, and the last has 1000 workers, scores for all three cells will be equal, for many accessibility models.

interpretation so it becomes difficult to determine if a particular index is high because of opportunity magnitude, population magnitude, or effective transport. Too often, one unit of opportunity substitutes directly for one unit of travel cost – each component is given equal weighting. These problems can limit the effectiveness of accessibility measures for informing urban planning policy. Accessibility scores based solely on job counts, for example, will assign a high accessibility score to job-rich regions. Regions such as Downtown Los Angeles have extremely high job counts and will, therefore, receive high accessibility scores. Does this mean that all of our communities should be encouraged to replicate the Downtown patterns? It seems more appropriate to define accessibility in terms of *level-of-service* – in terms of how well a given location serves surrounding populations. In the case of accessibility to employment, for example, if the jobs provided at a given location effectively match worker demand, and the linkages connecting them have sufficient capacity, a high accessibility score is appropriate.

Modeling Accessibility

A number of researchers have suggested using GIS to model intra-metropolitan accessibility (Kwan 1998, Miller 1991, Arentze, Borgers, and Timmermans 1994, Geertman and Ritsema Van Eck, 1995). Couclelis cautions, however, that the GIS environment presents a very specific representation of space (1991). GIS are geared toward spatial objects and an absolute or *container* view of space; they emphasize *site* characteristics over *situation* characteristics[5]. Consequently, the end result of any combination of GIS operations will typically involve information relating to specific locations or to specific spatial objects (points, lines, areas, or volumes). GIS are limited in their ability to represent non-localized spatial processes inherent in spatial organization, configuration, pattern, spatial dynamics, restructuring, transformation, or change (Couclelis 1991, 15).

At a more general level, Sheppard (1996) discusses the concepts of site and situation in relation to *local context*. He notes that the concept of local context is critical for explaining why seemingly similar processes may lead to very different outcomes in different places. Unfortunately, when places are modeled as discrete containers, and the concept of local context is narrowly defined in terms of site characteristics only, the possibility for *action at a distance* – the possibility for objects or phenomena, such as information flows, to effect local context – is ignored. Sheppard (1996) suggests a broadening of the concept of local context to incorporate both site and situation characteristics.

The merging of both site and situation characteristics within a GIS environment, however, is not a straightforward undertaking. At issue is the fundamental conflict between two very different conceptualizations of space: absolute space and rela-

[5] Where site characteristics reflect the attributes and qualities of particular locations or spatial objects, situation characteristics reflect each location's embeddedness within a broader spatial structure involving other locations and other spatial objects.

tive space[6]. Couclelis (1991; 1997) has developed the idea of *proximal space* as a way to bridge these two concepts. Where absolute space emphasizes the locational coordinates and attribute characteristics of site, and relative space emphasizes the spatial relations associated with situation, the key notion in proximal space is the neighborhood (Couclelis 1997, 170). This notion of neighborhood – neither static nor restricted to physical contiguity – reflects Sheppard's extended concept of local context.

Proximal space emphasizes spatial dependence in the form of local interactions where local interactions may be relevant from three different perspectives: spatial proximity, functional proximity, and statistical spatial dependency. Spatial proximity reflects the physical spatial relationships associated with a local site and its neighbors. Functional proximity reflects relationships based on influence or interaction. Statistical spatial dependence reflects the cohesion and homogeneity of a site and its neighbors.

Most spatial interaction accessibility models incorporate both a site and a situation component (typically a spatial opportunities variable and a distance or impedance variable). They are inadequate for representing intra-metropolitan accessibility and the proximal space construct, however, on at least two grounds:

(1) Scale. As with GIS operations, most spatial interaction models present intra-metropolitan accessibility as a single score or indicator for each location of interest. This score reflects a location's absolute site attributes on the one hand, and its situation relative to other locations and their attributes on the other. The concept of accessibility, however, is more realistically represented as a process or function of space, time, and technology. Individuals may trade-off time and distance, or may utilize telecommunications technologies, to gain access to spatial opportunities at a distance. Space, time and available technology, therefore, function as structures, which both constrain and enable human activities. With spatial interaction accessibility models, these structures are implemented by *discounting* spatial opportunities by an impedance factor. For any given location i, the result is a single indicator representing some quantity of *whole* opportunities associated with location i itself, plus some quantity of *partial* opportunities distributed various distances away. Interpretation is awkward. This strategy of discounting spatial opportunities is not entirely effective for representing the spatial dynamics of intra-metropolitan accessibility with changes in scale or with movements through space.

---

[6] From the absolute or Newtonian perspective, space is represented as a distinct entity with characteristics similar to a system of pigeonholes or containers (Lawton 1983, 197). It is conceptualized as emptiness – an entity with existence independent of matter, possessing the structure to hold or to individuate phenomena – a universal receptacle in which objects exist and events occur (Smith 1984; Harvey 1973). From the relative or Liebnitz perspective, on the other hand, space is an abstract concept reflecting the spatial relationships between perceived objects, endowed with structure and properties that are intimately tied to process (Lawton 1983, 197; Couclelis 1992, 221).

(2) Proximity. Most spatial interaction accessibility models include an impedance component, typically reflecting the friction of time, distance, or some other travel cost. With rapid developments in a broad range of spatial technologies[7], however, spatial relations become much more complex and these impedance functions become rather limiting. Gatrell (1983, 5) reminds us that despite a proliferation of urban models with a time or distance component in geography, time or distance themselves do not have causal properties. Instead, it is the *implications* of time and distance that have consequences, and it is these implications that are being modified by rapid developments in transportation and telecommunications technologies.

Gatrell (1983, 7) notes that there is an almost infinite number of relations associated with any given set of spatial objects – physical distance is only one, fairly constrained type of relation. He refers to these relations as *proximities*, distinguishing between attribute proximities and interaction proximities. Attribute proximities define relations between two or more locations based on attribute profiles (income, racial, or occupational profiles, for example). Defining relations using attribute proximities, location itself and the distances separating different locations are immaterial. In contrast, interaction proximities define relations between two or more locations on the basis of spatial interaction volumes (social networks, journey-to-work trips, commodity flows, for example). Here, location and distance are central components of the relations defined for a set of spatial objects. Within the proximal space construct (Couclelis 1997), these relations may be operationalized as a series of functions, heuristics or other symbols linking each location to every other location under study. This same flexibility would be difficult to attain using traditional spatial interaction accessibility models.

Context

It is important to recognize that many of the specific operational details (definitional, representational, and methodological) associated with measuring intra-metropolitan accessibility are necessarily influenced by the scope and overall objectives of the research study at hand. For the purposes of this chapter, I am specifically interested in understanding how broad spatial processes of economic restructuring, suburbanization, and rapid technological developments are impact-

[7] Couclelis (1994) coined the term *spatial technologies* in reference to the bundle of transportation, communication, and information technologies that, in conjunction, modify spatial relations.

ing urban spatial structure[8] and in how these impacts affect intra-metropolitan accessibility. The proposed analytical framework, therefore, presents an aggregate-level analysis rather than focusing on impacts associated with any one individual (disaggregate-level analysis). These broad structural changes have important implications for urban and transportation planning. In addition, results from this type of aggregate-level analysis can provide an effective springboard for more detailed, disaggregate-level analyses of intra-metropolitan accessibility.

3.3 Analytical Framework

This section of the chapter presents an analytical framework for measuring intra-metropolitan accessibility. The framework is based on a *level-of-service* definition of accessibility, the Couclelis (1997) proximal space construct, and the Getis/Ord G_i^* statistic (Getis and Ord 1992, Ord and Getis 1995).

Conceptualizing Accessibility

While accessibility need not be considered a physical entity – it is a concept, a perception, something each of us experiences, evaluates, or judges differently – the proposed analytical framework gives the concept of accessibility substance by defining accessibility to be a characteristic, or attribute, of proximal space. From this perspective, accessibility is a multi-dimensional attribute (see Hanson, Chapter 16) of the proximal spaces defined for any given urban activity system. It is multi-dimensional because the concept of accessibility comprises both structural and functional elements, encompassing both potential accessibility and realized accessibility. The structural elements of accessibility comprise the spatial distribution of people and opportunities, as well as the transportation and communications infrastructure connecting them. Functional elements of accessibility comprise the variety of attributes associated with different groups of individuals (their resources, aptitudes, constraints, preferences, ingenuity, etc.) that lead to different patterns of realized accessibility. These structural and functional dimensions of accessibility operate over multiple spatial scales through processes inherently tied

[8] In theory, urban spatial structure may be defined in terms of the spatial relationships linking a region's urban activities (employment, schools, medical facilities, public services, and/or recreational activities) to the spatial distribution of a region's inhabitants (Pred 1977, 10; Simpson 1987, 120). In the examples presented in the next section, however, I limit this definition of urban spatial structure to a focus on the spatial relationships between employment opportunities and workers by place of residence. I emphasize accessibility to employment opportunities because employment represents a fundamental component of the urban landscape and has been directly impacted by the broad spatial processes of economic restructuring, suburbanization, and rapid technological developments.

to space, time, and available technology. The empirical analyses presented in the next section, therefore, define local interactions – the proximal space construct – using a variety of travel costs. This approach is adopted in order to highlight the spatial dynamics of intra-metropolitan accessibility associated with changes in scale, and to reflect the idea that individuals may trade-off time and distance to gain access to opportunities at a distance. The statistical framework used to measure both structural and functional accessibility, and to model the proximal space construct is described next.

The Getis/Ord G_i^* Statistic

The Getis/Ord G_i^* statistic measures the degree of association or spatial clustering associated with a single variable (X) distributed over a spatial surface (Getis and Ord 1992; Ord and Getis 1995). Consider an urban landscape divided into n regions, $i = 1, 2, \ldots, n$, where each region is identified with a location whose Cartesian coordinates are known. Each location i has associated with it a value x_i of the variable X. Each value, x_i, is affiliated with a set of neighbors – a set of x_j values – where the *neighborhood* is defined by a spatial weights matrix (typically a binary, one/zero matrix). The spatial weights matrix is square, with row and column entries for each pair of locations i and j. As one example, a binary spatial weights matrix may define local interactions – the neighborhood – in terms of physical distance: a 2-mile-distance radius, for example. Each row, then, would be associated with column entries of either one or zero: values of one would reflect membership in the neighborhood, and would be associated with all locations j within a 2-mile-distance radius of location i, including the diagonal where $j=i$ (footnote[9]). The G_i^* statistic may be written as follows (Ord and Getis 1995):

$$G_i^*(d) = \frac{\sum_j w_{ij}(d)\, x_j - W_i^* \bar{x}}{s\{[(n s_{1i}^*) - W_i^{*2}]/(n-1)\}^{1/2}}$$

where:

$G_i^*(d)$ = the G_i^* score for a particular distance (or other neighborhood membership criteria), d.

$w_{ij}(d)$ = the spatial weights matrix for a particular distance (or other neighborhood membership criteria), d

x_j = the spatial variable, X, being measured at location j, where j may equal i

W_i^* = the sum of column entries for row i of the spatial weights matrix

$\bar{x}$ = the mean for all x_j observations

[9] The G_i^* and G_i formulations differ in the way the neighborhood is defined. With G_i^* j may equal i; for the G_i formulation, however, the spatial weights matrix has a zero for each diagonal entry. This distinction results in slight differences in the way the G_i^* and G_i formulations are calculated (see Ord and Getis 1995).

s = the square root of the variance for all x_j observations
n = the total number of observations
s_{1i}^* = the sum of squared column entries for row i of the spatial weights matrix[10]

The G_i^* statistic compares local neighborhood averages to the global average for a given variable X in a defined study region. Where high values of X (high in relation to the global mean) cluster together, G_i^* scores will be positive; where low values of X cluster together, G_i^* scores will be negative. These G_i^* scores may be interpreted as standard normal deviations where the expectation is zero and the variance is one (Ord and Getis 1995). To understand why the G_i^* statistic is effective for measuring intra-metropolitan accessibility, however, it is necessary to consider how *perfect* accessibility might express itself on the urban landscape, and further, to consider how specific aspects of the urban landscape may serve to detract from this notion of perfection.

Whenever we deal with accessibility, we are implicitly confronting a situation involving both supply and demand. When measuring accessibility to employment opportunities, however, this distinction becomes blurred. From the employers' perspective, demand is reflected by a need for workers to fill particular jobs, and the number of workers available represents supply. From the workers' perspective, demand reflects a need for jobs, while the number of jobs available represents supply. One approach to resolution is to measure accessibility by level-of-service. Suppose we define a spatial opportunities variable (X) to reflect both jobs and workers[11].

Let:

e_i = the number of jobs (employment opportunities) at location i
E = the total number of jobs in the entire study region
u_i = the number of workers, by residence, at location i
U = the total number of workers in the entire study region.

Now define each x_i as follows:

$$x_i = (e_i / E) - (u_i / U)$$

Notice that if location i contains 10 percent of the region's jobs and 10 percent of the region's workers, x_i will equal 0. If, however, location i contains a larger proportion of jobs than workers, x_i will be positive. Positive x_i values reflect loca-

[10] The spatial weights matrix need not be binary. When it is binary, s_{1i}^* is equal to W_i^*.

[11] This 'level-of-service' definition for accessibility is appropriate because the relationship between jobs and workers is represented as a simple one-to-one correspondence. Measuring accessibility to other types of services (retail or medical facilities, for example), may require different level-of-service formulations.

tions where employers have less than perfect accessibility to potential employees. Similarly, negative x_i values reflect locations offering workers less than perfect accessibility to potential employment opportunities. By defining accessibility in terms of how well a particular location serves surrounding populations, both the employers' and the workers' perspectives on accessibility are represented[12].

This notion of *perfect accessibility* needs further clarification, however, as it provides a primary justification for using the G_i^* statistic – a measure of spatial association – to measure intra-metropolitan accessibility. Imagine a highly urbanized hypothetical landscape exhibiting the following, quite unrealistic, characteristics: (1) all jobs and all workers are evenly distributed throughout the study region; (2) each location i is connected to every other location j by an equivalent and effective transportation network; and (3) all jobs and workers within the study region have identical characteristics: workers have identical preferences, levels of mobility, skills, etc., and jobs offer equivalent working environments and wages. In this hypothetical landscape, no matter what our scale of analysis, or how space is partitioned into discrete zones (census tracts, for example), worker proportions will match job proportions and the spatial opportunities variable (X), as defined above, will consistently tend toward zero. This hypothetical landscape reflects a notion of 'perfect' accessibility defined in terms of *level-of-service* or *social equity*; each worker in the study area has similar accessibility to the region's employment opportunities. Now let us relax each of the assumptions given above. Urban landscapes, generally, are not associated with evenly distributed jobs and workers. Instead, they are typically structured into a mosaic of residential and industrial land uses. Where job-rich (worker-poor) sites intermingle with worker-rich (job-poor) sites, worker shortages at one location may balance job shortages at locations nearby, so that accessibility is maintained. Where job-rich or worker-rich sites cluster together, however, worker/job shortages are additive. Scale of analysis is important here. In addition, transportation and communications networks develop unevenly over time and space. Where residential neighborhoods and employment centers are efficiently connected via transportation or telecommunications networks, costs[13] to access opportunities at a distance will be small; where there are physical or *functional* barriers restricting movement or communications, opportunity costs will be large. Finally, employment outcomes (a firm's decision to hire an individual or a worker's decision to accept a job offer) are influenced by a large number of complex individual characteristics and behavioral factors associated with worker or firm preferences, motivations, aspirations, objectives, social networks, search strategies, and sometimes just plain dumb luck. One

---

[12] This level-of-service variable does not include information regarding commuting patterns.

[13] There are, of course, costs associated with operating a vehicle or using public transportation to overcome distance. Other, traditional costs, may involve the time needed to transcend space (walking to work). Even when telecommunications technologies are employed, however, there are costs associated with gaining access to appropriate equipment, education, and authority to use these technologies. Many employees, who choose to telecommute to the workplace, may pay for this convenience in the form of reduced opportunities for promotion or salary increases as a result of spending large amounts of time off-site.

approach to understanding how these complex individual factors play out in the aggregate, however, is to examine observed journey-to-work travel behavior.

The G_i^* statistic presents an effective approach for evaluating intra-metropolitan accessibility by considering each location *i* within the context of its proximal locations *j*, and determining the degree of spatial clustering associated with job and worker spatial patterns at multiple scales of analysis. The proximal space relations reflecting the complex physical and functional connections among sites in the study area are modeled via the G_i^* statistic spatial weights matrix. In the empirical analyses presented in the next section, the spatial weights matrix is modeled to reflect Euclidean distances, travel time costs, and functional time costs at multiple scales of analysis. Functional travel times, derived from actual journey-to-work travel patterns, are compared to actual travel times, providing insight regarding the impact changes in transportation and telecommunications technologies are having on urban spatial structure. These statements are best clarified through the examples presented in the next section.

3.4 Application

Greater Los Angeles (Figure 3.1) is generally viewed as a region of endless urban sprawl with widely dispersed population and employment patterns (Giuliano and Small 1991); the region has been described as *the* prototypical example of urban restructuring (Soja, Morales, and Wolff 1983). This study area, therefore, provides an especially appropriate context for investigating questions relating to intra-metropolitan accessibility. In this final section of the chapter, the proposed analytical framework is applied to 1990 employment data for the five county Greater Los Angeles region[14]. Three analyses are presented. While each is performed using the framework outlined above, the method used to model the relations connecting origins (workers by place of residence) and destinations (jobs) varies.

[14] The study area encompasses most of the urbanized portions of the five-county Greater Los Angeles region covered by the Southern California Association of Governments (SCAG): Los Angeles, Orange, San Bernardino, Riverside and Ventura counties. Employment data reflecting the number of jobs and the number of resident workers for each census tract in 1990 were kindly provided by SCAG. Travel-time data and journey-to-work travel-flow data were obtained from the 1990 Census Transportation Planning Package (CTPP) CD-ROM distributed by the U.S. Bureau of Transportation Statistics.

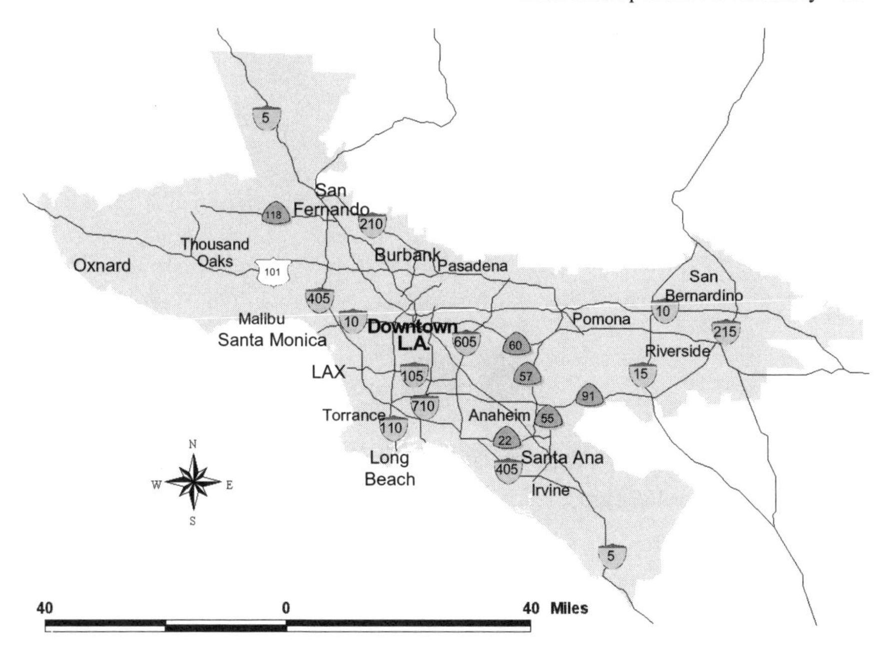

Figure 3.1. The Greater Los Angeles study area.

Spatial Distributions

One approach to evaluating intra-metropolitan accessibility to employment opportunities is to examine spatial relationships between job locations and resident workers. In Figures 3.2, 3.3, and 3.4, the G_i^* statistic has been applied to the level-of-service spatial opportunities variable (described in the previous section) using a spatial weights matrix based on Euclidean distance. In Figure 3.2, each location (census-tract centroid) is evaluated within the context of its neighbors, where the neighborhood is defined as those census tracts within a 5-mile distance radius. In Figure 3.3, the neighborhood is extended to encompass neighbors within a 10-mile radius. Similarly, the context for analysis in Figure 3.4 is defined using a 15-mile distance radius. The G_i^* statistic measures the statistical spatial dependence of each neighborhood, reflecting its cohesion and homogeneity. Tracts with positive G_i^* scores are associated with job-rich regions, which, while they may offer good access to job opportunities for local workers, provide poor access to employers needing to fill those jobs. Similarly, tracts associated with negative G_i^* scores reflect job-poor regions providing insufficient access to employment opportunities for local workers. Tracts with scores near zero represent regions offering effective accessibility to both employers and workers[15].

Figure 3.2. Greater Los Angeles, 1990. Accessibility scores based on the spatial distribution of jobs and workers. Scale of analysis: 5 miles.

In Figure 3.2, using a distance radius of 5 miles, four significant job-rich (worker-poor) clusters are apparent: the Downtown Los Angeles area, the Westwood and West Los Angeles area, the Torrance area, and the Santa Ana/Anaheim area. A number of worker-rich (job-poor) clusters are also discernible, including the residential communities west of San Bernardino, west of Riverside, and north of San Fernando. Interestingly, at this scale of analysis, a number of tracts in South Central Los Angeles are associated with job-poor regions, adjuring further research addressing spatial mismatch hypotheses.

When the scale of analysis is increased to 10 miles, these spatial patterns become more consolidated. Two job-rich (worker-poor) regions stand out: the Downtown Los Angeles area and the Santa Ana/Anaheim area in Orange County. Dominant worker-rich (job-poor) regions are diffuse to the north, east, and south of these two job-rich clusters. At the 15-mile scale of analysis, however, while the Downtown Los Angeles job-rich cluster continues to expand, the Orange County job-rich cluster has disappeared almost completely: the physical spatial relationships between jobs and workers in the Santa Ana/Anaheim area are fairly balanced

[15] In Figures 3.2, 3.3, 3.4, and 3.6, accessibility designations of *very poor* are associated with G_i^* scores greater than +2, or less than -2, standard deviations; designations of *poor* are associated with G_i^* scores ranging from +1 to +2 or from -1 to -2. *Effective* accessibility is associated with G_i^* scores between -1 and +1.

Figure 3.3. Greater Los Angeles, 1990. Accessibility scores based on the spatial distribution of jobs and workers. Scale of analysis: 10 miles.

Figure 3.4. Greater Los Angeles, 1990. Accessibility scores based on the spatial distribution of jobs and workers. Scale of analysis: 15 miles.

if workers are willing and able to endure 15-mile commutes. At the same time, however, worker-rich (job-poor) clustering in the eastern and northern portions of the study region have become more intense: the job deficiencies in these regions are additive with increases in scale of analysis.

Analysis is performed at 5, 10, and 15 miles to emphasize the idea that accessibility is not a static score, but a function, or process of space, time, and technolgy. Consider, for example, the two tracts labeled *1* and *2* in Figure 3.4. Both tracts are associated with similar accessibility scores when evaluated for a 15-mile distance radius. Considering only this one score, however, obscures important details about variations in accessibility with changes in scale. Figure 3.5 graphs the G_i^* scores at 5, 10, and 15 miles. For the Orange County tract (labeled *2*), accessibility to employment opportunities increases rapidly if one is willing and able to travel 10 or 15 miles. For the Los Angeles County tract (labeled *1*), however, increased travel offers diminishing returns. Getis (1994) suggests one approach to capturing these scale-related variations in G_i^* scores is to report not only the G_i^* scores at multiple spatial scales, but to also report the slope associated with these scores.

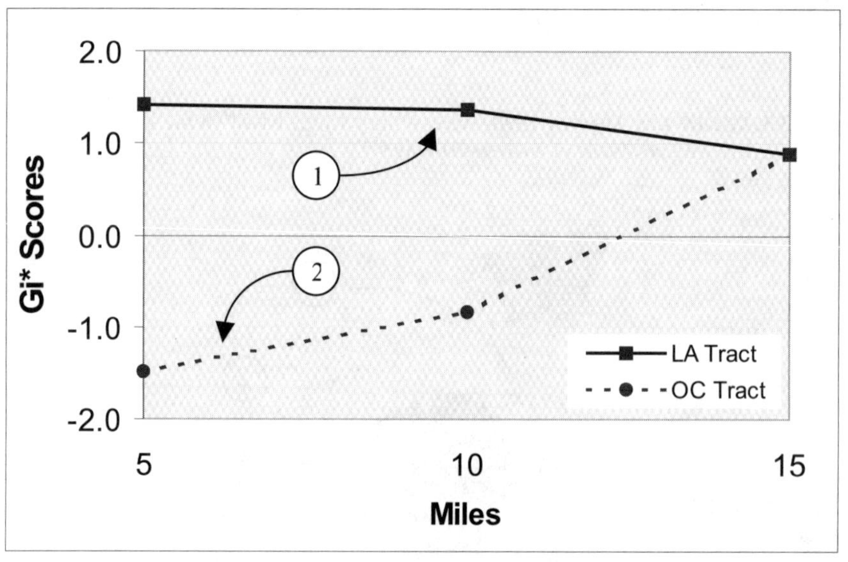

Figure 3.5. Variations in accessibility with changes in scale.

Defining the *neighborhood* by physical distance, as demonstrated in Figures 3.2, 3.3, and 3.4 above, allows us to focus on the spatial distributions of origins and destinations. This type of analysis may be useful for addressing research questions dealing with jobs/housing balance or spatial mismatch issues. In addition, this type of analysis may be extended by disaggregating both jobs and resident workers into different occupational, racial/ethnic, or gender categories in order to evaluate how sensitive these different categories (groups of individuals) are to employment proximity. Other research questions may be addressed by re-defining the linkages connecting origins and destinations; this possibility is discussed next.

Transportation Infrastructure

A second approach to evaluating intra-metropolitan accessibility is to examine the transportation infrastructure connecting origins and destinations. In Figures 3.2, 3.3, and 3.4, the spatial weights matrix was based on Euclidean distance. In Figure 3.6, local interactions are structured as travel-time isochrones.[16] The median travel time for all known 5-, 10-, and 15-mile commutes is 18, 25, and 30 minutes, respectively. To demonstrate analysis using travel-time costs, each tract in Figure 3.6 is evaluated for all neighboring tracts within a 30-minute travel time. Comparing Figure 3.6 to Figure 3.4, we see definite similarities: job-rich clusters associated with the central study area and worker-rich clusters in the eastern counties.

Figure 3.6. Greater Los Angeles, 1990. Accessibility scores based on travel-time costs. Scale of analysis: 30 minutes

[16] Median tract-to-tract travel times were extracted from the CTPP data set. Unfortunately, these data are of poor quality. Not only did a large number of missing time costs need to be estimated, but approximately 10 percent of the time values had to be discarded altogether because they were clearly erroneous (indications that individuals driving alone in a car could travel approximately 90 miles in 15 minutes, or required 99 minutes to travel approximately 0.5 miles). Where necessary, travel time estimates were constructed from *similar*, plausible commute times (similar distances and proximal origins and destinations). The large number of estimates needed, however, limits the extent to which analysis is possible. These estimates are sufficient, however, for demonstrating the potential, given accurate data, to address a broad range of urban research questions using the methods outlined in this chapter.

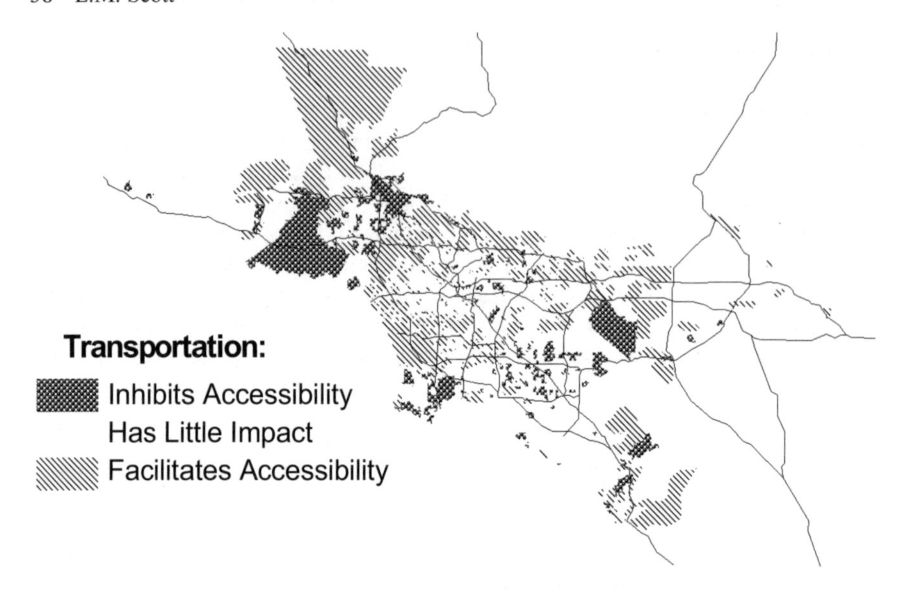

Figure 3.7. Greater Los Angeles, 1990. Transportation network impacts on accessibility. Scale of analysis: 30 minutes.

Striking differences are found in the northern portion of the study area, where accessibility based on distance indicates poor accessibility for *workers*, but based on travel times indicates poor accessibility for *employers*. In this region, effective transportation networks allow workers in the San Fernando Valley access to job-rich tracts to the south. Figure 3.7, based on differences in accessibility scores using both the 15-mile-distance radius and the 30-minute-time-cost isochrone, indicates how the transportation network impacts intra-metropolitan accessibility at this scale of analysis. Tracts with a block fill pattern indicate regions with congestion or insufficient transport capacity, reducing accessibility potential. Tracts with a striped fill pattern indicate regions where transportation infrastructure is facilitating accessibility. The accessibility scores associated with tracts shown using a smooth fill pattern did not change as a result of factoring in transportation infrastructure influences.

Analyses using travel-time costs allow evaluation and isolation of the impacts on intra-metropolitan accessibility imposed by transportation infrastructure, and offer some very interesting possibilities for extended analysis. Figure 3.7 indicates where insufficient infrastructure is limiting accessibility potential. An alternative to using travel-time costs is using actual road networks and their associated capacity constraints in constructing the spatial weights matrix. With this type of model, it would be very interesting to test whether the congestion depicted in Figure 3.7 would be most effectively relieved by building more roads or, alternatively, by encouraging flex-time employment opportunities[17]. In addition, incorporating capacity constraints into the structure of the spatial weights matrix adds a temporal dimension to the G_i^* analysis, offering potential for a dynamic model of intra-metropolitan accessibility.

Some may argue that in the *information age*, evaluation of transportation infrastructure ought to include data reflecting communications infrastructure. For the majority of individuals, however, telecommunications infrastructure is not the limiting factor inhibiting access. More substantial limitations are associated with lack of technical skills, poor access to appropriate equipment, or lack of opportunity (telecommuting is not an option for some occupations, for example). These limitations are addressed in the next section, focusing on functional components of intra-metropolitan accessibility.

Functional Proximity

A final example of the G_i^* analytical framework represents the relations between origins and destinations as functional time costs. Functional time costs are one instance of what Gatrell (1983) refers to as *interaction proximities*. The concept of functional time costs or functional distances is based on the idea that two places with high rates of spatial interaction are functionally 'closer' than two places with very little interaction. Participation in place-based activities promotes familiarity, which may be expressed through extended social networks, development of strategic contacts, and/or expanded knowledge about a place or region (see Hanson, 1998). In a recursive manner, participation in place-based activities promotes familiarity, while familiarity increases the likelihood of participation in place-based activities. Using a similar logic, two places with very limited spatial interaction may be represented as being functionally distant.

To operationalize the concept of functional travel times, I adapt a method developed by Tobler and Wineburg (1971) in an article titled *A Cappadocian Speculation*. The adapted method involves inverting the doubly constrained gravity model with known journey-to-work flows to solve for travel time. The doubly constrained gravity model may be written as:

$$F_{ij} = A_i B_j \frac{O_i D_j}{t_{ij}^{b}}$$

and may be interpreted as follows:

[17] Flex-time refers to a relaxation of the 8-to-5 workday. Employers allow employees to begin and end work over a broader time frame, as long as they fulfill a required number of working hours.

F_{ij} = the number of journey-to-work flows from site i to site j
O_i = the number of workers at site i (origins)
D_j = the number of jobs at site j (destinations)
t_{ij} = the travel time from site i to site j
b = an exponent reflecting the concept of a *time decay*
A_i = a vector of scaling factors to ensure $\sum_j F_{ij} = O_i$, for every O_i
B_j = a vector of scaling factors to ensure $\sum_i F_{ij} = D_j$, for every D_j.

Inverting the model and using known values for F_{ij}, we may solve for t_{ij} to obtain functional time:

$$t_{ij} = \left(A_i B_j \frac{O_i D_j}{F_{ij}} \right)^{1/b}$$

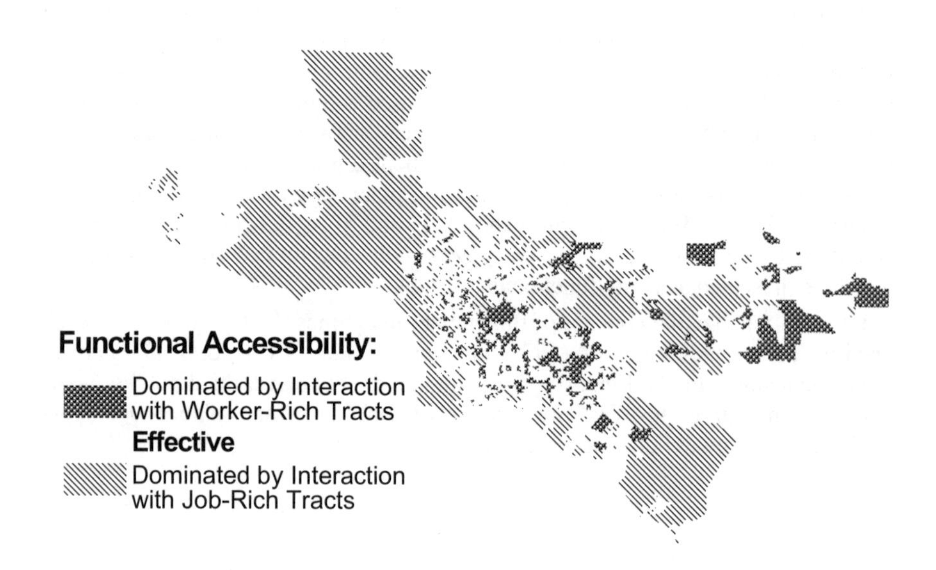

Figure 3.8. Greater Los Angeles, 1990. Accessibility scores based on *functional* travel times. Scale of analysis: 30 minutes.

Figure 3.9. Greater Los Angeles, 1990. Comparison of actual travel times to functional travel times.

Where actual spatial interaction rates are higher than predicted by the gravity model, functional travel times will be shorter than actual travel times; where spatial interaction rates are lower than predicted, functional times will be longer than actual travel times. Figure 3.8 maps the G_i^* accessibility scores calculated using a spatial weights matrix based on a 30-minute *functional* time-cost isochrone. This map is quite different from the others. Notice that many of the tracts near Downtown Los Angeles, while physically closer to job-rich (worker-poor) tracts, are *functionally* closer to worker-rich tracts. Similarly, the job-poor tracts in south Orange County are functionally closer to job-rich tracts to the north than to the worker-rich tracts physically nearby. Workers in south Orange County are, therefore, exposed to a large number of jobs because of their commuting behavior. Notice, also, that Figure 3.8 is *not* a *complete* reversal of Figure 3.6. The worker-rich tracts in the eastern-most portion of the study area, for example, continue to be associated with worker-rich clusters. One interesting extension of this analysis would be to actually map the stretching and pulling of space depicted by the functional travel times used in the Figure 3.8 analysis. An alternative analysis is presented in Figure 3.9. Here the average functional travel times for all journey-to-work flows are compared to the average actual travel times for these same flows. Where functional travel-time averages are similar to actual travel-time averages, we may conclude that accessibility is primarily determined by structural components of accessibility: the spatial distribution of jobs in relation to resident workers, and the impacts imposed by transportation infrastructure. However, tracts associated with much shorter functional travel-time averages, indicate that non-structural influences are impacting accessibility. These non-structural influences might include higher incomes, effective access to private vehicles, association with occupations allowing telecommuting, indifference to long-distance commutes, along with a host of other possibilities.

While this aggregate-level analysis of functional travel times cannot explain the behavior of any one individual, *by applying this method to longitudinal data, we may begin to address questions concerned with how broad spatial processes associated with economic restructuring, suburbanization, and rapid developments in transportation and telecommunications technologies, impact the urban landscape from both a structural and non-structural perspective.* This aggregate-level analysis may also provide a springboard for more detailed disaggregate-level inquiries. Regression analysis, for example, could be used to measure how well variations in income, race/ethnicity, gender, or occupational categories explain the functional vs. actual travel-time differences depicted in Figure 3.9. The CTPP data set includes many of these variables both by place of residence and by place of work.

As a final look at accessibility in the Greater Los Angeles region, Figure 3.10 identifies tracts with consistent G_i^* accessibility scores across all three of the analyses presented: the spatial distribution of jobs and workers evaluated within the context of a 15-mile distance radius, the distribution of jobs and workers in relation to transportation infrastructure at a 30-minute scale of analysis, and the functional relationships between jobs and workers based on actual travel flows and a 30-minute-functional-time isochrone. Tracts consistently associated with worker-rich (job-poor) local interactions represent a high priority for implementation of urban planning strategies promoting job development. Tracts consistently associated with job-rich (worker-poor) local interactions represent priority areas for urban planning promoting housing development. Finally, careful study of the tracts consistently associated with effective accessibility may provide guidelines for promoting more efficient, equitable, and sustainable communities.

Figure 3.10. Greater Los Angeles, 1990. Accessibility score stability. Based on: 15-mile distance, 30-minute travel time, and 30-minute functional time analysis.

3.5 Conclusions

This chapter has reviewed some of the definitional and operational challenges that must be met if we are to effectively examine impacts on intra-metropolitan accessibility associated with the broad spatial processes of economic restructuring, suburbanization, and rapid developments in transportation and telecommunications technologies. It has outlined an analytical framework structured around a level-of-service definition of accessibility, the Couclelis proximal space construct, and the Getis/Ord G_i^* statistic. Future research will consider use of actual road networks, incorporation of highway capacity constraints to develop a dynamic accessibility model, exploration of the efficacy of defining the relations between origins and destinations using other attribute and interaction proximities, and consideration of other methods for deriving functional travel times. Taking analysis of intra-metropolitan accessibility in the Greater Los Angeles region to the next level, will require accurate data over a series of time periods. While the challenges are daunting, understanding how broad spatial processes in the *information age* shape urban spatial structure represents one of the most interesting and exciting tasks facing urban geographers at this time.

References

Arentze, T.A., Borgers, A.W.J., and Timmermans, H.J.P. 1994. Multistop-based measurements of accessibility in a GIS environment. *International Journal of Geographical Information Systems* 8:343-56.

Breheny, M.J. 1978. The measurement of spatial opportunity in strategic planning. *Regional Studies* 12:463-79.

Burns, L.D. 1979. *Transportation, temporal, and spatial components of accessibility*. Lexington: Lexington Books.

Chinitz, B. 1991. A framework for speculating about future urban growth patterns in the US. *Urban Studies* 28:939-59.

Couclelis, H. 1991. Requirements for planning-relevant GIS: a spatial perspective. *Papers in Regional Science* 70:9-19.

Couclelis, H. 1992. *Location, Place, Region, and Space*. In Abler, R.F., Marcus, M.G., and Olson, J.M. (eds.) *Geography's Inner Worlds: Pervasive Themes in Contemporary American Geography*. New Brunswick, NJ: Rutgers University Press, 215-33.

Couclelis, H. 1994. Editorial: Spatial Technologies. *Environment and Planning B: Planning and Design* 21:142-43.

Couclelis, H. 1997. From cellular automata to urban models: new principles for model development and implementation. *Environment and Planning B: Planning and Design* 24:165-74.

Gatrell, A.C. 1983. *Distance and Space: a Geographical Perspective*. Oxford: Clarendon Press.

Geertman, S.C.M. and Ritsema Van Eck, J.R. 1995. GIS and models of accessibility potential: an application in planning. *International Journal of Geographical Information Systems* 9:67-80.

Getis, A. 1994. *Spatial Dependence and Heterogeneity and Proximal Databases*. In Fotheringham, S. and Rogerson, R. (eds.) *Spatial Analysis and GIS*. London: Taylor & Francis, 213-29.

Getis, A. and Ord, J.K. 1992. The analysis of spatial association by use of distance statistics. *Geographical Analysis* 24:189-206.

Giuliano, G. 1995. The weakening transportation-land use connection. *Access*, 6:3-11.

Giuliano, G. and Small, K.A. 1991. Subcenters in the Los Angeles Region. *Regional Science and Urban Economics* 21:163-82.

Gould, P. 1969. *Spatial Diffusion: Commission on College Geography*. Washington DC: Association of American Geographers.

Graham, S. and Marvin, S. 1996. *Telecommunications and the City: Electronic Spaces, Urban Places*. London: Routledge.

Handy, S.L. 1994. Highway blues: nothing a little accessibility can't cure. *Access* 5:3-7.

Hanson, S. 1998. Off the road? Reflections on transportation geography in the information age. *Journal of Transport Geography* 6:241-49.

Hanson, S. and Schwab, M. 1987. Accessibility and intraurban travel. *Environment and Planning A* 19: 735-48.

Harvey, D. 1973. *Social Justice and the City*. Baltimore: The Johns Hopkins University Press.

Haynes, K.E. and Fotheringham, A.S. 1984. *Gravity and Spatial Interaction Models*. Beverly Hills: Sage Publications.

Helling, A. 1996. *Why We Should Care About Intra-Metropolitan Accessibility and How We Measure It*. In Couclelis, H. (ed.) *Spatial Technologies, Geographic Information, and the City Proceedings* (Baltimore, Maryland, Sep 9-11). Santa Barbara CA: National Center for Geographic Information and Analysis.

Helling, A. 1998. Changing intra-metropolitan accessibility in the U.S.: evidence from Atlanta. *Progress in Planning* 49:55-107.

Ingram, D.R. 1971. The concept of accessibility: a search for an operational form. *Regional Studies* 5:101-7.

Jones, S.R. 1981. *Accessibility Measures: a Literature Review*. Transport and Road Research Laboratory Report 967. Crowthorne, Berkshire: TRRL.

Knox, P.L. 1980. Measures of accessibility as social indicators: a note. *Social Indicators Research* 7:367-77.

Koenig, J.G. 1980. Indicators of urban accessibility: theory and application. *Transportation* 9: 145-72.

Kwan, M. 1998. Space-Time and integral measures of individual accessibility: a comparative analysis using a point-based framework. *Geographical Analysis* 30: 191-216.

Lawton, R. 1983. Space, place and time. *Geography* 68:193-207.

Miller, H.J. 1991. Modelling accessibility using space-time prism concepts within geographical information systems. *International Journal of Geographical Information Systems* 5: 287-301.

Morris, J.M., Dumble, P.L., and Wigan, M.R. 1979. Accessibility indicators for transport planning. *Transportation Research A* 13A: 91-109.

Ord, J.K. and Getis, A. 1995. Local spatial autocorrelation statistics: distributional issues and an application. *Geographical Analysis* 27: 287-306.

Pirie, G.H. 1979. Measuring accessibility: a review and proposal. *Environment and Planning A* 11:299-312.

Pred, A. 1977. *City-Systems in Advanced Economies*. London: Hutchinson.

Sheppard, E. 1996. Commentary: Site, situation, and social theory. *Environment and Planning A* 28:1339-44.

Simpson, W. 1987. Workplace location, residential location, and urban commuting. *Urban Studies* 24:119-28.

Smith, N. 1984. *Uneven Development: Nature, Capital, and the Production of Space*. Oxford: Basil Blackwell.

Soja, E., Morales, R., and Wolff, G. 1983. Urban restructuring: an analysis of social and spatial change in Los Angeles. *Economic Geography* 59:195-230.

Tobler, W.R. and Wineburg, S. 1971. A Cappadocian speculation. *Nature* 231:39-41.

Wachs, M. and Kumagai, T.G. 1973. Physical accessibility as a social indicator. *Socio-Economic Planning Sciences* 7:437-56

4 Transportation, Telecommunications, and the Changing Geography of Opportunity[1]

Qing Shen

Department of Urban Studies and Planning, Massachusetts Institute of Technology, Cambridge MA 02139, USA. Email: qshen@mit.edu

4.1 Introduction

New telecommunications – digital information and communication technologies that support interaction and transaction over long distances – are emerging as a primary force in shaping cities. And more fundamentally, they are becoming one of the most important variables in defining spatial relationships among people and organizations located in metropolitan areas. Manifested by the rapid growth of the Internet, ATMs, and mobile phones, telecommunications are permeating the physical structure, the economic production, and the social life of cities. Visibly and invisibly, these technologies are creating new spatial paths and barriers that will profoundly affect people's access to economic opportunities and social services. Therefore, one of the most important tasks for urban researchers in the information age is to help policy makers and the general public to understand, monitor, predict, and respond to spatial consequences resulting from a massive-scale deployment of new telecommunications.

Despite their pervasiveness, spatial effects of telecommunications are largely uncharted (Atkinson 1998, Batty 1996). Part of the explanation for this can be found in the dynamic nature of these new technologies. Because the synthesis of digital information and communication systems is still at an early stage, the technologies as well as the effects they generate are changing rapidly. Under these circumstances, empirical data often become obsolete quickly, and forecasts of future outcomes tend to be speculative. However, perhaps the more important reason is found in the limitation of existing analytical approaches. Specifically, few studies attempt to incorporate relationships between transportation and telecommunications, or more generally relationships between the physical space and the virtual space, explicitly into the analytical framework. Therefore, analyses of spatial consequences of the new technologies are often incomplete and potentially biased. Also, they often fail to address systematically and rigorously important issues of social equity related to the spatial consequences. Undoubtedly, there is

---

[1] Reprinted with permission from *Urban Geography*, Vol. 20, No. 4, pp. 334-355.

an urgent need for new concepts and methods that will allow urban researchers to examine the new geography as an interdependent whole (Couclelis 1996).

This chapter aims to achieve two objectives. The first is to develop an analytical framework that integrates location, transportation, and telecommunication variables into a unified representation of spatial relationships. The second objective is to use the new analytical framework as a tool to undertake structured explorations of spatial and social implications of telecommunications. The focus of this study is on the changing geography of opportunity in American metropolitan areas.

The next section presents the new analytical framework, which is developed based on literature in three areas of urban research. The first is spatial representation using accessibility measures (Morris, Dumble, and Wigan 1979, Shen 1998a; Weibull 1976), the second is urban industrial restructuring (Kasarda 1995, Office of Technology Assessment, U.S. Congress 1995), and the third is telecommuting (Mokhtarian 1990, Salomon and Mokhtarian 1997). The third section will apply the new analytical framework to an analysis of the changing geography of opportunity. The discussion starts with observations of spatial reconfigurations that have taken place in metropolitan areas that have been shaped by the automobile. It is then expanded by taking into consideration new options of access, new compositions of economic opportunities, and new spatial patterns of metropolitan growth that are emerging in the information age. The fourth section will discuss some findings of a case study of employment accessibility in the Boston Metropolitan Area. The analysis of existing patterns and the simulation of future change both shed light on the understanding of the spatial and social effects of new telecommunications. In the concluding section, some policy implications will be drawn and directions for future research will be identified.

4.2 Spatial Technologies, Accessibility, and Geography of Opportunity

Basic Concepts

The discussion in this chapter involves several important concepts. One such concept is *spatial technologies*, which is defined by Couclelis (1996) as the complex of transportation, communication, and information technologies that together modify spatial relations. Another concept is *accessibility*, which is a measure of the strength and extensiveness of spatial relationships between opportunity seekers and relevant opportunities. Accessibility is a basic concept in geography, because it indicates the potential for spatial interaction. A third concept is *geography of opportunity*, which can be defined as a collection of people, economic opportunities, and the spatial relationships between them. In many studies, including this one, the metropolitan area is considered as the territorial extent of the geography of opportunity.

The general relationships among these three concepts are clear. Spatial technologies modify spatial relations, and the effects can be measured concretely by examining changes in accessibility. Changes in accessibility, in turn, imply possible reconfigurations of the geography of opportunity. These relationships become specific and meaningful when examined in some real context, such as a metropolitan area. They become much more complex once the social dimension is added to the analysis.

Transportation and Accessibility

Over the decades, urban researchers have developed accessibility measures to help understand the complex relationship between transportation technologies and the changing geography of opportunity. Their efforts have resulted in a large volume of literature. Most researchers agree that the level of accessibility for an opportunity seeker is a function of (4.1) the total number of relevant opportunities, (4.2) the spatial distribution of these opportunities, (4.3) the spatial location of the individual, and (4.4) the individual's ability to overcome spatial separation. When opportunities are relatively scarce, the competition with other opportunity seekers must also be taken into account. Many mathematical formulations have been proposed for the measurement of accessibility (Morris, Dumble, and Wigan 1979). In the following formulation, first suggested by Weibull (1976) and recently generalized by Shen (1998a), each available opportunity is weighted by a demand factor, expressed as the denominator, which represents the competing demand generated by other opportunity seekers:

$$A_i^v = \Sigma_j \frac{O_j f(t_{ij}^v)}{\Sigma_m \Sigma_k P_k^m f(t_{kj}^m)} \tag{4.1}$$

where:

A_i^v is the accessibility for opportunity seekers who live in zone i and travel by mode v

O_j is the number of relevant opportunities in zone j

$f(t_{ij}^v)$ is the impedance for travel from i to j by mode v

$f(t_{kj}^m)$ is the impedance for travel from k to j by mode m

P_k^m is the number of opportunity seekers who live in zone k and travel by mode m

For a metropolitan area with N zones, $i, j, k = 1, 2, \ldots, N$

For a metropolitan area with M modes, $v, m = 1, 2, \ldots, M$

As an accessibility measure, equation 4.1 has several desirable properties. One important property is that it provides a consistent framework for incorporating geographic locations and transportation modes into the measurement, and therefore allows direct comparison of accessibility across modes as well as across locations. Because modal choices are related to individuals' socioeconomic conditions – for example, low-income people are much more likely to travel by public transportation, this measure can reveal accessibility differentials among social groups. It can help understand how geographic locations and transportation technologies, which are highly correlated with socioeconomic variables, jointly determine individuals' relative positions in the geography of opportunity. Another important property, mathematically proved by Shen (1998a), is that its expected value always equals the ratio of the total number of opportunities to the total number of opportunity seekers, and hence is predetermined. Therefore, for each mode of transportation, *accessibility-poor* zones can be distinguished easily from 'accessibility-rich' zones.

For a community, represented by a geographic zone, the overall position in the geography of opportunity is determined by the levels of accessibility for its residents. Shen (1998a) proposes the following general accessibility measure to specify this relationship:

$$A_i^G = \Sigma_v \, (P_i^v / P_i) \, A_i^v \tag{4.2}$$

where:

A_i^G is the general accessibility for all opportunity seekers who live in zone i

P_i^v is the number of opportunity seekers who live in zone i and travel by mode v

P_i is the total number of opportunity seekers in zone i

Equation 4.2 essentially calculates a weighted score of accessibility for each zone. The numbers of people traveling by the different modes are used as the weights. The higher the percentage of a zone's residents that use advanced transportation, the higher a zone's general accessibility will be. The expected value of general accessibility also equals the ratio of the total number of opportunities to the total number of opportunity seekers (Shen 1998a). Hence zones in which the residents are on average *accessibility-poor* can be distinguished easily from zones in which the residents are on average *accessibility-rich*. Therefore, this measure depicts an overall picture of the geography of opportunity of a metropolitan area by defining the relative position of each zone in terms of accessibility.

Telecommunications and Accessibility

Telecommunications are a type of spatial technology that has some distinctive characteristics. These characteristics fundamentally determine how telecommunications modify spatial relationships and reconfigure the geography of opportunity. First of all, they are a means of access that is of a very high speed. In fact, when human interaction takes place over telecommunication networks, the time cost is so small that it is practically invariant with physical distance. Therefore, access in the virtual space follows logical links rather than physical paths, resulting in *despatialization of interaction* (Mitchell 1995). For opportunities that belong exclusively to the virtual space, measurement of accessibility through the specification of some time or distance impedance function is neither practical nor meaningful.

Secondly, telecommunications are useful for accessing only certain kinds of opportunities – or more precisely, certain elements of certain kinds of opportunities – which do not require the physical presence of the opportunity seekers. The development and deployment of digital information and communication technologies is associated with a fundamental restructuring of the economy, which is characterized by a rapid increase of opportunity elements that are accessible through telecommunications. An important indication is that information workers, whose primary economic activity involves the creation, processing, or distribution of information, now constitute about 50 percent of the labor force of the United States (U.S. DOT 1993). Therefore, any representation of the geography of opportunity will be incomplete if telecommunications are excluded. On the other hand, relatively few opportunities can be accessed fully through electronic means. Transportation and face-to-face interaction are still important, although the frequency of certain types of travel is reduced.

Some of the more thorough studies of the relationship between physical and electronic means of spatial interaction are those by researchers in the area of telecommuting (Mokhtarian 1990, Salomon 1986, Salomon and Mokhtarian 1997, U.S. DOT 1993). They found that telecommuters generally reduce, rather than eliminate, commuting trips. Most telecommuters still make several work-related trips every week for the purpose of maintaining a certain frequency of face-to-face interactions with co-workers and clients. This is essential for establishing and sustaining good working relationships with them, which enables telecommuters to function effectively while working at home. Furthermore, many telecommuters must travel to workplace from time to time because some of their tasks require special facilities or close interactions with co-workers, and therefore cannot be done remotely. The relationship between commuting and telecommuting is partly substituting and partly complementary.

Thirdly, telecommunications not only provide an important means of accessing opportunities, but also are themselves an indispensable component of an increasing number of opportunities. We have seen parallel situations for transportation; for example, operating motor vehicles is an essential part of the jobs of police officers and bus drivers. But such a dual function is much more common for telecommunications. To understand how this will potentially affect accessibility and the geography of opportunity, it is useful to adopt the concept of *skill mismatch*, which appears frequently in the literature on industrial restructuring (Kasarda 1995, Office of Technology Assessment, U.S. Congress 1995). The ability to work with colleagues and clients remotely through information and communication systems will be required as a basic skill for a higher and higher percentage of jobs in the information age. People who do not have such a skill will be excluded from these economic opportunities.

Taken together, these characteristics suggest that it is both possible and necessary to develop a unified framework – which incorporates both transportation and telecommunications – for the measurement of accessibility. A starting point for developing such an operational measure is to redefine the components: opportunities, opportunity seekers, and impedance functions.

It is a reasonable simplification to define three categories of opportunities:[2]

- One category consists of opportunities that are accessed through telecommunications, at least in part. Telecommunications are an indispensable component of these opportunities. This category will be denoted as C in mathematical expressions. Assuming that it constitutes ϕ proportion of the total opportunities O, there will be ϕO opportunities in this category. They are accessible only to people with telecommunications capabilities.
- A second category consists of opportunities that can be accessed through either transportation or telecommunications. Telecommunications are the preferable means of accessing these opportunities, but are not an indispensable component of these opportunities. Assuming that this category constitutes λ proportion of the total, there will be λO opportunities in it. They are accessible to everyone, through transportation or telecommunications.
- A third category consists of traditional types of opportunities that belong exclusively to the physical space, and therefore are accessed through transportation only. Obviously, there will be $(1-\phi-\lambda)O$ opportunities in this category. They are also accessible to everyone, because everyone has certain transportation capabilities.

Opportunity seekers can be divided into two broad categories based on whether or not they have telecommunications capabilities:

[2] It is important to note that the classification presented here is different from what was described in an earlier paper by this author (Shen 1998b). The old classification included only two categories of opportunities – opportunities that can only be accessed through telecommunications and opportunities that can only be accessed through transportation. The new classification has an additional category, which consists of opportunities that can be accessed through either transportation or telecommunication. The new scheme is more sophisticated and realistic. Because of the difference in classification, the resulting formulations and measurements of accessibility are also different in some ways.

- One category is comprised of opportunity seekers who have telecommunications capabilities. Assuming it constitutes δ proportion of the total opportunity seekers P, there will be δP people in this category. These people can access all three categories of opportunities.
- The other category is comprised of opportunity seekers who do not have telecommunications capabilities. Obviously, there will be $(1-\delta) P$ people in this category. These people cannot access the first category of opportunities, but they can access the other two.

If there are M transportation modes, there will actually be $2M$ groups of opportunity seekers, which is the total number of possible combinations of telecommunications status and transportation modes. The term *telecommunications capabilities* is purposefully left unspecified here. It is intended to be a broad concept that can accommodate a wide range of combinations of technologies and skills for using the technologies. In some contexts, it may simply indicate the capability to use the telephone for telemarketing; in other contexts, it may describe the capability to use sophisticated digital information and communication systems for Internet commerce and service. This concept carries more specific meanings once the problem to be studied is decided. In this research, *telecommunications capabilities* mean some combination of available equipment and skills that enables a worker to access and distribute information remotely using digital information systems and communication networks.

The reconceptualized accessibility measures can be generally expressed by the following two equations:

- Accessibility for seekers who do not have telecommunications capabilities

$$A_i^{v} = A_i^{v(1)} + A_i^{v(2)} \tag{4.3}$$

- Accessibility for seekers who have telecommunications capabilities

$$A_i^{cv} = A_i^{v(1)} + A_i^{cv(2)} + A_i^{cv(3)} \tag{4.4}$$

where:

$A_i^{v(1)}$ is accessibility to opportunities that are accessed through transportation only;

$A_i^{v(2)}$ is accessibility to opportunities that can be accessed through either transportation or telecommunications, as measured for people who can use transportation only;

$A_i^{cv(2)}$ is accessibility to opportunities that can be accessed through either transportation or telecommunications, as measured for people who can use telecommunications;

$A_i^{cv(3)}$ is accessibility to opportunities that are accessed through telecommunications.

To make these general formulas specific and operational, impedance functions must be defined. The usual approach is to determine the zone-to-zone travel times (or travel distances) for each means of access. For the transportation component, travel time by mode is used straightforwardly. For the telecommunications component, however, travel time can be defined only if people who use telecommunications to access opportunities still make at least some complementary trips. The literature on telecommuting suggests that this is indeed the case. Assuming that these people typically travel to a certain type of opportunities once every τ days, instead of everyday, we can calculate a *perceived average daily travel time* for them:

$$t_{ij}^{cv} = \sigma (1 / \tau)\, t_{ij}^{v} \tag{4.5}$$

In other words, there are two variables that determine the impedance for an opportunity seeker who has telecommunications capabilities. One is the frequency of using transportation for opportunity-seeking trips, and the other is the zone-to-zone travel times for the transportation mode used. Note that σ is a parameter that converts an *actual* average daily travel time into a *perceived* average daily travel time.[3] There are M different specifications of impedance for people who have telecommunications capabilities, each corresponding to a transportation mode. To keep the discussion simple, it is assumed that the travel frequency for people who use telecommunications to access opportunity does not vary across geographic locations or types of opportunities.

There are many possible mathematical formulations for specifying equations 4.3 and 4.4. Here the discussion is based on the assumption that equation 4.1, described previously, is a proper representation of accessibility in the physical space, and can be extended to represent more completely accessibility in the emerging geography of the information age.

Accessibility for different groups of opportunity seekers can be measured using the following equations, which are the extended versions of equation 4.1:

---

[3] The value of σ should normally be determined through calibration. But because the empirical data required for calibration are not currently available, σ is given the value of 1 by assuming that the perceived average daily travel time equals the actual average daily travel time.

- For opportunity seekers who do not have telecommunications capabilities

$$A_i^v = \Sigma_j \frac{(1-\phi_j-\lambda_j)O_j\, f(t_{ij}^v)}{\Sigma_m \Sigma_k\, P_k^m\, f(t_{kj}^m)} + \Sigma_j \frac{\lambda_j O_j\, f(t_{ij}^v)}{\Sigma_m \Sigma_k\, [(1-\delta_k)P_k^m\, f(t_{kj}^m) + \delta_k P_k^m\, f(t_{kj}^{cm})]} \quad (4.6)$$

- For opportunity seekers who have telecommunications capabilities

$$A_i^{cv} = \Sigma_j \frac{(1-\phi_j-\lambda_j)O_j\, f(t_{ij}^v)}{\Sigma_m \Sigma_k\, P_k^m\, f(t_{kj}^m)} + \Sigma_j \frac{\lambda_j O_j\, f(t_{ij}^{cv})}{\Sigma_m \Sigma_k\, [(1-\delta_k)P_k^m\, f(t_{kj}^m) + \delta_k P_k^m\, f(t_{kj}^{cm})]}$$

$$+ \Sigma_j \frac{\phi_j O_j\, f(t_{ij}^{cv})}{\Sigma_m \Sigma_k\, \delta_k P_k^m\, f(t_{kj}^{cm})} \quad (4.7)$$

Although these two equations appear quite complicated, they are in fact rather straightforward to understand. The accessibility for opportunity seekers who do not have telecommunications capabilities, which is represented by equation 4.6, has two components. The first component is essentially the same as the right side of equation 4.1, except that only $(1-\phi_j-\lambda_j)O_j$ opportunities in each zone are now accessed exclusively through transportation. These opportunities are accessible to all opportunity seekers $(\Sigma_m \Sigma_k P_k^m)$. Hence the denominator, which represents the demand factor, includes competing demands generated by all of them. The second component calculates accessibility attributed to $\lambda_j O_j$ opportunities in each zone that are accessible both to people with telecommunications capabilities $(\Sigma_m \Sigma_k \delta_k P_k^m)$ and people with only transportation $(\Sigma_m \Sigma_k (1-\delta_k) P_k^m)$. The demand factor of this component, which is the denominator, consists of competing demands generated by these two broad categories of opportunity seekers. It is important to note that the two impedance functions contain different travel times: travel time is t_{kj}^{cm} is for people with telecommunications; it is t_{kj}^m is for people with only transportation.

The accessibility for opportunity seekers who have telecommunications capabilities, which is represented by equation 4.7, has three components. The first component is identical to the first component in equation (4.6), because they can also use transportation to access the $(1-\phi_j-\lambda_j)O_j$ opportunities in each zone. The second component is also essentially the same as the second component in equation 4.6, except that t_{ij}^{cv} instead of t_{ij}^v is used to specify the impedance function in the numerator. The third component calculates accessibility attributed to the $\phi_j O_j$

opportunities in each zone that require telecommunications capabilities, and therefore are accessible only to the $\Sigma_m \Sigma_k \delta_k P_k^m$ opportunity seekers who have such capabilities. The denominator, which represents the demand factor, includes only competing demands generated by these people.

This extended accessibility measure – the combination of equations 4.6 and 4.7 – has properties similar to the simpler measure expressed by equation 4.1. It is a unified framework that allows for meaningful comparison across transportation modes, telecommunications capabilities, and geographic locations. Because telecommunications capabilities, like modal choices, are related to individuals' socioeconomic conditions, this measure can indicate accessibility differentials among social groups in the information age. It can help understand how telecommunications, together with transportation technologies and geographic locations, determines individuals' relative positions in the new geography of opportunity. The other important property – that its expected value equals the ratio of the total number of opportunities to the total number of opportunity seekers – can also be proved. The mathematical procedure is tedious, and is omitted here. Because of this property, *accessibility-poor* zones can be distinguished easily from *accessibility-rich* zones for each group of opportunity seekers, which is defined by a combination of the transportation mode and the telecommunications status.

Notice that when ϕ_j and λ_j, the proportions of opportunities that are accessed exclusively and partly through telecommunications, are both equal to zero, equations 4.6 and 4.7 become one single equation identical to equation 4.1.

For a community, the varying levels of accessibility for its residents determine the community's overall position in the geography of opportunity. This relationship can also be described by a composite measure of general accessibility:

$$A_i^G = \Sigma_v [(P_i^v - P_i^{cv}) / P_i] A_i^v + \Sigma_v (P_i^{cv} / P_i) A_i^{cv} \qquad (4.8)$$

Notice that although all opportunity seekers in zone i can use transportation as the only means of access, only $\Sigma_v(P_i^v - P_i^{cv})$ of them actually have to do so. This weighted measure, similar to equation (4.2), shows that the higher the percentage of a zone's residents that use advanced spatial technologies, the higher the zone's general accessibility will be. In addition, as in equation 4.2, its expected value equals the ratio of the total number of opportunities to the total number of opportunity seekers. Hence, zones in which the residents are on average *accessibility-poor* are easily distinguished from zones in which the residents are on average *accessibility-rich*. This general accessibility measure depicts an overall picture of the emerging geography of opportunity, by defining the relative position of each zone.

4.3 New Technologies and the Changing Geography of Opportunity

With properly constructed accessibility measures, one can think through more clearly how spatial technologies – transportation and telecommunications – may separately and jointly reconfigure spatial relationships in a metropolitan area. One can effectively explore future changes in the geography of opportunity and address some 'what-if' questions systematically and rigorously.

Spatial and Social Effects of the Automobile

The automobile has been the most important technological force in shaping American metropolitan areas in the twentieth century. It has fundamentally changed the geography of opportunity. Using equation 4.1 described in the previous section, spatial and social effects of the automobile can be characterized in terms of increases and decreases in accessibility. The spatial effects of the automobile are largely attributed to four basic changes brought directly or indirectly by this great invention: speed of access, spatial distribution of jobs and services, spatial distribution of people, and industrial restructuring.

Speed of Access. The automobile has provided the great majority of households with a fast and flexible means of transportation for daily activities. This great benefit, however, does not go to the relatively small percentage of households that cannot travel by car due to financial, physical, or other constraints. These people are still dependent on public transportation. Unfortunately, increased usage of personal motor vehicles has often been associated with decreased quality of service of public transportation. Therefore, the automobile has caused a widening gap among people's ability to overcome spatial separation. It has reduced drivers' zone-to-zone travel times, but has often increased transit passengers' zone-to-zone travel times. Everything else being equal, the accessibility for the former would increase, whereas the accessibility for the latter would decrease.

Spatial Distribution of Jobs and Services. The automobile has enabled urban employment to decentralize, and has therefore facilitated a spatially dispersed pattern of metropolitan growth. Consequently, a high percentage of private enterprises and governmental agencies are now located in suburban areas not served by public transit. In the accessibility measure, this change is depicted by reduced opportunities in zones located within or near the central city. In the meantime, zones that are difficult to access by public transit, and therefore have long transit travel times from other zones, gain a large share of the redistributed opportunities. Assuming all other factors were constant, accessibility would overall decrease in the center but increase at the periphery. The greatest decrease in accessibility would occur for transit-dependent groups living in the central city.

Spatial Distribution of People. The automobile has also effected the decentralization of urban population. In particular, people who own a private automobile have gained more freedom in choosing home and work locations, as well as the freedom to trade accessibility for better housing. The transit-dependent population groups, however, are confined to residential and employment locations connected to the public transportation network. For low-income minorities subject to discrimination in the housing and labor markets, location choices are even more limited. These changes can be properly described in terms of accessibility only if measurement is made separately for the privileged and the disadvantaged. For the former, population decentralization has partly offset employment decentralization, and therefore has reduced the magnitude of accessibility loss in the center and accessibility gain at the periphery. For the disadvantaged groups, such offsetting effect is more limited.

Industrial Restructuring. Decentralization has been especially prevailing for industries – such as manufacturing and wholesale – that require more land per unit of output but less face-to-face interaction with customers. Thus, an increasingly large share of business establishments located in the central city are *information-processing industries*. Employment decentralization has changed occupational compositions both in the central city and in the suburbs. These changes can also be properly described in terms of accessibility if measurement is made separately for different occupation groups. For those workers whose primary economic activity involves collection and processing of information, industrial restructuring has partly offset the effects of employment decentralization. For the rest, industrial restructuring has reinforced the effects of employment decentralization.

Overall, transit-dependent groups located in the central city have been worse off in terms of accessibility. They have been worse off not only relatively in comparison with the great majority of the population who can drive their own car, but also in absolute terms. The absolute decline in accessibility is due to the fact that while their travel speed has not increased, job opportunities and social services have become more dispersed. On the other hand, the overall level of accessibility for the auto-driving population located in the suburbs has increased both relatively and absolutely. This group has a fast means of access and is located in areas with increasing opportunities brought by employment decentralization. Because low-income and minority groups constitute disproportionately high percentages of the transit-dependent population in the central city but low percentages of the auto-driving population in the suburbs, the spatial consequences imply increased gaps between the privileged and the disadvantaged. Furthermore, because transit-dependent people constitute a large proportion of the central-city population, from equation 4.2 one can see that the central city's position in the geography of opportunity has declined.

This general discussion is useful for outlining some major changes in urban spatial structure during the age of the automobile. However, there are many questions that cannot be answered in abstract and general terms. They require empirical analysis of specific cases. For example, has the location advantage of the central city totally disappeared as far as low-skilled workers' residential choice is concerned? Is good residential location still an effective substitution for automobile ownership? Some of these questions will be addressed later with findings of a case study of employment accessibility in the Boston Metropolitan Area.

Spatial and Social Effects of Telecommunications

Telecommunications, especially the new digital information and communication technologies, can generate spatial effects that are even more revolutionary than those generated by the automobile. The effects can also be characterized in terms of increase and decrease in accessibility, measured using equations 4.6 and 4.7 described in the previous section. The analysis can, once again, proceed by examining four basic changes – speed of access, spatial distribution of jobs and services, spatial distribution of people, and industrial restructuring – which have, in this case, been brought directly or indirectly by telecommunications.

Speed of Access. For people who use these new technologies to access opportunities, the saving in time spent on travel can be enormous. For example, if workers telecommute two days every week, they will reduce their average daily commuting time by forty percent. Since there are still many households in the United States who do not have a telephone, it is most likely that a considerable percentage of the population will not have adequate telecommunications capabilities in the near future. Therefore, there will be further widening in the gap among people in terms of ability to overcome spatial separation. Travel time differentials across population groups will increase. Everything else being equal, the level of accessibility for people who have telecommunications capabilities will increase, whereas the level of accessibility for the others will decrease relatively.

Spatial Distribution of Jobs and Services. New telecommunications are likely to cause further decentralization of employment. Over the past three decades, many urban researchers have attempted to address the question of how telecommunications will affect the spatial distribution of urban activities and, ultimately, the future of the city (Berry 1973, Castells 1996, Gottmann 1983, Hall 1996, Meier 1962, Mitchell 1995, Webber 1964; 1996). Despite many differences among these researchers' approaches and views, they generally agree that information and communication technologies create a possibility for certain types of economic activities to become footloose. In all likelihood, new telecommunications will reinforce the existing trend of employment decentralization. Their spatial effects will have a similar pattern: accessibility will overall decrease in the center but increase at the periphery; the greatest decline in accessibility will occur for those people who live in the central city and do not have access to telecommunications. However, the magnitude of the impacts is likely to be greater.

Spatial Distribution of People. There will likely be a parallel decentralization of population. Leading the future wave of decentralization will be telecommuters who are equipped with both automobiles and telecommunications. However, this group will be relatively small compared to the auto-driving population. Some researchers reported recently that only about a third of employees in the United States are potential telecommuters (Salomon and Mokhtarian 1997). But overall, the existing pattern of population decentralization will be reinforced. Therefore, the offsetting effect, which applies mainly to the economically and socially privileged group, will also occur: population decentralization will reduce the magnitude of loss of accessibility in the center and gain of accessibility at the periphery.

Industrial Restructuring. This process now consists of three elements. First of all, occupational compositions in the central city and in the suburbs will both continue to change. In particular, the percentage of employment in traditional industries that always require the physical presence of the workers will decrease. Secondly, a large and increasing percentage of job opportunities and social services will be accessible through telecommunications as well as transportation. And finally, a considerable percentage of job opportunities and social services will require telecommunications capabilities for access. The emerging Internet commerce and tele-medicine are examples. These three elements are, respectively, represented in the three components of equation 4.7 described earlier. It is easy to see that everything else being equal, the level of accessibility for people who have telecommunications capabilities will increase, whereas the level of accessibility for the others will decrease.

In general, central-city residents who do not have telecommunications capabilities will be worse off in terms of accessibility. They will be worse off relatively in comparison with the people who can access opportunities through telecommunications. They will most likely be worse off in absolute terms, because an increase in opportunities that require telecommunications for access, and hence not accessible to them, will almost certain be accompanied by a corresponding decrease in opportunities that can be accessed fully through transportation. On the other hand, suburban residents who have telecommunications capabilities will have an overall higher level of accessibility than before, measured in absolute terms.

Many scholars suggest that *new classes* of privileged and disadvantaged may emerge in the information age. However, research findings indicate that the 'new classes' are in many ways reassembling the old classes (Hoffman and Novak 1998). Therefore, the spatial reconfigurations caused by telecommunications can also be translated into social effects. Undoubtedly, low-income and minority groups will constitute disproportionately high percentages of those central-city residents who do not have telecommunications capabilities but low percentages of those suburban residents who have such capabilities. Without strong policy intervention, a new form of marginalization will be in place. Furthermore, from equation 4.8 one can see that the central city's overall position in the geography of opportunity may continue to decline in the future, since a high percentage of its residents will be unlikely to have telecommunications capabilities.

Because of the increased complexity of spatial configurations in the information age, there are more issues that cannot be addressed without examining empirical data for specific contexts. In particular, it is not easy to see clearly the relative contributions of geographic locations, transportation modes, and telecommunications capabilities in determining the accessibility differentials among individuals and communities. The case study of employment accessibility in the Boston Metropolitan Area, reported in the next section, will help obtain additional insights into such complex relationships.

4.4 Employment Accessibility in an American Metropolitan Area

The Boston Metropolitan Area, which covers more than two thousand square miles of land and accommodates more than 4 million people, was selected for a case study. The case study was focused on employment accessibility because commuting and telecommuting data were more available in comparison with data for other types of travel or telecommunication usage. It consisted of two parts. The first part examined existing patterns of variations of employment accessibility across geographic locations and travel modes. It was an empirical analysis of 1990 data using equations 4.1 and 4.2. The second part explored future changes in employment accessibility caused by a large-scale deployment of telecommunications. It was a simulation of some future scenario using equations 4.5, 4.6, 4.7, and 4.8. Each part was implemented with a program written in the C language.

Auto Ownership and Polarization in Employment Accessibility

This analysis was based on demographic and employment data originating from the 1990 Census and peak-hour travel time matrices for auto and transit obtained from the Central Transportation Planning Staff (CTPS) in Boston. The metropolitan area was described by data associated with 787 traffic analysis zones (TAZs). The measurement of accessibility was made with alternative impedance functions – the exponential function and several travel-time-threshold functions, which generate similar results. The following findings were obtained consistently, no matter whether accessibility was measured for all workers or only for low-skilled workers:

First, the central location of the inner city still gives the residents advantage in accessing spatially distributed jobs. For workers who use a given transportation mode, those living near the central business district (CBD) have relatively higher employment accessibility. In general, workers would not increase their employment accessibility by moving to the suburbs.

Second, although central location of residence offers an advantage, auto ownership is the key determinant. More than half of the TAZs are *accessibility-rich* for workers commuting by auto, whereas only a very small proportion of the TAZs are *accessibility-rich* for those commuting by public transit. In fact, most of the workers who live in the suburbs and commute by auto have higher employment accessibility than those who live in the central city but commute by public transportation.

Third, for a large proportion of the TAZs located in the poor neighborhoods of the central city, the general employment accessibility is rather low because the level of auto ownership is low. The location advantage of these TAZs is in most cases more than offset by the low level of auto ownership.

Therefore, in a metropolitan area that is shaped by the automobile, geographic location is of decreased significance in defining spatial relationships, whereas travel mode has become critical. Regardless where they live, transit-dependent groups are faced with a major obstacle in accessing job opportunities and social services. This conclusion is consistent with the empirical finding that the lack of auto ownership significantly reduces the chance for low-skilled workers to find and keep jobs (Ong 1996). It is also broadly consistent with the widely accepted view that changes in the geography of opportunity have aggravated the economic and social problems of the central city (Kasarda 1995, Teitz 1997, Wilson 1996). However, it provides a new insight by showing clearly that the problem is not location *per se*, but the increasing spatial barrier for a high percentage of low-skilled workers who are dependent on the limited service of public transportation. It suggests that planning and policy making should focus on interaction instead of location, because even if there were better residential locations for the poor, the location advantage would be very limited, compared with the disadvantage associated with transit dependence.

Telecommuting and Changes in Employment Accessibility

A number of assumptions were made in order to construct a scenario of the future of commuting and telecommuting in the Boston Metropolitan Area. They were made based in part on empirical studies of telecommuting in the United States conducted by a number of researchers (Handy and Mokhtarian 1996, Mokhtarian 1990, Salomon and Mokhtarian 1997, U.S. DOT 1993). Specifically, it was assumed that approximately 45 percent of workers in the metropolitan area would be potential telecommuters. The percentage varies from one zone to another, depending on income and other factors. In addition, it was assumed that approximately 12 percent of jobs in the metropolitan area can be accessed through either transportation or telecommunications, and that roughly 2 percent of jobs are accessible only to people who have telecommunication capabilities. These percentages also vary from one zone to another, depending on the occupational composition of employment in each zone. Furthermore, it was assumed that on average, telecommuters take two job-related round-trips every week. Finally, it was assumed that telecommuters have the same transportation mode split as the rest of the labor force. These assumptions enabled variables in the accessibility measures to be estimated.

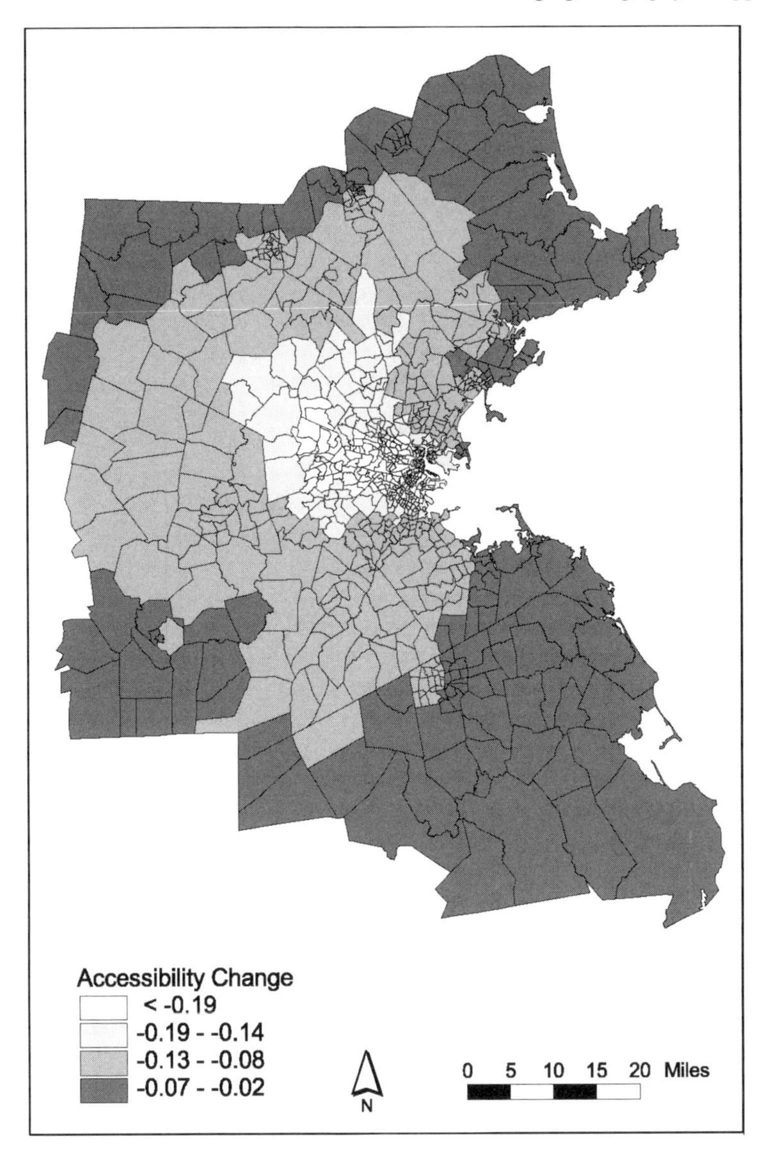

Figure 4.1. Decrease in employment accessibility for workers who travel by auto and who are not potential telecommuters.

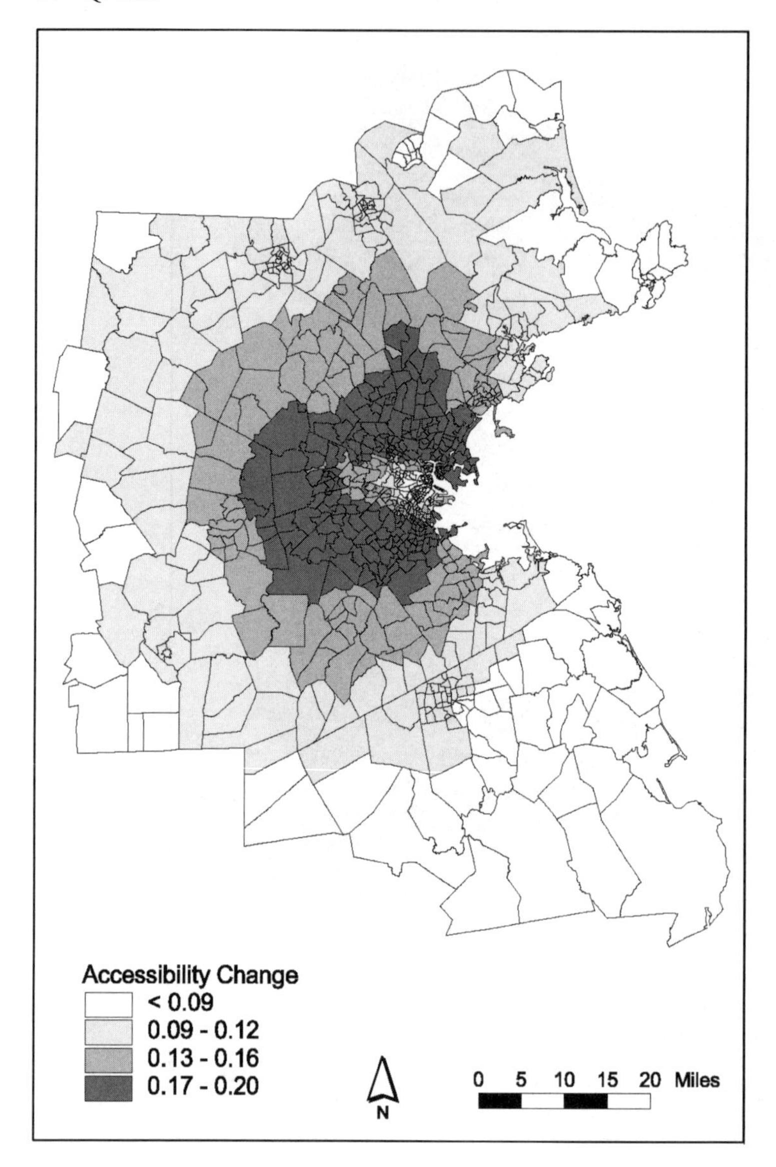

Figure 4.2. Increase in employment accessibility for workers who travel by auto and who are potential telecommuters.

Not surprisingly, the first finding of this part of the study is that for workers who travel by a given transportation mode, their employment accessibility will decrease if they do not have telecommuting capabilities. Of course, the effect is generally the opposite for those who have such capabilities.

The second finding, which is more intriguing, is that the magnitude of accessibility change is related to the pre-existing geographical configuration of the metropolitan area. Figure 4.1 depicts the pattern of accessibility decrease for workers who commute by auto and who are not potential telecommuters.[4] The amount of loss is positively correlated with the pre-existing level of accessibility, as the greatest losses occur in the central-city area and the smallest at the periphery.[5] Figure 4.2 shows the pattern of accessibility increase for workers who commute by auto and who are potential telecommuters. It also exhibits a distinctive, although more complicated, spatial pattern. In this case, the greatest gains occur in a suburban ring and the smallest gains both at the center and on the edge of the metropolitan area. Similar relationships between the magnitude of accessibility change and the pre-existing geography are observed in the results for workers who commute by public transit. The same findings can also be obtained from Figure 4.3 and Figure 4.4. These scatter charts show the relationships between the magnitude of accessibility change and the pre-existing level of accessibility for different categories of workers classified by travel mode and telecommuting capability.

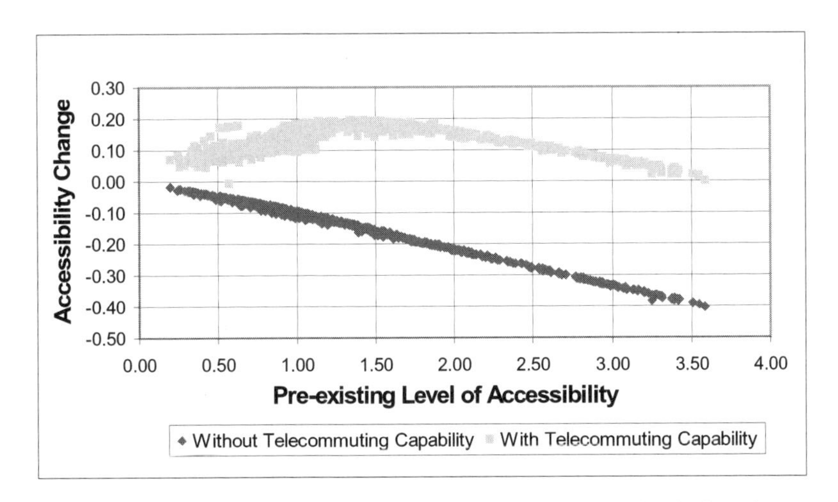

Figure 4.3. Relationship between the change and the pre-existing level of employment accessibility for workers who travel by auto.

[4] In Figure 4.1, Figure 4.2 , and Figure 4.5, the equal interval scheme is used to classify the zones into four groups based on the magnitude of accessibility change.

[5] Readers who are interested in seeing maps of existing levels of employment accessibility for workers who travel by different modes in the Boston Metropolitan Area are referred to an earlier paper by this author (Shen 1998b).

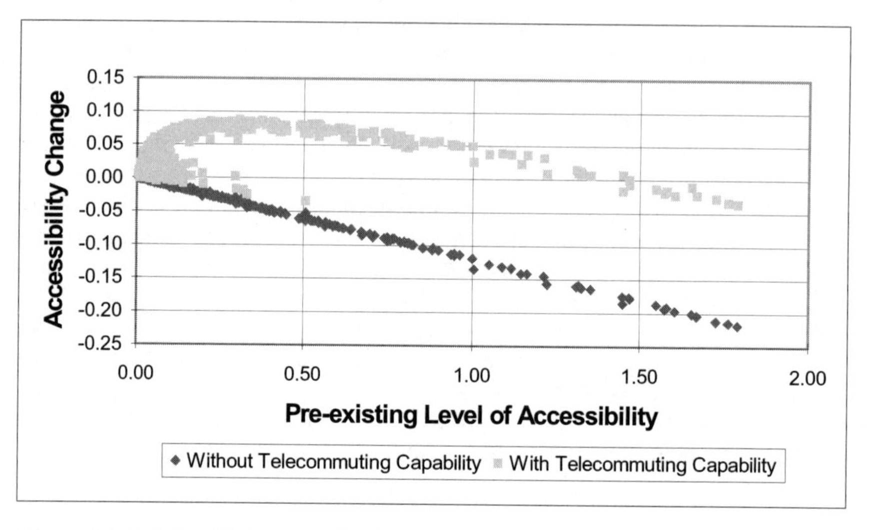

Figure 4.4. Relationship between the change and the pre-existing level of employment accessibility for workers who travel by public transit.

The third finding is that telecommuting increases the overall level of employment accessibility for zones where there are higher percentages of wealthier households, but reduces the overall level of employment accessibility for zones where poorer households predominate. The primary reason is that low-income, low-skilled workers are less likely to be potential telecommuters. This outcome is displayed in Figure 4.5. Geographically, zones that show a substantial decline in the general level of accessibility are highly concentrated in the central city.

The fourth finding is that the magnitude of losses and gains in the general level of employment accessibility is usually quite modest. The largest percentage change is only 15 percent. Approximately three-quarters of the zones either decrease or increase their general level of accessibility by less than 5 percent. Therefore, if telecommunications only partially substitute transportation, as it is assumed in this study, the spatial structure of the metropolitan area will certainly not be dissolved.Overall, the deployment of telecommunications causes the location advantage of the urban center to decrease. In fact, the importance of geographic location is generally reduced. In that sense, advanced spatial technologies lead to location equalization. This process is accompanied by further polarization in accessibility along the social dimension. The most distinctive social consequence is that the wealthier and more educated residents who can use the advanced spatial technologies become better off, whereas the poorer and less educated residents who cannot use these technologies become worse-off. The social distribution of accessibility gains and losses is interrelated with, rather than independent of, the pre-existing geography, which supports the argument made by

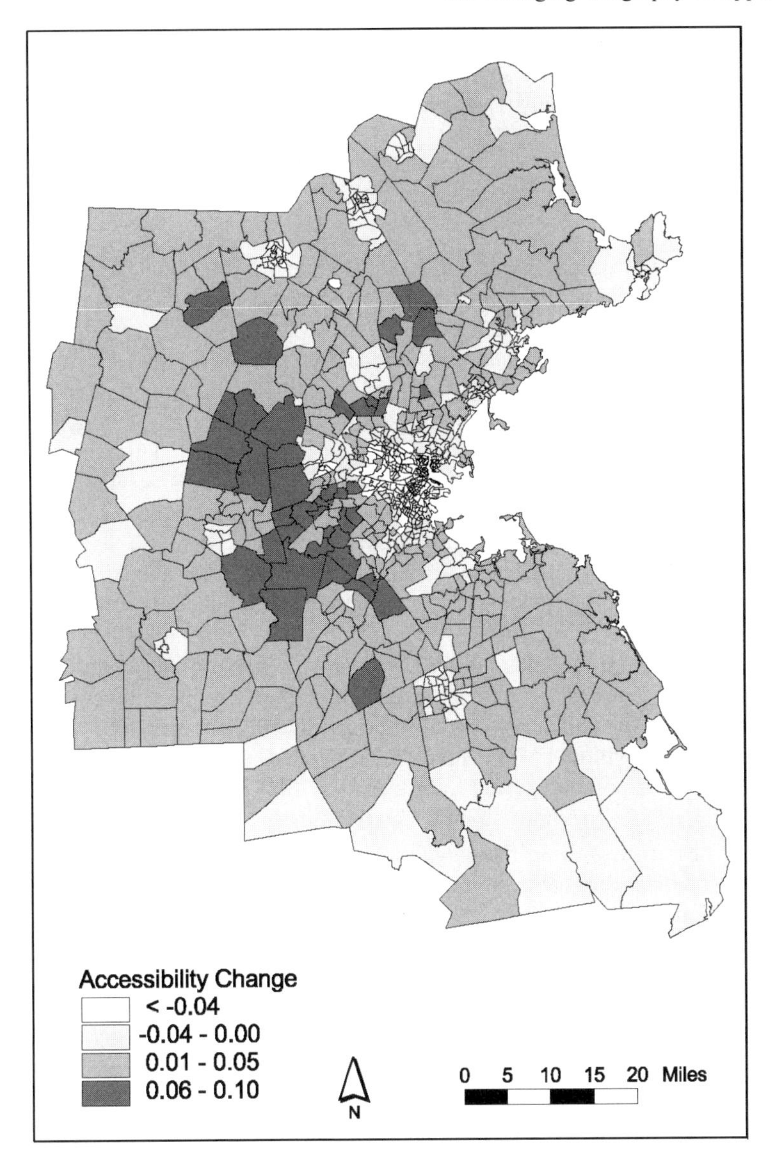

Figure 4.5. Change in the overall level of employment accessibility when telecommuting becomes an option for many workers.

a number of researchers (Gillespie and Robins 1989, Wheeler *et al* 1998). At the community level, telecommunications weaken the position of the central city and strengthen the position of the suburbs in the geography of opportunity. On the other hand, it is also important to note that the magnitude of changes in the general level of employment accessibility is quite modest. Therefore, despite significant impacts of telecommunications, the spatial structure of the metropolitan area in the foreseeable future will inherit many basic characteristics of the one existing today.

Discussion

The case study of employment accessibility in Boston Metropolitan Area has demonstrated the usefulness of the approach described in this chapter. However, it has also revealed a number of important issues. Some of these issues are related to data. Empirical data on which communities and individuals have what kinds of telecommunications capabilities are not currently available in any systematic form. Therefore, variations among communities and individuals in terms of availability and quality of technologies cannot be realistically captured. This kind of variation can be substantial, because adequate telecommunications service may not be available in the near future to some communities, especially those located far away from the metropolitan center. Similarly, empirical data on location choice and travel behavior of telecommuters are not available. Consequently, it is impossible to estimate accurately some of the variables and parameters of the extended accessibility measure, such as telecommuters' mode split and travel frequency. The lack of empirical data often forces researchers to make various assumptions, including ones that are rather simplistic. Obviously, it is imperative to explore mechanisms for collecting spatially disaggregated data on telecommunications and related activities. For example, governmental agencies in charge of surveys and censuses of population and industries may find it both desirable and feasible to ask respondents questions about telecommuting.

Some other issues are regarding the extended accessibility measure itself. It is not clear whether embedding telecommunication-based activities into a spatial interaction formulation is always appropriate. While equations 4.6 and 4.7 are basically sound for representing the potentials for interaction in the context of commuting and telecommuting, their applicability to other contexts remains a question. Furthermore, even in a context where a spatial interaction formulation is appropriate, the specification of impedance based solely on the time variable may limit the analytical power of the extended accessibility measure. It is conceivable that in some cases the main question is under what conditions people will choose to activate their telecommunications potentials given alternative modes of access, alternative goods for consumption, and time and budget constraints. Therefore, it is important to explore the possibility to construct alternative impedance functions that include these economic factors. Almost three decades ago, Wilson (1970) developed a budget version of a spatial interaction model in which the tradeoff between housing rent and transportation costs is specified. His approach can perhaps be extended to include the telecommuting variable.

4.5 Conclusion

This chapter has presented a unified analytical framework for understanding the spatial and social effects of both transportation and telecommunication technologies. This analytical framework characterizes these effects by measuring changes in accessibility. Its key component is a set of accessibility measures that incorporate not only the traditional land use and transportation components but also the telecommunications component. Travel impedance is specified for the telecommunications component on the basis of empirical evidence that shows partial substitution of these technologies for travel. The resulting framework allows consistent and meaningful comparison of accessibility levels across geographic locations, travel modes, and telecommunications capabilities.

With this unified analytical framework, one can examine more rigorously the ways in which the automobile and telecommunications interact with location and social variables in creating a new geography of opportunity in American metropolitan areas. In particular, the new framework can serve as a useful tool for understanding the likely consequences of a large-scale deployment of digital information and communication technologies. Structured explorations were undertaken to understand how different socioeconomic groups, defined by their available options of spatial technologies as well as by their residential locations, fare in the emerging geography of opportunity. Effects on communities were also examined, by aggregating the effects on the individuals. As an effort to go beyond abstract discussion and generalization, the analytical framework was applied to a case study of employment accessibility in the Boston Metropolitan Area. Existing spatial configurations were analyzed, and likely future changes were simulated.

Some of the research findings are not surprising. As it is commonly believed, advanced spatial technologies reinforce polarization among socioeconomic groups. The level of accessibility generally increases for those who have telecommunications capabilities, but decreases for those who depend on more traditional transportation means. Wealthier communities, including most of those in the suburbs, are better off because their residents tend to have telecommunications capabilities. In contrast, poorer communities, especially those in the central city, are worse off. Unless effective policies are implemented to make new telecommunications a viable option for disadvantaged groups, the resulting spatial and social effects will undoubtedly aggravate the problems of the central city.

There are important new insights generated by this research. One useful insight is that geographic location is overall of reduced importance in determining the potential for spatial interactions in the information age. A good location is no longer an effective substitution for access to advanced spatial technologies. Spatial relationships are increasingly defined by technological and fundamentally social factors. This suggests that in order to improve the situations of disadvantaged groups, planning and policy making should focus on interaction rather than narrowly on location. Another important insight is that the distribution of accessibility benefits generated by new technologies is dependent on the existing spatial configurations of a metropolitan area. This suggests that it is essential to examine carefully the interaction between telecommunications and location, transportation, and social factors and to identify potential synergies.

For the purpose of identifying directions for future research, it is important to point out that the practical value of this analytical approach is currently limited by the lack of empirical data on telecommunications and related activities. More specifically, some of the variables and parameters of the extended accessibility measure are not accurately estimated. Consequently, application of this framework necessarily involves many assumptions. Where to obtain the data and how to calibrate the parameters are acute questions that need to be addressed.

It will also be crucial to answer the question of how generally applicable this analytical approach is. In particular, the spatial interaction formulation is based on commuting and telecommuting behaviors. It is not currently clear whether this formulation is appropriate in other contexts.

In addition, the research described in this chapter represents only one possible way to measure the effects of new telecommunications. The approach is based on concepts and methods developed in geographic sciences. Its focus is on characterizing and interpreting the changing spatial relationships in metropolitan areas as measured by changes in the potential for spatial interaction. Economic factors such as budget constraint and tradeoff between alternative goods have not been directly included in the analytical framework. Needless to say, advanced spatial technologies come with both benefits and costs. It will be an excellent complement if some future research can develop alternative versions of accessibility measures that are useful for rigorous economic analysis of the changing urban geography.

Furthermore, this research examines only the spatial effects of telecommunications as measured by changes in the potentials for direct interactions between opportunity seekers and opportunities. These spatial effects can be considered as only part of the total (Niles 1994). Another substantial aspect of the effects can be indirectly measured by indicators such as changes in the cost, quality, choice, and convenience of virtually all types of goods and services we consume. Is the distribution of these indirect benefits much more equal geographically and socially? This is another important question for future research.

References

Atkinson, R.D. 1998. Technological change and cities. *Cityscape*, 3:129-70.

Batty, M. 1996. Access and information: How the new electronic media changes everything geographical but perhaps changes nothing! In Couclelis, H. (ed.) *Spatial Technologies, Geographic Information, and the City*. Technical Report 96-10, pp. 42-4. National Center for Geographic Information and Analysis, University of California, Santa Barbara.

Berry, B. 1973. *The Human Consequences of Urbanization*. New York: St. Martin's Press.

Castells, M. 1996. *The Rise of the Network Society*. Oxford: Blackwell.

Couclelis, H. (ed.) 1996. *Spatial Technologies, Geographic Information, and the City*. Technical Report 96-10, National Center for Geographic Information and Analysis, University of California, Santa Barbara.

Gillespie, A. and Robins, K. 1989. Geographical inequalities: The spatial bias of new communications technologies. *Journal of Communication,* 39:7-18.

Gottmann, J. 1983. *The Coming of the Transactional City*. Institute of Urban Studies, University of Maryland, College Park.

Hall, P. 1996. Revisiting the nonplace urban realm: Have we come full circle? *International Planning Studies* 1:7-15.

Handy, S. and Mokhtarian, P. 1996. Forecasting telecommuting: An exploration of methodologies and research needs. *Transportation* 23:163-90.

Hoffman, D.L. and Novak, T.P. 1998. Bridging the racial divide on the Internet. *Science* 280:390-1.

Kasarda, J. 1995. Industrial restructuring and changing location of jobs. In Farley, R. (ed.) *State of the Union: America in the 1990s, Volume I: Economic Trends*. New York: Russel Sage Foundation, 215-67.

Meier, R.L. 1962. *A Communications Theory of Urban Growth*. Cambridge MA: MIT Press.

Mitchell, W.J 1995. *City of Bits: Space, Place and the Infobahn*. Cambridge MA: MIT Press.

Mokhtarian, P.L. 1990. A typology of relationships between telecommunications and transportation. *Transportation Research* 24:231-42.

Morris, J.M., Dumble, P., and Wigan, M. 1979. Accessibility indicators for transportation planning. *Transportation Research A*. 13:91-109.

Niles, J.S. 1994. *Beyond Telecommuting*. Washington DC: Office of Energy Research, US Department of Energy.

Office of Technology Assessment, US Congress 1995. *The Technological Reshaping of Metropolitan America*. Washington, DC: US Government Printing Office.

Ong, P. 1996. Work and car ownership among welfare recipients. *Social Work Research* 20: 255-62.

Salomon, I. 1986. Telecommunications and travel relationships: A review. *Transportation Research A* 20:223-38.

Salomon, I. and Mokhtarian, P.L. 1997. Why don't you telecommute? *Access* No. 10:27-29.

Shen, Q. 1998a. Location characteristics of inner-city neighborhoods and employment accessibility of low-wage workers. *Environment and Planning B* 25:345-65.

Shen, Q. 1998b. Spatial technologies, accessibility, and the social construction of urban space. *Computers, Environment and Urban Systems* 22:447-64.

Teitz, M.B. 1997. American planning in the 1990s: Part II, the dilemma of the cities. *Urban Studies* 34:775-95.

U.S. DOT 1993. *Transportation Implications of Telecommuting*. Washington DC: US Government Printing Office.

Webber, M.M. 1964. The urban place and the nonplace urban realm. In M. M. Webber, editor, *Explorations into Urban Structure*. Philadelphia PA: University of Pennsylvania Press, 79-153.

Webber, M.M. 1996. Tenacious cities. In Couclelis, H. (ed.) *Spatial Technologies, Geographic Information, and the City*. Technical Report 96-10, pp. 214-18. National Center for Geographic Information and Analysis, University of California, Santa Barbara.

Weibull, J.W. 1976. An axiomatic approach to the measurement of accessibility. *Regional Science and Urban Economics* 6:357-79.

Wheeler, J.O., Muller, P., Thrall, G., and Fik, T. 1998. *Economic Geography*, Third edition. New York: John Wiley & Sons, Inc.

Wilson, A.G. 1970. *Entropy in Urban and Regional Modelling*. London: Pion Limited.

Wilson, W.J. 1996. *When Work Disappears*. New York: Alfred A. Knopf.

5 Space, Time and Sequencing: Substitution at the Physical / Virtual Interface

Pip Forer and Otto Huisman

Spatial Analysis Facility, Department of Geography, University of Auckland,
10 Symonds St., P.B. 92019, Auckland 1001, New Zealand.
Email: p.forer@auckland.ac.nz; o.huisman@auckland.ac.nz

5.1 Introduction

This chapter is concerned with methodologies for determining accessibility at an individual and aggregate level, both from the perspective of what the individual can access and of the degree to which many individuals can access a location. Throughout this chapter, however, the authors view accessibility as a time-space phenomenon, both in terms of how accessibility should be conceived and of how it should be reported. In essence, we attempt to take the space-time view of Hägerstrand (1970; 1975) and build from it a framework for defining accessibility in an enhanced way, making that definition operational for large numbers of people and extracting new forms of expression and query from it along the way.

The specific focus of the chapter concentrates on issues surrounding virtual (multi-modal communications) technologies and their influence on accessibility and its definition. This is a complex area where rampant technical innovation and unconstrained speculation on its impact go hand in hand. To allow the chapter to evaluate some ideas in this area as rigorously as possible we have sought to find a context where the issues are relatively simplified, and to draw examples from this domain as we develop our ideas. The domain chosen in this case is University education and the access needs of students. The structured nature of student lives, the growing pressures on their daily schedules and the potential significance of virtual technologies in the delivery of aspects of University learning provide the justification for this.

The chapter develops through six further sections. The next one considers the general nature of accessibility as embodied in the space-time view of Hägerstrand, and how virtual activities can alter the definition of accessibility and the way that altered accessibility reflects altered spatio-temporal constraints. The third describes the basis of current student lives, and reviews a specific implementation of a cellular, three-dimensional model of accessibility and interaction based on physical presence and time-geographic concepts. The fourth addresses issues of aggregation, elaborates practically on this by developing aggregate measures of accessibility from individual student lifelines. Section five discusses how virtual technologies in education may alter the parameters for activity scheduling through total or partial substitution, with significant implications for accessibility patterns. The sixth section works through an example of the possible impact of 'virtualization' of a course using the original cellular model with refined scheduling constraints. The chapter concludes with a brief discussion on the wider application of the concepts and techniques presented.

5.2 Some Reflections on Accessibility in Space-time

The concepts of access and accessibility are used widely within geography and owe their origin to early quantitative and theoretical work in the 1960s (Bunge 1966, Warntz 1967, Garrison 1960, Kissling 1966). In a spatial sense, accessibility is both a specific and aggregate property. As a specific condition it describes individual circumstances, i.e., either whether an individual can access a point, or whether they can access a facility at a point. The basic question is *can A get to XY*? As an aggregate property most working definitions seek to provide a generalized statement of what can be reached. This often involves showing the relative advantage of one location over another across what is usually presented as a smooth surface but in reality is usually a derivative from point values on a network. An accessibility surface often answers the question of *how many Bs can get to place XY, for all XY*?

Accessibility is generally about physical movement and physical access. The simplest spatial accessibility measures use isotropic space and Euclidean distance, but most of the definitions used in research employ a model of space derived from the context of local transport systems and the realities of travel and mobility. Cost and time measures of distance, defined through a transport network model, are the most common definitions of separation used (Weibull 1976; 1980; Pirie 1979). For aggregate descriptions of access, the measure of distance is frequently modified using a non-linear transformation or a threshold distance value for access. When travel time is used as a distance measure in the spatially conceptualized view of access, then mobility and accessibility are clearly twinned concepts (Dijst and Vidakovic 1997). Pooler (1995) and Handy and Niemeier (1997) provide reviews of spatial accessibility measures.

There are a number of reasons for questioning spatial measures of accessibility, some of which relate to their failure to acknowledge non-spatial constraints on access, such as the use of authority to ban entry to places. Another critique is to note that accessibility, from a process perspective, is more fundamentally about whether (or how fully, or at what marginal cost) person A can be at location XY *for a given purpose*. The italicized component often requires a facility constraint to be met (for instance, a golf course to play golf on). It almost always requires a minimum temporal stay, so that the individual query above is rewritten as *can A get to XY for duration t?* The same is true for the aggregate queries. Inevitably, the duration requirement is specific to the task to be undertaken, which is itself constrained within a regime of required, and competing, personal activities. This is in turn mediated by mobility and changes in spaces defined by parameters such as travel time (Janelle 1968, Chapin 1974, Kwan 1998, Miller, Wu, and Ming 1999).

This latter, dynamic conceptualization of space and accessibility has major implications. First, the notion of a generic, purpose-independent accessibility surface or definition of accessibility becomes untenable. Access varies depending on the process being considered (the conditions that a particular activity requires to be met: for instance timing or minimum duration). Second, space and time (whether modified by transport or not) are no longer an independent metric for directly mediating accessibility, since the criteria by which a place is deemed accessible may be predominantly constrained by scheduling options. These are themselves created by the available options for scheduling other activities (which are in turn constrained by different, process-specific time and space issues). In a sense, scheduling is itself a spatial agent in modifying accessibility for the individual, since the ability to interact with others and make use of facilities or services is conditional upon the existence of, or the ability of individuals to create, appropriate sequences of time slots to perform activities. The intermeshing of these demands throughout the day already creates quite distinct diurnal geographies in the city (Janelle, Klinkenberg, and Goodchild 1998).

Third, in terms of virtual technologies, we see a new partitioning between what activities can be undertaken virtually through communications media and what can only be done through physical presence. This is only in part a reclassification of *entire* activities into the virtual sphere (see Chapter 14 by Kwan and Chapter 17 by Occelli). We also see the emergence of a continuum of optional modes for undertaking specific activities, which allow choice-driven substitution of virtual for physical presence, or vice versa. Central to this partitioning and choice are issues related to the economics of delivery and the (actual or perceived) quality of the virtual versus physical experience.

We can expect that these trade-offs vary greatly between different activities and individuals. In either event, virtual technologies exercise two influences on issues of access. On the one hand, virtual technologies generally permit the actor greater flexibility in allowing the re-grouping of activities in a schedule, and so influence the possible schedule choices for the remaining activities. On the other hand, physical access to virtual technologies may not be spatially or individually ubiquitous (see Chapter 9 by Harvey and Macnab and Chapter 10 by Moss and Townsend; and see Salomon 1988). Thus, two geographies of access emerge within any individual's schedule. One is a geography based on the (internally aspatial) activity sequences of web-based contacts (within which time is the significant dimension to juggle), and the other is based on the needs of the physical-presence activities. Of course, these two geographies intermesh, and over time the nature and interaction of both is changing as technology changes access to, and capabilities of, the Web.

This chapter seeks to explore these ideas through considering a particular domain, in this case that of University students. University education is an excellent context in which to consider the application of new technologies. Student life styles are currently heavily structured by scheduled teaching situations such as lectures, laboratories and tutorials. New IT is providing a range of options that offer trade-offs in the way that these traditional learning experiences can be delivered (Forer 1998), This will impact equally on individual lifelines, and city-wide patterns of accessibility.

The next section juxtaposes issues of evolving student lifestyles with a cellular space-time approach to modeling individual accessibility (Forer 1993; 1995; 1997; Huisman, Forer, and Albrecht 1997).

5.3 A Space-time Approach to Modeling Accessibility with Students

The Elements of Current Student Life-Styles

Much of the empiric work on which the examples within this paper are based derives from a study of student life styles at the University of Auckland. The city of Auckland has over one million inhabitants, and, as in many larger cities, students must cope with commuting and parking restrictions to study at a central site, such as the University of Auckland's main campus, which hosts over 20,000 students. They must also juggle with the unallocated time in their day so that they are able to sustain paid work, either in the central area or elsewhere. Further details of these contextual issues can be found in Boswell (1995) and Forer (1998).

Student lifelines are well suited for analysis because it is possible to identify major components of their necessary daily activities from known course choices and their related formal teaching timetables. Having defined the suburb a notional student may live in during term time, we can add to this the location of study and the formal components of their study timetables to define the major marker episodes in their working day. (i.e., specified activities of which the location and starting and ending times are known and largely defined (Lenntørp 1976, Hägerstrand 1982)). It is also possible to use some simplifying assumptions to move from records of location and timetable to the creation of a notional activity schedule for domestic chores, sleeping and study for the whole day, with associated space-time coordinates. This is discussed in Huisman and Forer (1998a). These mandatory episodes can in turn form the basis for analysis of individual accessibility to activities in the remaining periods of discretionary time.

Individual Space-time Paths and the Ability to Access

It is argued elsewhere (Forer 1997, and see Chapter 16 by Hanson) that Hägerstrand's conceptualization of lifelines achieved limited deployment because of the difficulty of making paths and prisms easily operational. These difficulties related to all facets of generating and modeling activity-based data, for both lifelines and prisms. In recent years a number of authors have noted the potential of GIS to cope with two key components in understanding space-time accessibility: the locational data of individuals, and the issues of movement through transport networks (Miller 1991, Mey and ter Heide 1997, Löytönen 1998). However, implementations of systems to model lifelines within the geometry of the space-time aquarium remained relatively few. In 1993 one of the authors developed a framework for implementing lifelines and prisms in a cellular way, using a combination of raster and network methods to populate a three-dimensional matrix with binary values denoting absence or potential presence (Forer 1993; 1995; 1997). This work also proposed defining other actual or potential space-time *objects*, for instance an open library, in the same way within the cellular framework. Figure 5.1 illustrates the kind of representation developed, as well as an example of how object types could be combined to answer spatio-temporal queries through Boolean operations on object masks.

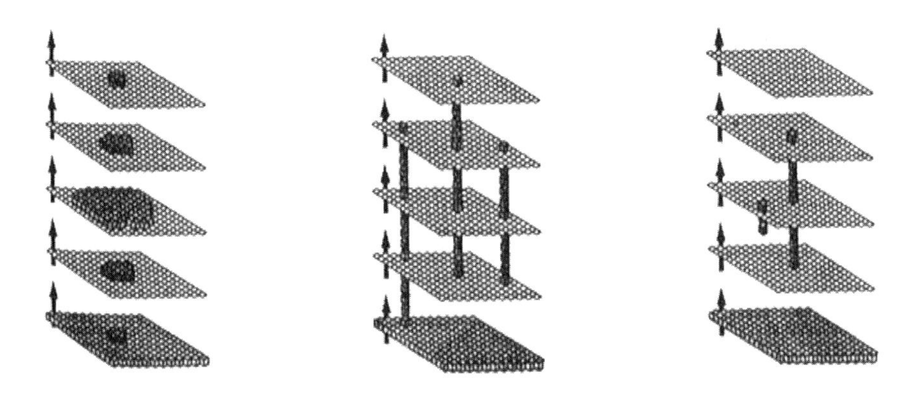

Figure 5.1. An exploded view of three binary masks for a rectangular area over five time periods. The left-hand mask displays an individual's prism. The central mask represents three facilities that are open at different times. The right hand mask is a mask of the other two derived from a simple AND operation, and represents the times that the individual can be at the facilities while they are open. *(Source: Forer 1993).*

A *mask* in this respect is simply a set of identically coded cells in a three-dimensional space-time grid associated with an individual or object over a specified time, usually a day. The basic individual mask is binary, includes both lifelines and prisms, and has cells containing a '1' wherever an individual or object is present or could be present, and otherwise a zero. Figure 5.1 shows two basic masks, and a derived accessibility measure of the library access options in space and time for that individual. The major limitation in deriving such masks remains the ability to move from coping with clearly defined marker events to coping with activity schedules where only a desired activity mix, rather than precise event timing and location, are known.

The construction of prisms is normally based around gaps between known marker events. For the present implementation, standard allocation procedures and network data structures are used, in normal and reverse modes, to identify locations accessible to individuals given a specific travel time from a marker event and current mobility conditions for the individual. In some circumstances raster processing is also used to identify travel time across open spaces, such as large parks. These data are then used to populate a three dimensional binary matrix. The process for static objects such as libraries is much simpler.

A matrix of space-time object types was subsequently proposed (Huisman, Forer, and Albrecht 1997), which has been extended to include virtual events and activities (Table 5.1). Intermittent events are defined by their spatial extent and location, and the periods of time for which they are *accessible*, i.e., their opening or functioning hours. Continuous events represent those objects and events that exist at any point in time (such as the actual or possible location of an individual). In the virtual realm these events and activities represent locations in virtual space (servers or websites) that can be physically (though remotely) accessed within temporal constraints.

Table 5.1. Representing *presence* of objects and events in space-time

	Actual presence	Potential presence	Virtual presence
Intermittent events	Static facilities and amenities	Mobile services and facilities	Online services/ resources
Continuous events	Individual life-lines	Physical action spaces	Virtual action spaces

Source: Modified from Huisman, Forer, and Albrecht (1997)

Overall, it is therefore possible to routinely generate space-time masks in complex real-world situations that can show actual and potential physical presence for an individual or other object, over a specified time. Much of what can then be achieved stems from a threefold extension of the concept of individual binary masks. The first of these is the creation of syntax for the investigation of interaction between individuals and/or objects through the combination of binary masks. The second is the population of the masks with non-binary values, which could represent a number of concepts, such as possible presence in a cell or remaining available time in that cell. A third is to derive various spatial or volumetric outcomes from the masks, which provide a direct statement about accessibility or potential interaction. A map of maximum duration for an activity available at a point would be such an example, and others are suggested below.

The establishment of this rather extended view of Hägerstrand's concept is shown in diagrammatic form in Figure 5.2. The main strengths of this approach include an ability to generate space-time measures of accessibility, to model impacts of shifts in marker events, and to cope with some of the issues of virtualized access, at least conceptually.

5.4 Generalized Volumes and What or Where is Accessible

Individual accessibility *to* a point is easily modeled by individual masks, which can also reveal aspects such as length of access, timing of access, and existence of multiple access opportunities. However, the accessibility of any space-time point to a group of individuals, or the likely number of members of a group at a space-time point, is a matter of aggregate behavior. These aggregate patterns, too, can be derived from individual masks, through their redefinition and combination.

In this section, we consider accessibility as involving the ability of an individual to be at a location for a minimum time needed to undertake an activity. Implicit in this is that the individual has the discretion to move to certain points. Aggregate accessibility is consequently derived from a combination of numerous individual lifelines and prisms. The simplest question we can derive involves identifying space-time locations where the most people can simultaneously meet, however briefly. Simply adding together $\boldsymbol{N}$ binary masks representing the lifelines of $\boldsymbol{N}$ individuals will provide an aquarium in which cell values reflect accessibility, in these terms. The values will potentially range from N (all present) to $\boldsymbol{0}$ (inaccessible to all). If the requirement is to know how many people can visit a place for a set duration, say 30 minutes, then individual masks can be trimmed down to include only prism volumes that extend through cells that represent duration of at least 30 minutes, and then be added together.

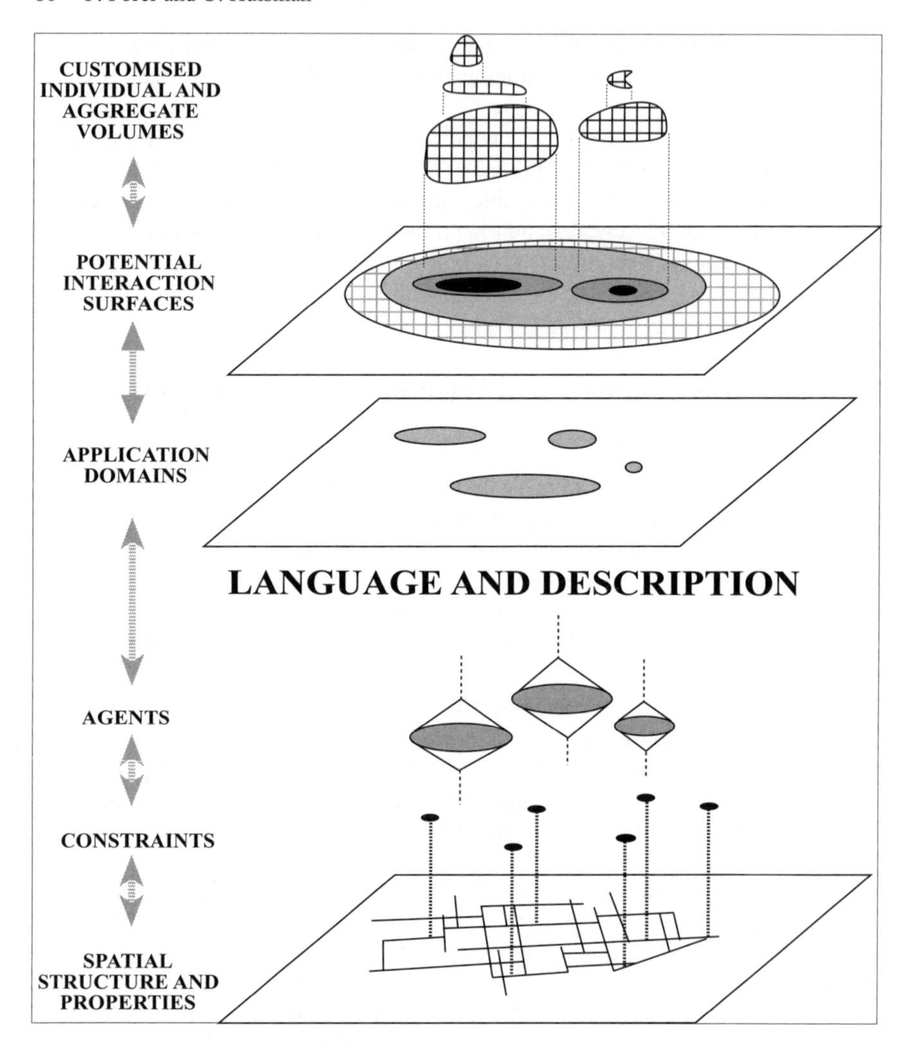

Figure 5.2. Extending time-geographic concepts. *Source: Huisman and Forer (1998a)*

If what is required is a measure of likely interaction, then the contents of the prism cell can be weighted to reflect some model of likely presence based on the spatial behavior of an individual given a particular window of available time. Once again, the masks can be combined to provide an aggregate picture of accessibility. The most powerful artifacts to emerge from this are the resultant space-time volumes, which show details of both spatial and temporal variations in accessibility or likely interaction.

We have discussed elsewhere a number of derivatives from single and aggregate masks (Huisman and Forer 1998a). Figure 5.3 shows one example created from analysis of 100 students. This surface shows the number of discretionary people/minutes that can be spent at a location by the target group between the hours of 12 noon and 2:00 p.m. of a given day. Essentially it is one measure of the accessibility of each point to the student body, or of possible student interaction over space during this time period.

In general, then, masks can be derived to show both actual presence (lifelines) and potential presence (prisms) in a cellular form. These can be combined in various ways to answer queries and to yield aggregate patterns of accessibility. The main required ingredient is an ability to specify the space-time location of marker events. The next section briefly discusses the challenge of applying this in an operational context where marker events may need to be derived, and leads on to consider how virtual meetings could be modeled and integrated into models of accessibility.

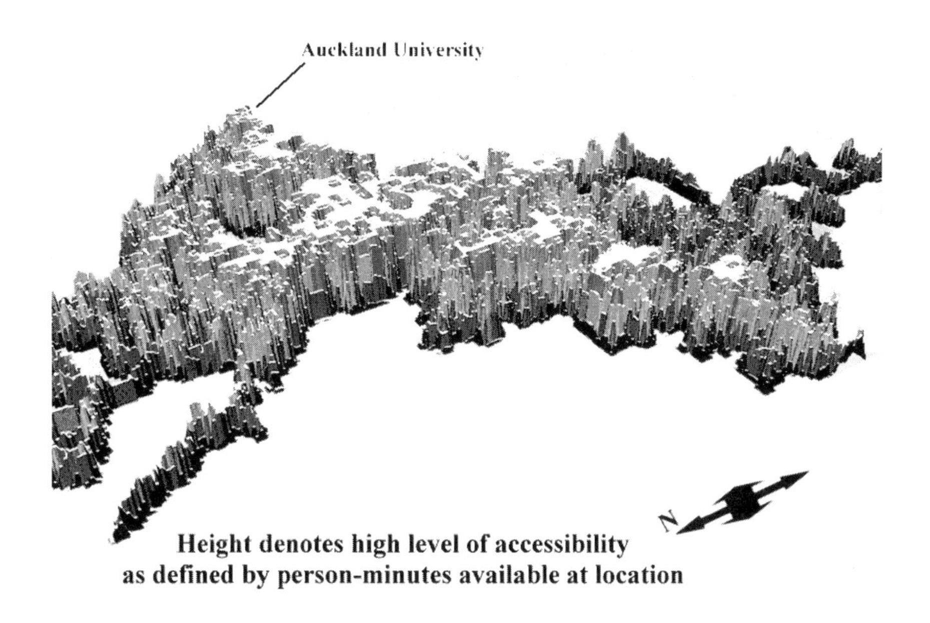

Figure 5.3. Potential access and interaction in Auckland. Accessibility surface showing the number of person/minutes that can be spent at a given location between the hours of 12 noon and 2 p.m. by the sample of 100 students. The height of each grid cell is proportional to the number of person-minutes for which it can be occupied.

5.5 The Changing Nature of Student Life-Styles

University timetables do show some evidence of greater flexibility through the use of short courses and self-paced methods, such as Open Learning. However, for most students a fairly rigid timetable based on a campus location still dictates their days, as noted above. The growing need to work for money to pay tuition fees places greater pressure on finding a way, and place, to undertake paid work (Boswell 1995, George 1998). In this respect students face a challenge in scheduling that is more structured than for many groups, but in generic terms their case is typical of a large number of different groups in society who seek to restructure lifestyles in response to new pressures and opportunities.

However, students may see a greater immediate spatial effect from Cyber or Virtual technologies than most, because these technologies offer very significant possibilities to re-schedule their days. Amongst the benefits offered by Virtual or Cyber technologies in education are the ability to substitute virtual and physical experiences at will, and the ability to divorce service delivery from the need to be at a specific location. Examples would include the optional delivery of video versions of lectures using on-demand video servers, and web-based tutorials. These can be independent of a campus location, as can be the delivery of digitally based labs and other Intranet and library resources that can employ virtual Intranets. The nature of these technologies in education, and their impact on learning options, is widely debated (Peters and Roberts 1998), but one clear implication of their successful implementation is that learning options for mainstream students become more flexible. One possible outcome of these technologies is not a wholesale shift into a virtual or distance learning mode, but rather a modification of mainstream tuition to embody aspects of both social learning and distance methods in so called *high tech/high touch* scenarios (Forer 1997, 1998). This implies a trade-off between virtual and real learning environments, a situation that may represent the ways in which virtual services or activities come to integrate with traditional processes in the wider economy.

Current physical access to University results in a student day being composed of starting and ending times dictated by formal timetable requirements, coupled with a series of continuous commitments to lectures and laboratories. Discretionary time between these is often filled with local *informal* activities, such as library visits. Timetabling options (in terms of lecture streams) are limited, and hence it is not uncommon for the timetable to be unhelpful, for instance by containing only two lectures, one at either end of the day. Anecdotal evidence of students choosing courses so as to minimize such problems is quite widespread.

A more virtual campus would provide flexible access to both formal and discretionary learning resources. Faculty would probably respond to the technology by adjusting some aspects of timetabling, and students would react by scheduling activities differently. For them, virtual technologies could decouple many aspects of learning delivery from a need to be on campus, and so free up their control over daily schedules. Library access for many functions could be undertaken as effectively at home, and could be undertaken when desired. Certain lectures at key times could be viewed from video downloaded to home at the student's discretion. In short, students could exercise a choice to substitute virtual for real presence, and staff might reduce and consolidate the periods of mandatory real presence. In general, the key times when substitution occurred might well be when physical attendance was problematic because of the need to keep a block of time clear for other activities, or when a day contained large amounts of dead time. Students would likely seek efficiencies in time use by having fewer but larger blocks of time on campus. *Efficiencies* in this context could mean simply saving unused time between an early and late lecture, but more likely might mean the ability to accommodate a major commitment such as work along with substantial study.

Given this kind of change, if we thought in traditional spatial accessibility terms, the changes in the flexibility of learning delivery would show little effect in terms of accessibility measures, individual or aggregate. However, in a space-time definition we would expect that the changing life styles would result in a changing balance between the formal and discretionary parts of the day, and that this might result in major shifts in discretionary time and spatial opportunity. The next section details some basic investigations in this area.

5.6 Varying Virtuality and Changing Accessibility

Extreme scenarios of real/virtual substitution are unlikely to prove either popular or sustainable for the population at large. Even for mundane actions, such as shopping it is likely that the best cyber-market experience that could be contrived will fail to provide the social, tactile and environmental stimuli of the best shopping experiences imaginable. In almost every area of activity, the most likely impact of virtualization will be to see virtual services emerge that cater for minority groups or act to complement the experience of the majority. In every sphere of activity this will modify the frequency and nature of physical trips, as well as lead to changes in the nature of the physical process (as it comes to embody certain virtual aspects to it). In shopping, the nature of Malls will change to emphasize the competitive edge of *real* shopping, but malls and issues of their accessibility will remain. This simply restates more generally the specific impacts that are expected for virtual technologies in education.

Given this eventuality, two things become clear. One is that people will recast their time use in new ways, taking into account the ability to flexibly trade and aggregate different activities, particularly the ability to clump many virtual substitute activities, which are largely independent of space and time constraints. The 'real' component of their activity schedules will continue to dominate their movement decisions, and so space-time constraints will not disappear as a factor in determining accessibility. However, the nature of the scheduling will be harder to predict, as the relationship between doing an activity and the need for a physical presence will be determined by individual decisions on the trade-offs of using a virtual option.

If we focus back on accessibility as we have treated it in this paper, then defining space-time accessibility becomes much harder because the marker events that define discretionary periods for access are more difficult to formalize, because at an individual level the real or virtual option could be selected. The real option is tightly location constrained, the latter is either loosely or minimally constrained. Furthermore, the link between a location and a function may become much less deterministic. Real activities frequently rely on the nature of the place that hosts them; virtual activities largely import their own cyberplace to wherever they are being undertaken. Presence at home may mean university work is being done, for instance, as may presence in a Cyber-café in a local mall. In general, we can expect that even a space-time accessibility measure will become a more generalized representation of aggregate access, because of the mutant nature of the possible virtual/real substitutions. We can also expect that to deploy the space-time framework to understand the operation of virtual/real systems will require a much greater understanding of the activity processes involved, both in isolation and in the specific contexts of the individuals.

In spite of these caveats, it is worth speculating on how we might view the impact of a restricted version of virtualized education as discussed previously. An implementation of an accessibility surface for one hundred student lifelines is described below. Issues of access to cyberspace are addressed in a second, shorter section. The section also looks at how some of these substitutions could be modeled in the cellular space-time model described earlier. It identifies issues in including virtualized activities in the model, notes the centrality of scheduling in generating new activity patterns in these circumstances and demonstrates the impact of virtualizing courses under simple assumptions of schedule choices.

Virtual Lectures and Relaxed Scheduling

This analysis works on the known timetables of one hundred notional commerce students in 1996. Their lifelines were developed, and general accessibility surfaces deriving from their discretionary time were produced. This process is described in Huisman and Forer (1998a). In the analysis below, the following assumptions have been made in respect of modified learning delivery using virtual technologies:

- Lecture hours are halved due to greater use of Web resources.
- Activities, whether virtual or actual, are still required to be performed within the same 12-hour period (i.e., the same day).
- Students still have access to the same range of transport alternatives as before, with the addition of a virtual resource.

- Each student in the sample is assumed to have access to a terminal, irrespective of financial costs.

The students have then been allocated new schedules based on these changes and their lifelines and prisms have been calculated. From this, new aggregate surfaces of student accessibility have been derived (Figure 5.4). The left column shows the original access surface, the right column the surfaces resulting from physical/virtual substitution.

The analysis largely shows that discretionary time is increased, which should free up student choice considerably and allow easier access to paid work or other activities. It also shows the spatial pattern of this, and demonstrates a simple way to identify changes that might spring from a given change in learning delivery. The basic idea could be deployed with more sophistication, for instance by stochastically modeling virtual/real substitutions for particular activities. This could be traced through their resultant impact on the overall individual mask and on the aggregation of masks.

Access to Terminals

A significant tenet of activity modeling in the past has been a relatively tight mapping of activity to location. In many cases the mapping has been one-to-one: for instance University classes are only at the University campus. In others a one-to-many relationship has held, as with shoe shopping being possible at several locations. Occasionally one location has supported many kinds of activities, but this is sometimes due to a scale issue, as when a Mall is treated as a single entity but supports multiple kinds of shopping activity.

Virtual technologies are unlikely ever to be ubiquitously available, in the sense of there being no restriction on where high-speed links could be made, or where high quality screens were available. In the short term (a few techno-nomads aside), getting on to the Web for virtual services will remain tied to specific nodes. While these will proliferate very quickly, access to them will remain restricted by a range of constraints, notably the fact that most will be within private property and subject to log-on restrictions. In the short term the nodes into cyber-space are likely to be quite specific. For individuals these could be home links, links from educational institutions, links from cyber-cafes or other pay-for-access locations, or links from work. Some of these locations will be more or less suitable for certain kinds of virtual links. Surfing or e-mailing a help desk could be done from a cyber-café, but undertaking a demanding laboratory exercise or downloading video might require a fast link and a quiet work area. From a modeling point of view, locating the ports to cyber-space would be simple, and these possible activity points could be used as part of a student's lifeline and mask. Once again, however, the specific nature of particular ports brings us back to a detailed consideration of the student's constraints and activity needs.

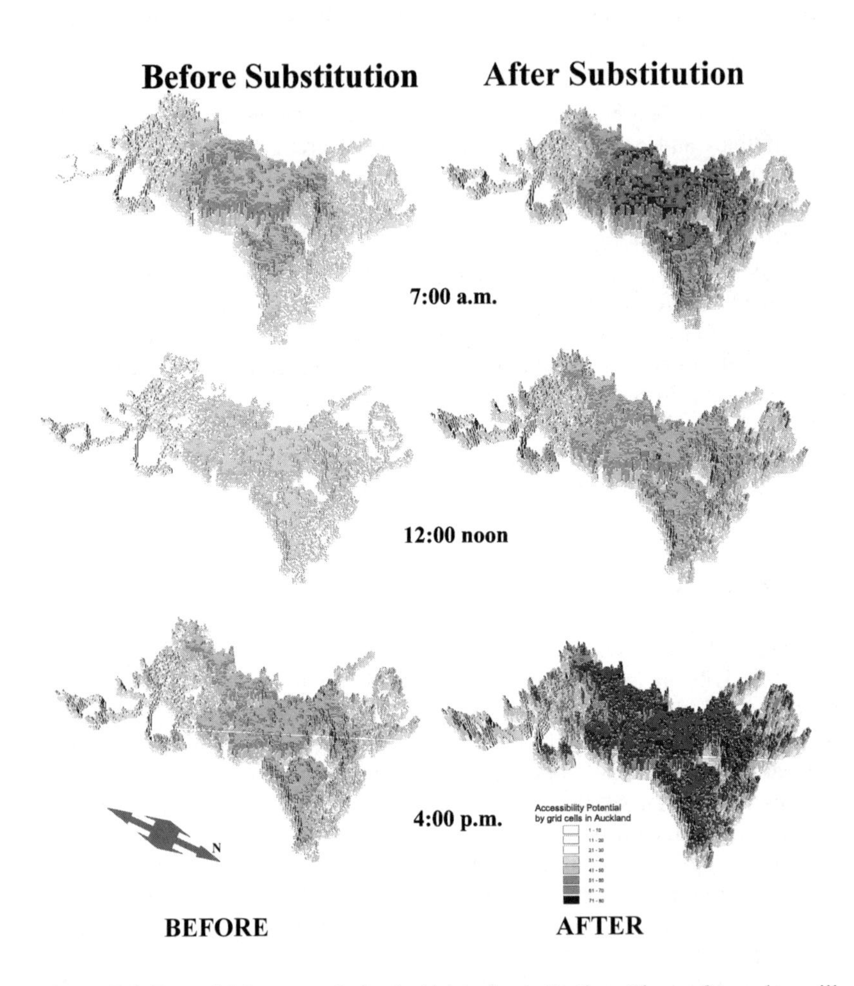

Figure 5.4. Potential impacts of physical/virtual substitution. The surfaces above illustrate the impact of physical/virtual substitution on accessibility patterns at different times of the day. Each surface illustrates a ten-minute time interval. These figures cover the same extent as Figure 5.3. Once again, height represents the possible duration of presence (physical access). Lightness is used here to illustrate the results better. Physical access values above range from 0 (flat) to 1000 person/minutes (highest) for the sample of 100 students.

An interesting side issue for universities is whether a virtual component of learning delivery would encourage more distributed delivery through secondary campus sites or cyber-ports. At the heart of the impact of virtual technologies is an unresolved issue of how and when people need to physically congregate to pursue learning to best effect. That issue is both pedagogic and practical, in the sense that economics often dictates the provision of centralized specialist facilities. The possibility of new arrangements of campus cells in an urban area, which better meet student life-style needs though a real/virtual blend of approaches, would be an interesting situation to model for real and physical time-budgets and accessibility.

5.7 Summary: Alternative Concepts and Realities

The discussion above has sought to explore the properties of accessibility that can be derived from Hägerstrand's dimensional view of human activity. Such a view is based on atomistic assumptions about an individual's ability to be at a point (or a point's ability to be visited), and an aggregation of individual properties into systematic patterns. It yields a more specific view of accessibility that is inevitably more closely linked to process and to the scheduling and sequencing of activities. This is at the same time more powerful but less general than the traditional spatial formulations of accessibility.

One advantage of this approach is that it implicitly embodies questions that arise when cyber-activities come into consideration, in that, in general, such activities are likely to be embedded in a wider process where the virtual/real trade-off is decided. Many activities, and certainly education, will retain a mix of virtual and real, where the real component is significant and will retain the space-time constraints of physical accessibility. The issue in determining how virtual/real substitution takes place is one that is built around scheduling of the physical activities within a much less bounded set of virtual activities.

In one respect, virtualizing activities throws the entire issue of accessibility open, in that the one spatial constraint remains the constraint on accessing a terminal. In another it simply redefines the scheduling task, which as we noted is in fact a key aspect of defining the *marker* episodes and thus the residual areas of discretionary use of time. Aspects of process may define what virtual activities are best aggregated and best undertaken, and these will then need to be integrated into a wider context of overall activities. When the necessary virtual and physical *markers* have been established, a rather different set of accessibility measurements will emerge for the student's discretionary time.

For universities, and for their communities, these changes may have significant impacts on the geography of tertiary learning, although that is not the key issue under discussion here. The indications are that the ideal form for the university, and the likely interactions with its hinterland that are possible, will change significantly as real and virtual accessibility adjusts to new technologies. More widely, the same opportunities to interchange the virtual and real experience will be present in many other spatial processes, including shopping, working, tourism and socializing. In each case we can expect that physical constraints and access will remain important, but that there will be a greater emphasis on the nature of the specific processes as determinants of substitution and the process of activity choice and scheduling. As we know more about different dimensions of cyber-geographies, perhaps through concepts such as Heikkila's *fuzzy club* (see Chapter 6), we may be able to develop better ideas on modeling the two separate geographies (virtual and real) at a more sophisticated level, and eventually achieve a more satisfying integration of the real and virtual components of accessibility.

References

Boswell, S. 1995. *The Splintering of Student Geographies: Restructuring of the Tertiary Education Sector in New Zealand.* Unpublished M.A. Thesis. University of Auckland, New Zealand.

Bunge, W. 1966. Theoretical geography (2nd Ed.). *Lund Studies in Geography* Series C #1. Lund: Geerlup.

Chapin, F.S. 1974. *Human Activity Patterns in the City: What People Do in Time and Space.* New York: John Wiley.

Dijst, M., and Vidakovic, V. 1997. Individual action space in the city. In Ettema, D. and Timmermans, H. (eds.). *Activity Based Approaches to Travel Analysis.* London: Elsevier Press, 117-34.

Egenhofer, M.J. and Golledge, R.G. (eds.) 1997. *Spatial and Temporal Reasoning in Geographic Information Systems.* New York: Oxford University Press.

Forer, P.C. 1993. Geometric approaches to the nexus of time, space, and microprocess: Implementing a practical model for mundane socio-spatial systems. Paper presented to *Initiative 21, Symposium.* Lake Arrowhead, U.S.A.: September.

Forer, P.C. 1995. An implementation of a discrete 3-d space time model of urban accessibility. *Proceedings of the 1993 Conference of the New Zealand Geographical Society.* Wellington: NZGS, 51-7.

Forer, P.C. 1997. Geometric approaches to the nexus of time, space, and microprocess: Implementing a practical model for mundane socio-spatial systems. In Egenhofer, M.J. and Golledge, R.G. (eds.) *Spatial and Temporal Reasoning in Geographic Information Systems.* New York: Oxford University Press.

Forer, P.C. 1998. Cyberia and the premature death of distance. In Peters, M.A. (ed.) *Virtual Technologies in Education.* Wellington: Dunmore Press, 146-70.

Garrison, W.L. 1960. Connectivity of the inter-state highway system. *Papers of the Regional Science Association* 10:121-37.

George, G. 1997. Student misery should depress us all. *New Zealand Herald,* 8 October *A2.*

Hägerstrand, T. 1970. What about people in regional science? *Papers of the Regional Science Association* 24:7-21.

Hägerstrand, T. 1975. Space, time and human conditions. In Karlqvist A., Lundqvist L and Snickars F. (eds.) *Dynamic Allocation of Urban Space.* Farnborough: Saxon House, 3-14.

Hägerstrand, T. 1982. Diorama, path and project. *Tijdschrift voor Economische en Sociale Geografie* 73:323-39.

Handy, S.L., and Niemeier, D.A.. 1997. Measuring accessibility: an exploration of issues and alternatives. *Environment and Planning A* 29:1175-94.

Huisman, O. and Forer, P.C. 1998a. Towards a geometric framework for modeling space-time opportunities and interaction potential. Paper presented at the *IGU'98 Con-*

ference (Commission on Modeling Geographical Systems) Lisbon, Portugal: 28-29 August.

Huisman, O. and Forer, P.C. 1998b. Computational agents and urban life-spaces: A preliminary investigation of the time-geography of student life-styles. *Proceedings 3rd International GeoComputation Conference.* Bristol, UK: 17-19 September. (CD-ROM/Hypertext 23 pages).

Huisman, O., Forer, P.C., and Albrecht, J.A. 1997. A geometric model of urban accessibility. *Proceedings of the 25th International AURISA Conference.* Christchurch, New Zealand: 17-21 November (CD-ROM/ Hypertext 9 pages).

Janelle, D.G. 1969. Spatial reorganization: a model and concept. *Annals of the Association of American Geographers* 58:348-64.

Janelle, D.G. 1986. Metropolitan expansion and the communications-transportation trade-off. In Hanson, S. (ed.) *The Geography of Urban Transportation*, Guilford Press, 357-87.

Janelle, D.G., Klinkenberg, B., and Goodchild, M.F. 1998, The temporal ordering of urban space and daily activity patterns for population role groups. *Geographical Systems* 5:117-37.

Kissling, C.C. 1966. *Transportation Networks: Accessibility and Urban Functions, Empirical and Theoretical Analysis.* Doctoral Thesis in Geography, McGill University.

Kwan, M. 1998. Space-time and integral measures of individual accessibility: A comparative analysis using a point-based framework. *Geographical Analysis* 30(3): 191-216.

Lenntørp, B. 1976. Paths in space-time environments: A time-geographic study of the movement possibilities of individuals, *Lund Studies in Geography,* Series B #44. Lund: Geerlup.

Löytönen, M. 1998. GIS, time geography and health. In Gatrell, A., and Löytönen, M. (eds.) *GIS and Health.* (GISDATA VI). London: Taylor and Francis.

Mey, M.G., and ter Heide, H. 1997. Towards spatiotemporal planning: Practicable analysis of day-to-day paths through space and time. *Environment and Planning B* 24:709-23.

Miller, H.J. 1991. Modeling accessibility using space-time prism concepts within geographical information systems. *International Journal of Geographical Information Systems* 5:287-301.

Miller, H.J., Wu, Y., and Ming, C.1999. GIS-based dynamic traffic congestion modeling to support time-critical logistics. *Proceedings of the Hawaii International Conference on Systems Science*, Maui, Hawaii: January.

Peters, M.A. and Roberts, P. (eds.) 1998. *Virtual Technologies in Education.* Wellington: Dunmore Press.

Pirie, G.H. 1979. Measuring accessibility: A review and proposal. *Environment and Planning A* 11:299-312.

Pooler, J. 1995. The use of spatial separation in the measurement of transportation accessibility. *Transportation Research A* 29(6):421-27.

Salomon, I. 1988. Geographical variations in telecommunications systems: The implications for location of activities. *Transportation* 14(3):311-27.

Warntz, W. 1967. Global science and the tyranny of space. *Papers of the Regional Science Association* 17:7-21.

Weibull, J.W. 1976. An axiomatic approach to the measurement of accessibility. *Regional Science and Urban Economics* 6(4):357-79.

Weibull, J.W. 1980. On the numerical measurement of accessibility. *Environment and Planning A* 12:53-67.

6 The Fuzzy Logic of Accessibility

Eric J. Heikkila

School of Policy, Planning and Development, University of Southern California
Los Angeles CA 90098-0626, USA. Email: heikkila@usc.edu

6.1 Introduction

Accessibility is a measure of association linking people (or places) with some target destination or node. In its most general formulation, accessibility is not limited to a strictly geographic interpretation. Thus, instead of spatial proximity, accessibility may represent the ease with which one may gain entry to certain social or communications networks. The rapid advent of emerging information technologies increases the imperative task for geographers and other social scientists to develop models that are sufficiently robust to accommodate the concept of accessibility in its physical, social, and technological manifestations. To that end, this paper examines the potential to combine the fuzzy logic of Zadeh (1965) and Kosko (1992) with the club theory of Tiebout (1956) and Buchanan (1965) to model accessibility in both geographic and non-geographic contexts.

Club theory argues that individuals voluntarily form groups or clusters to derive tangible or intangible benefits through mutual association. The Tiebout (1956) model is an early example of this, where individuals voluntarily *vote with their feet* to join clubs called municipalities that in turn provide services in the form of local public goods to their members. Access to Tiebout's clubs was by virtue of a physical presence within the geographical boundaries of the local service area. Club theory extends the Tiebout model so that club members may derive benefits not only from physical services but also from the characteristics of fellow club members, as is the case in professional associations (Cornes and Sandler, 1986). This club theory extension is consistent with but not confined to the implied geographic setting of the original Tiebout model. More recently, using factor analysis and analysis of variance techniques, Heikkila (1996) has demonstrated empirically that municipalities within complex urban settings, such as Los Angeles, can be usefully modeled as *clubs*.

This chapter pushes this idea further, by modeling places (either geographic or not) as *fuzzy clubs* based on the concept of fuzzy sets, in which membership is a matter of degree rather than being restricted to a binary form ('is a member' vs. 'is not a member'). From this perspective, membership in a fuzzy club supercedes the concept of accessibility to a place – where greater accessibility corresponds to a higher degree of membership. Recasting the notion of accessibility in this manner allows us to tap into the rapidly growing literature on fuzzy logic, the logic of reasoning with fuzzy sets. This paper is an initial exploration of the implications and potential of this approach.

The next section reviews fuzzy set theory and fuzzy logic. Several ideas appear to be of fundamental importance in this regard. One is that fuzzy sets do not obey the classical 'laws' of non-contradiction and of the excludable middle that apply to non-fuzzy sets. This leads to characterizations of entropy that appear to have useful potential in the context of urban systems and other systems of accessibility. The fuzzy set approach also offers a graphical representation of fuzzy clubs that is highly intuitive and insightful. In the following section, these ideas are translated and applied more specifically to modeling the impact of information technology on systems of accessibility. Three fundamental dichotomies are identified: (1) participation-withdrawal, (2) entropy-order, and (3) strict-loose partitioning of clubs. The penultimate section of the paper sets out a strategy for measuring these phenomena empirically.

6.2 Review of Fuzzy Set Theory

Fuzzy Sets and Fuzzy Operators

In classical set theory all sets A are *crisply* defined, so that the membership in A of any element x_i of some referential or universal set X is unambiguously defined with respect to some bivalent membership function $m_A(x)$. Each element x_i either *is* or *is not* a member of the crisp set A:

$$\forall\ x_i \in X, \qquad x_i \in A \ \text{ iff } m_A(x) = 1 \qquad (6.1)$$
$$x_i \notin A \ \text{ iff } m_A(x) = 0$$

So the membership function in the classical case takes on one of two values, zero or one.

$$m_A(x): X \rightarrow \{0,1\} \qquad (6.2)$$

Lotfi A. Zadeh (1965) extended the membership function $m_A(x)$ to map from X to the entire unit interval.

$$m_A(x): X \rightarrow [0,1] \qquad (6.3)$$

The resulting sets are 'fuzzy' in the sense that the classical *yes-no* membership dichotomy now becomes a question of degree or extent. In graphical form, the extension from crispy to fuzzy sets is represented in Figure 6.1, where in this example the reference set is a real line representation of geographical space. Here, the urban-rural dichotomy is transformed into a fuzzier notion of urbanity, where the height of the membership function determines the extent to which a given location x_i is a member of the set of all urban locations. In a similar fashion, fuzzy sets might describe locations that are 'close to work' or 'located in good school districts'. Neither crispy nor fuzzy sets need be single peaked in general.

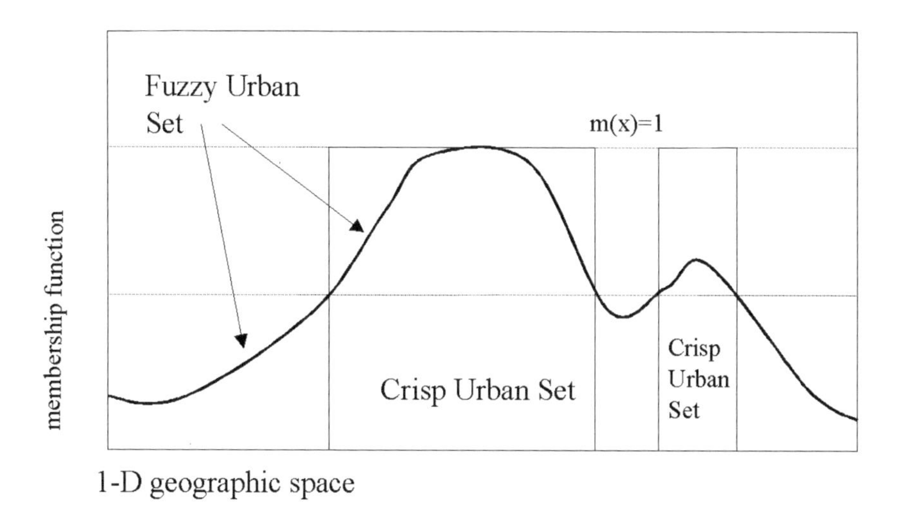

Figure 6.1. Crisp versus fuzzy urban sets.

Three important fuzzy-set operations proposed by Zadeh (1965) in the context of fuzzy set theory and by Lukasiewicz[1] in the 1920s in the context of continuous or fuzzy logics are fuzzy-set intersection (by pairwise minimum), union (by pairwise maximum) and complementarity (by order reversal). In symbolic form we have

$$m_{A\cap B}(x) = \min [m_A(x), m_B(x)] \tag{6.4}$$

$$m_{A\cup B}(x) = \max [m_A(x), m_B(x)] \tag{6.5}$$

$$m_A^{\ c} = 1 - m_A \tag{6.6}$$

The intersection and union of two fuzzy sets A and B are depicted in Figure 6.2.

[1] According to Kosko (1992), the attribution to Lukasiewicz is given by Rescher (1969).

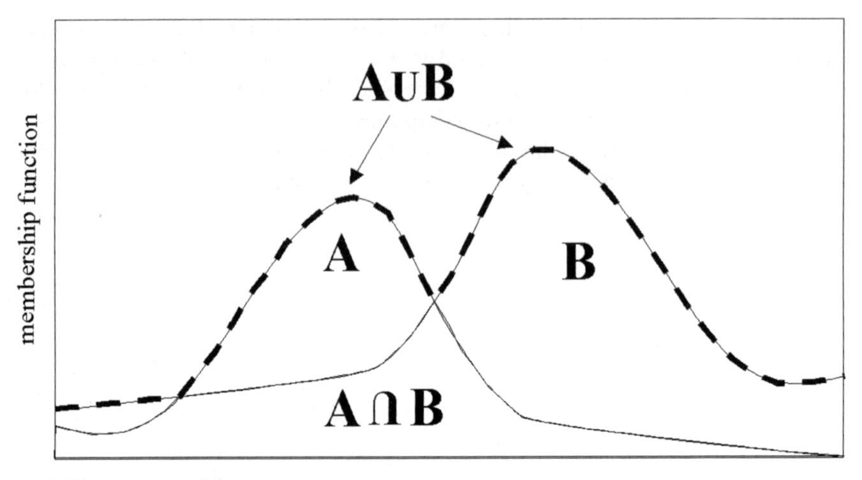

Figure 6.2. Intersection and union of fuzzy sets.

The fuzzy set paradigm abolishes two 'laws' of classical set theory. These are the *law of noncontradiction*,

$$A \cap A^c = \varnothing \tag{6.7}$$

which forbids any overlap between a set A and its complement A^c, and the *law of the excludable middle*,

$$A \cup A^c = X \tag{6.8}$$

which states that any set A and its complement A^c jointly comprise the universal or reference set X. In fuzzy set theory, these classical laws *only* hold for the extreme case that $m_A(x) \in \{0,1\}\ \forall x \in X$, that is, where the set in question is a crisp one.

To see that this is so, consider first the law of noncontradiction. By definition, $m_A{}^c = 1 - m_A$, so

$$m_{A \cap Ac}(x) = \min\,[m_A, (1\text{-}m_A)] \tag{6.9}$$

By definition, the measure of elementhood for the null set is zero. That is, $m_\varnothing(x_i) = 0\ \ \forall x \in X$. Thus, the law of noncontradiction can only hold if

$$\min [m_A, (1\text{-}m_A)] = 0 \quad \forall x \in X \tag{6.10}$$

This in turn can hold only if $m_A(x) \in \{0,1\}$ $\forall x \in X$, which corresponds to the classical definition of a (crisp) set. A similar line of reasoning applies to the law of the excluded middle. In this case we need

$$m_{A \cap Ac}(x) = \max [m_A, (1\text{-}m_A)] = 1 \quad \forall x \in X \tag{6.11}$$

This condition, too, can only hold if $m_A(x) \in \{0,1\}$ $\forall x \in X$, and so, as Kosko (1992) asserts, fuzziness begins where classical bivalent set theory leaves off.

Entropy and Subsethood

It is useful to introduce two more concepts pertaining to fuzzy sets, entropy and subsethood, before turning to issues of accessibility. Entropy measures the uncertainty or ambiguity of a system or message. The uncertainty in question need not be probabilistic; it may be fuzzy in nature. In the context of fuzzy sets, entropy is a measure of the fuzziness of a set, where crisp sets have an entropy measure of zero, and where the fuzziest set has an entropy measure of unity. Kosko (1992) answers the question, 'How fuzzy is a fuzzy set?' with the following entropy measure:

$$E(A) = M(A \cap A^c) / M(A \cup A^c) \tag{6.12}$$

where M(.) is a measure of the cardinality or size of a set:

$$M(A) = \Sigma_i \| m_A(x_i) \| \tag{6.13}$$

The entropy measure E(A) is given as the ratio of the cardinalities of the overlap (intersection) and underlap (union) of the set A with its complement A^c. From the preceding discussion regarding the laws of non-contradiction and of the excludible middle, we know that $M(A \cap A^c) = 0$ and $M(A \cup A^c) = 1$ for crispy sets, and so crispy sets have zero entropy: $E(A) = 0/1 = 0$. In contrast, maximum fuzziness occurs for sets with $m_A(x) = 0.5$ $\forall x \in X$, for in that case $E(A) = 0.5/0.5 = 1$. Most sets will lie somewhere between these two extremes.

Subsethood, S(A, B) measures the extent or degree to which one set, A, is a subset of another, B. Kosko (1992) operationalizes subsethood as:

$$S(A,B) = M(A \cap B) / M(A) \tag{6.14}$$

Note that in fuzzy set theory, it is possible for a set A to be a subset (to some degree) of another set B that is wholly contained within it, for in that case

$$M(A \cap B) = M(B) \rightarrow 1 > S(A,B) = M(B) / M(A) > 0 \tag{6.15}$$

6.3 Accessibility and Fuzzy Sets

Accessibility and Membership in Fuzzy Clubs

Club theory, in the tradition of Buchanan (1965) and Tiebout (1956), examines the economic rationale by which individuals voluntarily form groups or clusters in order to derive tangible or intangible benefits through mutual association. The club theory literature, which is summarized well by Cornes and Sandler (1986), is relevant to the formation of social organizations as diverse as municipalities, health maintenance organizations, and country clubs. The club metaphor is extended very easily to issues of membership, where the strength or degree of an individual's membership in, or association with, a club may vary according to circumstance. This notion lends itself very naturally to the formulation of membership in *fuzzy clubs*:

$$m_A(x) = \text{degree}\ (x \in A)\ \forall x \in X \tag{6.16}$$

where $m_A(x)$ measures the strength of association that any individual person or place x has with some club A for any individual x belonging to some referential or universal set X.

Accessibility is a kind of association, where the more accessible a person is to a place (or, more generally, to a node) the stronger is the implied association. As such, accessibility can be modeled using the same fuzzy set theoretical approach described above. Viewed from this perspective, accessibility is merely one manifestation of the more general notion of association. In a geographic sense, accessibility or *proximity* is an essential measure of the strength or degree of an association. In a transactional sense, *interaction* is an alternative measure of association. In a socio-political sense, *influence* is yet another. Taken together, these ideas suggest that traditional notions of spatial accessibility are misplaced. Geographic location within an urban area, for example, appears to be secondary to considerations of 'fuzzy club membership' where those clubs may or may not be geographically defined and where club membership need not be binary.

To develop this idea a bit further, we borrow again from Kosko (1992) to develop a geometric representation of fuzzy clubs. Consider the reference set $X = \{x_1, x_2, \dots x_n\}$. Its power set, denoted by 2^X, contains all crisp subsets of X. For example, in the two-dimensional case (n=2) which is depicted on the left-hand side of Figure 6.3, $2^X = \{\varnothing, \{x_1\}, \{x_2\}, X\}$. These elements of 2^X (which are themselves sets) constitute the vertices of an n-dimensional hypercube such as the one in Figure 6.3. This is the fuzzy power set of X. Using this representation, fuzzy *sets* can be depicted as *points* within a *cube.*[2]

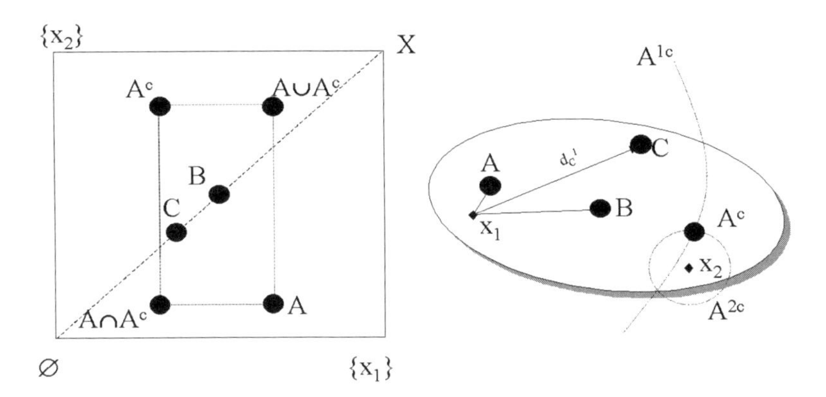

Figure 6.3. Fuzzy power sets derived from geographic space.

Now consider the stylized map of an urban area on the right-hand side of Figure 6.3 in which two locations x_1 and x_2 are situated with reference to three urban sub-centers or nodes A, B and C; and where the distance from x_1 to A, for example, is denoted by d_A^1. Membership of x_1 in the fuzzy club denoted by A increases monotonically with decreases in d_A^1, that is,

$$m_A(x_1) = f(d_A^1) \quad \text{where } f'(.) < 0 \tag{6.17}$$

For reasons that will become apparent below, it is useful and intuitively reasonable to impose the following boundary conditions on f(.):

$$f(0) = 1 \tag{6.18}$$

$$\lim_{x \to \infty} f(d) = 0 \tag{6.19}$$

which (when taken in conjunction with the strict monotonicity in equation 6.17) state that full membership occurs only when there is no distance separating a member from the center, and that any finite distance yields a positive degree of membership, however small. These conditions are satisfied by any function with the familiar negative exponential form

[2] We use the terms 'cube' and 'hypercube' interchangeably.

$$f(d) = \exp(-\alpha d) \qquad \text{where } \alpha > 0 \tag{6.20}$$

In Figure 6.3 the fuzzy set or fuzzy club A is represented as a point in the fuzzy power set of X, where its location within the cube indicates the degree of membership of each element of X in A. On the right-hand side of Figure 6.3, x_1 is located much closer to A than is x_2, and so its degree of membership in the fuzzy club A is higher. In contrast, the fuzzy club B is equally accessible from both x_1 and x_2, and so it falls on the 45°-line, as does fuzzy club C, although the membership levels of x_1 and x_2 in fuzzy club C are less than they are for B, reflecting the fact that C is less accessible to them than is B.

Three other fuzzy sets are depicted on the left-hand side of Figure 6.3, corresponding to A^c, $(A \cap A^c)$, and $(A \cup A^c)$, whose coordinates in the fuzzy power set of X are given by

$$A^c = (1 - m_A(x_1), 1 - m_A(x_2)) \tag{6.21}$$

$$A \cap A^c = (\min[m_A(x_1), 1 - m_A(x_1)], \min[m_A(x_2), 1 - m_A(x_2)]) \tag{6.22}$$

$$A \cup A^c = (\max[m_A(x_1), 1 - m_A(x_1)], \max[m_A(x_2), 1 - m_A(x_2)]) \tag{6.23}$$

These four points (including the fuzzy set A) are situated symmetrically with respect to the four vertices that comprise the (non-fuzzy) power set of X. The further these points are from the vertices, the fuzzier the set is. Maximum entropy occurs at the midpoint, which happens to be B in this case, where the four sets converge so that $B = B^c = B \cap B^c = B \cup B^c$.

The Fuzzy Logic of Accessibility

Thus far we have established that the language of fuzzy sets as developed by Zadeh (1965), Kosko (1992) and others accommodates rather effortlessly concepts of accessibility in geographic or non-geographic contexts. This is encouraging, but the real question is whether the analytical tools of fuzzy logic can add to our general understanding of accessibility. We begin this part of the inquiry by probing more deeply the interpretation of the three fuzzy sets most closely associated with A, beginning with A^c. We established already that there is a one-to-one mapping of fuzzy sets in urban space on the right-hand side of Figure 6.3 to points in the fuzzy power set on the left-hand side of Figure 6.3. What about the inverse mapping? Specifically, is there a geographic location on the right-hand side of Figure 6.3 that corresponds to the complement of the fuzzy club A, and if so, what is its interpretation?

To answer this question, consider the inverse function $f^{-1}(.)$ which maps from club membership levels to distance, $f^{-1} : m \rightarrow d$. By taking the inverse of both sides of equation 6.17 we have

$$d_A^1 = f^{-1}[m_A(x_1)] \tag{6.24}$$

Building from the definition of the complement A^c, we may define the 'implicit complementary distance'

$$(d_A^1)^c = f^{-1}[1-m_A(x_1)] \tag{6.25}$$

which denotes the distance of x^1 from the fuzzy complement set A^c. From this we can infer that A^c will be located somewhere on the set of locations on the right-hand side of Figure 6.3 that jointly form a circle of radius $(d_A^1)^c$ centered on x_1. We denote this circle by A^{1c}, where

$$A^{1c} = \{x \mid d_{A^c}(x) = (d_A^1)^c\} \tag{6.26}$$

Likewise, A^c must also be located somewhere on another circular set of locations A^{2c} centered on x_2, and so we can narrow our search to the intersection of these two hyperspheres. In the case of n=2 the intersection of two overlapping hyperspheres is given by two points, as shown on the right-hand side of Figure 6.3, although at this juncture we must also allow for the possibility that the two circles in question do not overlap at all, or that they do so only tangentially. The interpretation of A^c in this context is that of a location in the original geographic space whose relationship to the reference set $X = \{x_1, x_2\}$ is complementary to the relationship that the urban node A has with this same reference set. If membership in the fuzzy club A is based on accessibility, then each element of the reference set X is as *inaccessible* to the location(s) defined by A^c as they are *accessible* to A.

A similar technique can be used to derive the reverse mapping for the fuzzy overlap and underlap clubs $(A \cap A^c)$ and $(A \cup A^c)$, which in the crisp version of the world correspond to the empty and universal sets, respectively. It is interesting that these logical expressions exist in a geographic context, and so one might actually *live* in the realm of contradiction or the excludible middle! These terms have a nice intuitive interpretation in a geographic context, as can be seen by considering the limiting case of a crisp world. In that case, as we have seen, the overlap $(A \cap A^c)$ corresponds to the null set, or $m_A(x_1) = m_A(x_2) = 0$. Using the function f(.) from equations 6.17 and 6.20, which maps distances into membership levels, zero membership corresponds to infinite distance, which implies that the excluded members are literally banished from the map, and so we are left with the empty set. Similarly, in the extreme crisp case the underlap $(A \cap A^c)$ corresponds to the universal set, or $m_A(x_1) = m_A(x_2) = 1$, indicating that the locations of both x_1 and x_2 coincide exactly with that of the urban center A, and so the entire reference set sits squarely 'downtown', or stated another way, the universal set is entirely subsumed in A. The same kind of reasoning applies when we admit partial membership in fuzzy clubs, but we are no longer constrained to these corner solutions of extreme dispersion and extreme concentration in the urban spatial structure. This suggests that fuzzy logic is a useful analytical tool for probing the intimate relationship between accessibility and entropy in the geographic formation of cities and regions.

The preceding discussion used a concrete example to convey a general concept. It is useful to pause, therefore, to consider how the specific example may be generalized. In Figure 6.3 the reference set X has dimensionality n=2, which is useful for graphical exposition purposes. However, generalized measures of accessibility typically make reference to a host of locations, so in practice the dimensionality of X may become quite large. This poses no problem conceptually, as the fuzzy power set in Figure 6.3 simply assumes added dimensionality concomitantly. It is most likely, however, that in this case too many constraints will yield a null solution:

$$A^c = A^{1c} \cap A^{2c} \ldots \cap A^{nc} = \varnothing \tag{6.27}$$

Thus, the existence and dimensionality of A^c in the original geographic space depends critically on the dimensionality of the reference set X.

Another assumption from Figure 6.3 that can be relaxed is the explicitly geographical nature of the original space, with A, B and C representing urban nodes or subcenters in that example, and with membership or accessibility represented as a function of geographical distance. Although geographic accessibility is subsumed very easily within the fuzzy club theoretical framework, more abstract notions of association may also be accommodated. For example, the fuzzy clubs A, B and C in Figure 6.3 may be on-line chat rooms, political power centers or community.

6.4 Modeling the Impact of Information Technology on Accessibility

Theoretical Approach

To model the impact of information technology on accessibility we first introduce some additional terminology with reference to the fuzzy power set in Figure 6.4, which differs from its cousin in Figure 6.3 insofar as it is based on n=3 rather than n=2. The corners of this cube are labeled X_1 ($= \{x_1\}$) or X_{23} ($= \{\{x_2\}, \{x_3\}\}$), for example, to indicate which crisp subsets of X 'belong' to the vertex in question. We refer to X_1, X_2 and X_3 as strictly partitioned sets while X_{12}, X_{23}, and X_{13} are loosely partitioned sets. A set S belonging to the power set of X is *strictly partitioned* if

$$x_i \in S \rightarrow x_i \notin T \qquad \forall i = 1 \ldots n \tag{6.28}$$

where T is some other set also belonging to the power set of X. *Loosely partitioned* sets drawn from the power set of X allow some overlap, but they are still 'partitioned' in the sense that there are some elements of X that are clearly excluded altogether, and there is another crisp subset of X that forms the perfect complement to the set in question. Considering for the moment the case where n is large, we would say that the set S, whose coordinates in the fuzzy power set of X are (1, 0, 0…0, 0, 1), is *more strictly partitioned* than T, with coordinates (0, 1, 1…1, 1, 0), although neither is strictly partitioned. Likewise, we say that T is *less strictly partitioned* than S. In the extreme case, as T becomes less and less partitioned, it happens that we approach full convergence, corresponding to X itself. In a social setting, a club that is more strictly partitioned may cater to narrower interest groups with high degrees of involvement. A religious sect or paramilitary organization that maintains rigid control over its members, demands a high degree of commitment from them, and shuns 'outsiders' would be an extreme case of a strictly partitioned set. The set of people belonging to the Democratic Party, or subscribers to the *Los Angeles Times*, which have clearly defined memberships but that draw from a large cross-section of the population, are less strictly partitioned.

Now consider the diagonal connecting ∅ and X in Figure 6.4. Its midpoint M is also the midpoint of the fuzzy power set of X and, as discussed earlier, it corresponds to the point of maximum entropy of the system. Conversely, the endpoints of this diagonal, ∅ and X, correspond to zero entropy. In the case of X there is full convergence and maximum participation. All elements are fully engaged at X. In the case of ∅ there is complete divergence and maximum withdrawal, but there is no ambiguity or entropy.

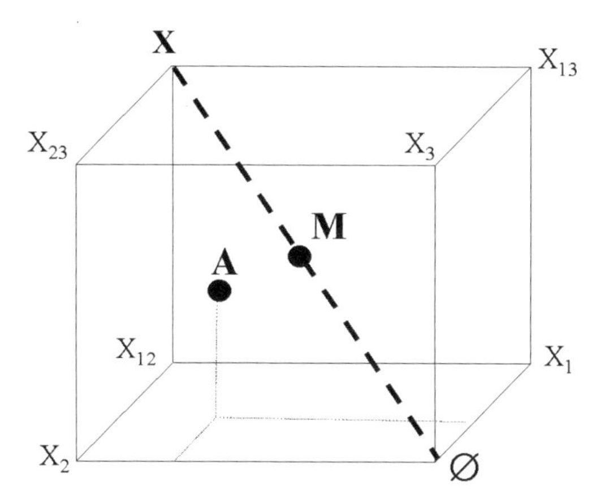

Figure 6.4. Fuzzy power set for n = 3.

To recapitulate, there are *three dichotomies* that are of interest here. One is the *participation-withdrawal* dichotomy, which is measured as distance in any direction from ∅. Another is the *entropy-order* dichotomy, which is measured as distance in any direction from the midpoint M. The third dichotomy is the *strictness-looseness* of partitioning, which can be measured as the angle away from the central diagonal. Each of these dichotomies is likely to be affected by our collective movement into the information age. Specifically, as we move from an era of geographic space to one of cyberspace, a number of changes are likely occurring, including:

- movement away from ∅ in response to increased levels of overall participation as technology provides more avenues for access to a wider range of fuzzy clubs;
- movement towards more loosely coupled partitions for the same reasons; and
- proliferation of more specialized clubs as like-minded individuals are able to congregate and find each other more easily.

These trends may often pull in opposing directions, thus making the overall impact on our three fundamental dichotomies unclear. For example, improved access may be expected to increase participation levels overall, but participation in some clubs may decline as competing opportunities for their members arise elsewhere. Likewise, the movement towards more loosely partitioned clubs may be offset by the proliferation of more strictly partitioned ones. The analytical framework is a theoretical one, but the question as to which of these trends is dominant must be answered ultimately by empirical means.

Empirical Approach

To implement this approach empirically requires data for some universal reference set X and the patterns of association each element x_i has with a set of identifiable clubs; A, B, C .... Z. Obvious candidates for study include (1) daily trip logs of individuals in an urban area and destination travel zones, (2) telephone records, or (3) Web browsing logs of a sample of users. The latter example is used in this discussion, where destination URLs [3] are the fuzzy clubs to which browsers belong, and where frequency and/or duration of visits measures strength of membership. Another advantage of this example is that the World Wide Web is undergoing very fast-paced change, and so the nature of its evolution for the three fundamental dichotomies is unfolding rapidly and requires shorter time frames. It may also offer insights for extending Bolton's (1997) notion of places as networks.

As before, X represents the universal reference set, in this case a sample of n Web users $X = \{x_1, x_2, \ldots x_n\}$, where n is large and where the survey records

[3] URL refers to the universal record locator that serves as a website's identifier and address on the World Wide Web.

cover two distinct periods, t=0 ('before') and t=1 ('after'). Any website that has been visited at least once by any user x_i in either period is a fuzzy club. The fuzzy power set defined by this data sample is an n-dimensional hypercube and each website is a fuzzy club located within this hypercube. For this example we define the membership value $m_A(x_i)$ for either period as the total time spent in that period by user x_i at fuzzy club A [4], where all such times are normalized so that the maximum membership value over both periods is unity. The result is a hypercube similar to the one depicted in Figure 6.4, but in n dimensions, and with numerous fuzzy clubs located throughout. The objective of this empirical approach is to determine how the system in question has changed with respect to the three fundamental dichotomies: (1) participation-withdrawal, (2) entropy-order, and (3) looseness-strictness of partitioning. In the context of this example it is reasonable to ascribe any such changes to the increasing availability of and advances in information technology, and so this provides a relatively 'pure' case study of how information technology impacts accessibility.

As noted above, the level of participation in a fuzzy club is measured by the distance of the club from ∅, where the maximum distance is √n, based on a Euclidean metric. More specifically, we have

$$P_{At} = \{ \Sigma_X [m_A(x_i)]^2 \}^{1/2} \quad \text{for } t = 0,1 \tag{6.29}$$

aggregating again over W, the set of all websites, yields an overall level of participation for period t, P_t, and the difference in the level of participation between the two study periods is given by

$$\Delta P = P_1 - P_0 = \Sigma_{A\in W} [P_{A1} - P_{A0}] \tag{6.30}$$

As for entropy, we may vary the notation from equation 12 slightly to conform to current usage:

$$E_{At} = M(A \cap A^c) / M(A \cup A^c) \text{ for } t = 0,1 \tag{6.31}$$

and then our measure of change in entropy between the two study periods for the entire system is

$$\Delta E = E_1 - E_0 = \Sigma_{A\in W} [E_{A1} - E_{A0}] \tag{6.32}$$

And finally, the degree of partitioning represented by any set A is given by the angle Θ_{At} formed between the central diagonal and the vector defined by A within

[4] In principle it should be easy to experiment with alternative specifications of the membership function that incorporate number of visits as well as time spent.

the fuzzy power set of X in period t. Let A^D denote the projection of the vector A onto the central diagonal. Then the degree of partitioning is given by

$$\Theta_{At} = \cos^{-1} [M(A^D) / M(A)] / (\pi/4) \quad \text{for } t = 0,1 \tag{6.33}$$

where the expression has been normalized by ($\pi/4$) so that the maximum degree of partitioning is one. Then, as with the preceding two cases, we have

$$\Delta\Theta = \Theta_1 - \Theta_0 = \Sigma_{A \in W} [\Theta_{A1} - \Theta_{A0}] \tag{6.34}$$

Equations 6.30, 6.32 and 6.34 measure the change in participation, entropy and degree of partitioning, respectively, of the system over the two time periods. As such, they are useful and practical means of assessing the impact of information technology on accessibility within a given system. More generally, we can think about the *process* by which one fuzzy state is mapped into another. Kosko (1992) refers to such mappings (from one fuzzy power set to another) as fuzzy systems, and argues persuasively that neural network techniques, which he terms fuzzy associative memories (FAM), are especially well suited to this task. Figure 6.5 illustrates the basic concept. Using fuzzy logic in this manner, neural networks can be trained computationally to reason with sets rather than with propositions regarding questions of accessibility. This idea is not pursued further in this paper, but it does suggest an intriguing avenue for developing predictive models about the manner in which accessibility systems unfold.

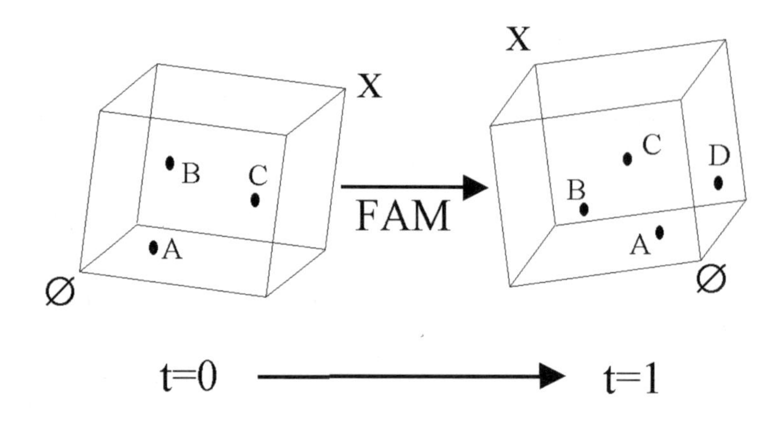

Figure 6.5. Fuzzy associate memories (FAM).

6.5 Fuzzy Conclusions

Fuzzy set theory and fuzzy logic is still relatively new and it is certainly not without its critics, some of whom argue that the introduction of multivalued membership functions to set theory adds little or nothing that could not be accomplished through other means. This paper does not seek to resolve that ongoing debate. However, the evidence presented here suggests quite strongly that the logic of fuzzy sets has much to offer geographers in their quest to model accessibility in a geographic and non-geographic context. The following conclusions appear to apply:

- The language of fuzzy sets can be adapted almost effortlessly to encompass a generalized notion of accessibility, and has considerable potential for integrating geographic, social, virtual, and other elements of accessibility or association;

- This approach also provides a natural extension to the club theory literature that grew from the Tiebout model, where municipalities or other nodes are modeled as 'fuzzy clubs';

- The fuzzy logical operators of complementarity, intersection, union, subsethood and entropy also have interesting and plausible interpretations in the context of accessibility, and so we may expect that as fuzzy logic continues to evolve, so too will the insights it offers to scholars concerned with issues of accessibility;

- Modeling fuzzy clubs as points in a cube leads us to a representation of three fundamental dichotomies in such systems: (1) participation-withdrawal, (2) entropy-order, and (3) looseness-strictness of partitioning; and

- We have demonstrated the feasibility in principle of implementing this approach to analyze empirically the impact of emerging technologies on a variety of systems in which accessibility issues are central.

Acknowledgements

This paper was written in response to the *Call for Participation* in the Varenius initiative on *Measuring and Representing Accessibility in the Information Age*. I am grateful to the organizers for prompting this inquiry, to my colleague Niraj Verma for helpful discussions generally, and to Qisheng Pan for research assistance. I am solely responsible for the paper's contents.

References

Bolton, R. 1997. 'Place' as 'network': Applications of network theory to local communities. Paper for Volume in Honor of T.R. Lakshmanan; Department of Economics, Williams College, Williamstown MA 01267; December.

Buchanan, J.M. 1965. An economic theory of clubs. *Economica.* 32:1-14.

Cornes, R., and Sandler, T. 1986. *The Theory of Externalities, Public Goods, and Club Goods*. Cambridge UK: Cambridge University Press.

Heikkila, E.J. 1996. Are municipalities Tieboutian clubs? *Regional Science and Urban Economics* 26:203-26.

Kosko, B. 1992. *Neural Networks and Fuzzy Systems: A Dynamical Systems Approach to Machine Intelligence.* Englewood Cliffs NJ: Prentice Hall.

Rescher, N. 1969. *Many-valued Logic*, New York: McGraw-Hill.

Tiebout, C. 1956. A pure theory of local expenditures. *Journal of Political Economy* 64:416-24.

Zadeh, L.A. 1965. Fuzzy sets. *Information and Control* 8:338-53.

7 The E-merging Geography of the Information Society: From Accessibility to Adaptability

Daniel Z. Sui

Department of Geography, Texas A&M University, College Station TX 77843, USA.
Email: D-sui@tamu.edu

7.1 Introduction

Issues related to the reconceptualization of access and accessibility in the information age have received significant attention from both policy makers and academic researchers in recent years (Couclelis 1996, Handy and Niemeier 1997, Kwan 1999, Litan and Niskanen 1998, Leebaert 1998, Miller 1999). This growing interest is attributed partly to the central importance of access and accessibility in geographic theories and models, and is caused partly by the extraordinary innovations in communication and transportation technologies in the late 20th century (Hodge 1997, Hanson 1998). From recent literature, one concludes that there is little consensus on how to define (much less how to measure) access and accessibility in the information age. Most authors do seem to agree that the traditional conceptualizations of access and accessibility are incapable of capturing the new reality and that we need to redefine and reconceptualize accessibility in light of the new Internet-led revolution in telecommunications (See Chapter 16 by Hanson and Chapter 17 by Occelli). However, few have asked the fundamental questions: why are we so obsessed with the access and accessibility issues, and what exactly are their roles in our social and economic lives? If the dazzling development in telecommunications has rendered the definition of access and accessibility so elusive that they defy conventional measurements, what kind of alternative questions can we ask to better understand the E-merging geography of the information society?

This chapter has two objectives. First, I present a philosophical critique on current research on access and accessibility. Second, I aim to examine the feasibility and utility of an adaptive perspective in studying the geography in the information age. The chapter discusses why accessibility becomes less critical in the information age and describes the need to focus more on adaptability. It traces the root metaphors embedded in accessibility and adaptability and provides an explanation of their syntax, semantics, and pragmatics. It also presents preliminary empirical results on the emerging geography of information society, and identifies future research needs.

7.2 Drawn to the Web: Accessibility or Adaptability?

Although it is generally agreed that accessibility is only a necessary condition for economic development or for improvement in the quality of life, accessibility is often treated as an end in itself. The literature is replete with mono-causal assertions that tend to regard accessibility as a panacea for all our social and economic problems. Without considering many other social, economic, and even cultural factors, policies designed solely to improve access and accessibility can be misleading and ineffective in accomplishing social goals. Hägerstrand (1975) observed that it is not as important to measure what people do as it is to measure what they are free to do. By the same token, I believe it is not as important to measure what people have access to as to measure the access of what people need (whether facilities, services, or information) at the right time and the right place. The relentless pursuit to improve accessibility in transportation in most major U.S. cities seems to be locked in a vicious cycle: the wider the roads become, the more crowded they seem to be; and, thus, making many places less accessible. The prerequisite to a mature geography of telecommunications, according to Abler (1991, 46) is

> the need to abandon the assumption that communications technology cause cultural, economic, political and social behavior and events and that behavior and events are consequences of the absence, presence or use of telecommunications hardware. Neither historical evidence, current events, nor prospects for the future support that assumption.

Abler's insight is particularly pertinent to our current discussions on access and accessibility. Indeed, it would be disastrous if we continue to let accessibility dictate our research and policy agenda in the information age. Many studies have indicated that having access to unlimited amounts of information is not necessarily beneficial to an individual or an organization. In fact, information overload (having access to too much information) may be equally or more harmful than having too little information. Davenport (1997) reports that information overflow is one of the main reasons for institutional dysfunction. The reason for the dysfunction is captured in Simon's law – the wealth of information creates a poverty of attention (Simon 1997). Economists even hotly debated the existence of the so-called productivity paradox, because many empirical studies show a decrease of productivity with increasing access to information technology (Brynjolfsson and Yang 1996). At the individual behavior level, the irony is even more disturbing. Several recent national studies indicate that the more time people spend on-line, the more likely they tend to be depressed and, thus, less productive in life (Schindler 1998). As a result, the University of Texas at Austin offers Internet addiction counseling services to students who are so addicted to surfing the Web that their normal course work is interrupted (Schindler 1998).

Perhaps, Brin (1990) best depicts the situation of information overload in his science fiction *Earth,* when everybody is wired to everybody else and to every conceivable information source via the global communication network. As a result, people are inundated with unwanted, incorrect, and irrelevant information. Consequently, most of them are confused, hallucinated, and depressed. The problem people in this kind of society have to deal with is not how to gain access to information, but precisely the opposite: how to filter out the unwanted information and retain only valuable information. In Brin's fictional world, the wealthy and powerful are differentiated from the poor and powerless by their ability to purchase artificially intelligent agent software capable of separating the wheat from the chaff. Although Brin's world has not yet arrived, Bednarz (1998) observes that the problem has. Mokyr (1997, 138) notes that

> of course we have access now to a much larger supply of information, but here we must always keep in mind that the human mind can process at most 50 bytes per second - and most of us probably do not do nearly as well. Few if any persons in our age, I venture, suffer from a lack of information – all have stacks of unread papers, unanswered e-mail messages, endlessly unopened Internet web pages, memos that are still in the their brown campus envelopes, videotapes of Public TV programs which we promise ourselves to watch 'when there is time.' The high priority is no longer in getting the information to us but in selecting and ranking, sorting the duplicative and false and the irrelevant from the information that we need. Like DNA, most of the information is junk.

According to a recent study in memetics, the most accessible information in our society is neither the best nor the most accurate (Lynch 1999). Instead, it is information that can replicate itself that is spread rapidly in various media. Unless we make judicious and persistent efforts to filter unnecessary information, the benefits gained from the telecommunication revolution can easily be canceled by the problems caused by information overload.

Furthermore, having access to the right information (or facilities, services, etc.) at the right time and the place is not as important as what people actually do with the information. One extreme example is that almost everybody has access to mud and clay and yet only a handful of artists can use this to make an artistic product. Our current obsession with accessibility issues might blind us to the real issue in the information society. The real issue we need to tackle is, in my opinion, adaptability: to study how people or organizations actually creatively use accessible information to their best advantage and to the benefit of society and the environment. To adapt in the information age means to learn and to translate what one learns into productive actions more quickly and effectively. Obviously, to study adaptability is much more complicated than measuring accessibility.

Undoubtedly, there are still lots of research and policy issues to be addressed to improve access and accessibility at the global, regional, and local levels. Accessibility issues are especially acute during the initial stage of a technological innovation. But the importance of accessibility declines as the technology evolves. History has shown that as a technological innovation matures, access to it will increasingly depend on one's ability to pay and on an equitable legal framework to guarantee such access (see Chapter 18 by Onsrud). Thus, the access or accessibility issue ultimately always becomes a socio-economic or political issue in disguise. In retrospect, the time required for universal access to a new technological innovation has shortened dramatically during the last 150 years. Electricity was first harnessed in 1831, yet it took 50 years to build the first power station in 1882 (Hughes 1983). It took another 50 years before electricity powered 80 percent of factories and households across the United States. Radio was in existence for 40 years before 50 million people tuned in. TV took 13 years to reach that benchmark. Sixteen years after the first PC kit came out, 50 million people were using one (Margherio 1997). Once it was opened to the general public, the Internet crossed that line in less than four years. Right now we have about 172 million Internet users worldwide (NUA 1998). Two recent national studies (Clemente 1998, Rogers 1998) show that the Internet has been increasingly accessed by women (40% of all users), the less affluent (the average annual household income with Internet access dropped below $41,000), and the less educated (about 50 % of the Internet users do not have college education). Based on the current pace of Internet growth, it is estimated that global Internet users will probably reach 1 billion by the year 2005 (Margherio 1997). So it is not unreasonable to assume that before long, the issue of accessibility may still exist but that it will be pretty trivial in the developed world, especially in the United States. Who would now conduct research on the accessibility of electricity or telephone or TV (although the issue might be interesting and legitimate during the initial stages of these technological innovations)?

With this historical background in mind, I believe that it will be more fruitful to shift our focus from accessibility to adaptability. By focusing on adaptability, we can better comprehend actual processes at the individual, organizational/institutional, and social levels, and appreciate more fully the E-merging geographical patterns that characterize the information age. Accessibility and adaptability entail different metaphors, but also different syntax, semantics, and pragmatics.

7.3 Understanding the Web: Mechanical (accessibility) versus Biological (adaptability) Metaphors

Human thinking and cognitive abilities are inherently metaphorical, and geographers have paid increasing attention to the role of metaphors in geographic theories and models (Sui 2000). Generally speaking, most models in social sciences are developed according to either a physical/mechanistic metaphor or a biological/organic metaphor. The root metaphor behind the various measures of accessibility is a mechanistic one, and the concept of adaptability is motivated by a biological metaphor. To shift the focus of research from accessibility to adaptability necessitates not only a change of metaphors but also different policy prescriptions about the geography of information societies.

The mechanistic metaphor embedded in various measurements of accessibility has a built-in deterministic and ahistorical ontology. The world is conceptualized more or less like a gigantic machine, composed of parts (bounded regions) and geography is about the interaction of those parts. Accessibility measures the reachability of various parts by people via certain kinds of impedance. All aspects of human life are determined and represented by the calculus of variation, and human agents have no choice but to behave in accordance with physical laws such as the laws of least effort. Such a worldview as ahistorical: reality is theorized in logical time, not historical time. Motion is completely reversible and gives rise to no qualitative changes (Barnes 1997).

The epistemology behind a mechanistic metaphor is based upon the Cartesian separation of object and subject and is, thus, inherently reductionistic. Social physics according, to the Newtonian mechanistic metaphor, always reduces the complexity of the real world to elemental components to facilitate analytical treatment. Implicit in the various accessibility measurements, the centrality of rational individuals is assumed and human irrationalities are not considered. Accessibility is operationalized predominantly from the perspective of supply-side economics although a few researchers have devised new measurements to take into account demand-side factors as well (Shen 1998). The mechanistic metaphor also imposes a strong sense of teleology to the system we model. The purpose of the system is to optimize or maximize over everybody's collective objective function. In fact, the meaning of life has been reduced to maximize certain utility functions according to the mechanistic worldview. Despite its analytical elegance and contributions to the development of neoclassical economic theories, more and more researchers realize that measurements and models developed according to a Newtonian mechanistic metaphor only present a very partial, limited view of the reality. In most cases, they've generated more heat than light in our quest to understand the world. Conclusions reflect more the internal logic of the mechanistic metaphor than that of the phenomenon to be modeled (Mirowski 1989, 1994).

The recent round of evolutionary thinking in general and the recent development of evolutionary economics in particular (Hodgson 1993; England 1994; Nelson 1995) inspired the idea of adaptability. The root metaphor of adaptability is a biological one. Its ontology rejects the Laplacian dream of full mathematical determinacy. Instead, the world is assumed to be a result of the interplay between chance and necessity. Simultaneous randomness and determinacy accentuate historical process and irreversible change. Reality is not only path-dependent but also nonlinear in its dynamics - small changes can provoke wider reverberations throughout the entire system. Its epistemology denies the understanding of the whole by analyzing its parts. The world is conceptualized as composed of an entangled, complicated web, and geography is about the evolution (the dynamics) of the web as manifested by the interaction among each strand in the web. The emergent behavior and spontaneous growth (order out of chaos) should be treated as the norm. The adaptability perspective assumes no teleology – the wider system is not necessarily improving or pursuing some ultimate goal of perfectibility. The best one can expect is a temporary sub-optimal result of a constantly evolving, unpredictable process.

Table 7.1. Accessibility versus. adaptability: syntax, semantics, and pragmatics

Versions	Accessibility	Adaptability
Syntax	Measurements are based upon analogies from Newtonian physics and Keyensian economics; deterministic gravity-type of models	Measurements are based on chaotic and self-organizing principles; in accordance with analogies from biology; fractals, cellular automata; neural computing, and genetic algorithms
Semantics	Realities are conceptualized as machines composed of simple systems; predefined spatial boundaries; aggregate; static; emphasis on employment, travel, and land use	Realities are conceptualized as organisms composed of complex systems; derived spatial units and grid cells; highly disaggregate; dynamic; emphasis on individuals and institutions
Pragmatics	Improvement of infrastructure (hardware and software); short-term, vicious cycle; seek to maximize; emphasis on behavior and maximum economic efficiency.	Improvement of learning and cognitive capabilities (humanware); long-term, virtuous cycle; seek adapt/satisfy; emphasis on cognitive capabilities and sustainable development.

To further differentiate accessibility and adaptability, we need to take a close look at their syntax, semantics, and pragmatics (Table 7.1). I want to emphasize their different pragmatics, which has profound policy and practical implications. Efforts on accessibility tend to emphasize the improvement of infrastructure (hardware and software) with the goal of maximizing economic efficiency, with less concern for social equity and environmental sustainability. In contrast, the adaptability perspective emphasizes the cognitive dimensions (humanware) of individuals, organizations, and regions. To adapt in an uncertain, unpredictable, and constantly changing world, agents must learn and adjust their tactics and strategies accordingly. Different learning schemes, such as learning by doing (Arrow 1962), learning by trying (Fleck 1993), and learning by building alliances (Mody 1993) are proposed as adaptive strategies in the information age. The adaptability perspective also mandates abandonment of the maximization of economic efficiency and the improvement of infrastructure as driving motivations for policy initiatives. Instead, the adaptive paradigm seeks to balance economic efficiency, social equity, and environmental sustainability so that society and environment can co-evolve in harmony and can avoid further tragedies for the commons (Sui 1998, see Couclelis in Chapter 20).

Just like social physics, various versions of social biology, especially social Darwinism, have been discredited and hotly contested. However, adaptability and other biological (or physical) principles can only be used in a strictly metaphorical sense. All metaphors are both liberating and constraining. I distance myself from the hardcore socio-biological thinking of Wilson's (1975) or Dawkins' (1976) depiction of society in *red in tooth and claw*. My line of reasoning falls squarely within the mainstream of evolutionary economics, which draws on the latest development of complexity theory and new computing theories. The evolutionary metaphor is invoked to stimulate us to ask new questions but not to seek answers. We must search our answers through meticulous study of the real world.

7.4 Caught in the Web: The Death of Distance or the Birth of a New Geography?

Theoretical speculations on the impacts of the Internet-led telematic revolution are rampant in the literature. Arguments range from the borderless world and the end of nation states (Ohmae 1990, 1995), space-time compression and distanciation (Harvey 1996), and the death of distance (Cairncross 1997), to the anywhere-anytime-anything paradigm for interpreting access to various social, economic, and cultural aspects of society (Mitchell 1995). And yet, rarely are these assertions supported by convincing empirical evidence because of the lack of relevant data and because of the enormous conceptual complexities of measuring access and accessibility in the information age. By shifting our perspective from accessibility to adaptability, I want to use a combination of several different data sources to shed light on the emerging geographic patterns in the new information age and to test the usefulness of the adaptability perspective in explaining these patterns.[1]

---

[1] Unless noted otherwise, data used in this paper come primarily from the following sources: (1). *Telegeography 1999: Global Telecommunication Traffic Statistics and Commentary*. Gregory Staple (ed.). Washington DC: Telegeography, Inc. (see http://www.telegeography.com, (2). Network Wizards (www.nw.com), RIPE, Inc. (www.ripe.net), Web21 (www.web21.com), (3). Clemente, P.C. 1998. *State of the Net: The new frontier*. New York: McGraw-Hill. Data used in this book are collected by FIND/SVP 1997 American Internet User Survey (www.cyberdialogue.com, (4). Matrix Information and Directory Services (MIDS), Inc. (www.mids.org, (5) CyberAtlas, Inc. (www.cyberatlas.com), and (6) U.S. Department of Commerce, National Telecommunications and Information Administration (NTIA) 1997. *Falling Through the Net II: New Data on the Digital Divide* (www.ntia.doc.gov/ntiahome/net2/falling.html).

Table 7.2. The most wired countries in the world: Top 15 nations by Internet population by the end of 1998

Nation	Internet Users (millions)
United States	76.5
Japan	9.75
United Kingdom	8.10
Germany	7.14
Canada	6.49
Australia	4.36
France	2.79
Sweden	2.58
Italy	2.14
Spain	1.98
Netherlands	1.96
Taiwan	1.65
China	1.58
Finland	1.57
Norway	1.34
Top 15 nations	129.9
Europe	36.02
Worldwide	147.8

Source: CyberAtlas, Inc.

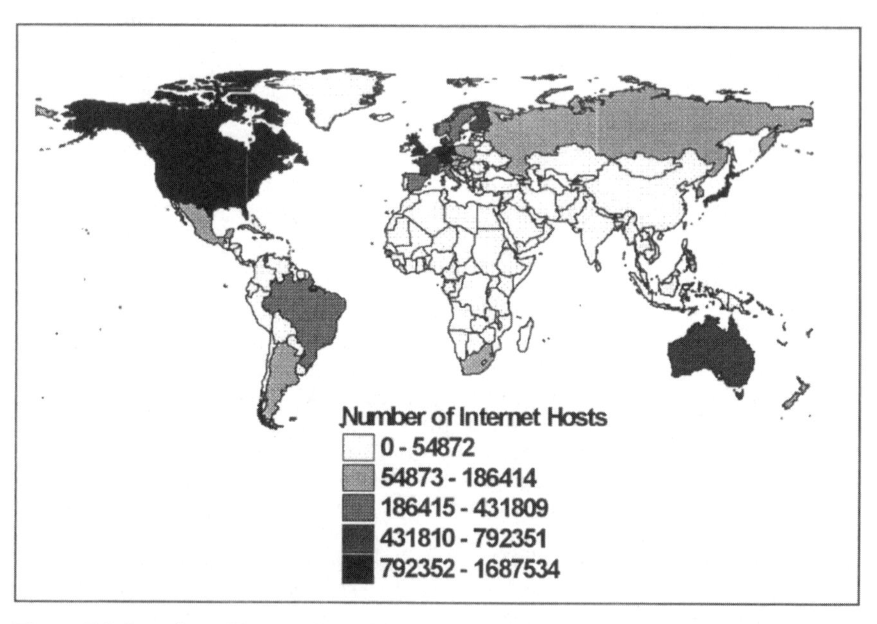

Figure 7.1. Location of Internet hosts by county, 1997 (Data source: TeleGeography, Inc.)

Despite the difficulties of collecting data on the new information infrastructure and their usage, several federal agencies and private companies have made some initial efforts to collect data to mirror spatial and temporal growth of the Internet. Although inconsistencies exist, these data sources do complement each other in many respects. By synthesizing different data sources currently available, we can gain a glimpse of the E-merging geography of the information society at the global, regional, and even local levels.

Table 7.3. The changing Internet user profile (1994-1997)

Demographics	**All U.S. Adults 1997**	**Internet Users 1994**	**Internet Users 1995**	**Internet Users 1996**	**Internet Users 1997**
Total Adults (millions)		3.5	8.4	22.5	36.3
Age	(Percentages)				
18-29	22	42	38	34	30
30-49	42	45	48	51	54
50 & Up	36	13	14	15	16
Average Age (Years)	44	35	36	37	38
Gender	(Percentages)				
Male	48	78	72	66	61
Female	52	22	28	34	39
Household Income ($)					
Average Income	44,900	66,300	61,500	56,700	51,900
Median Income	34,000	64,000	57,500	51,000	44,200
Education	(Percentages)				
College Graduate	21	51	48	45	42
Not College Graduate	79	49	52	55	58
Household Composition					
Married	57	66	66	67	68
Single	23	29	26	23	21
Divorced/Separated	20	5	8	10	11
Children Present	42	30	31	37	38
Occupation					
Knowledge Work	27	47	48	47	45
Other Employed	36	30	29	31	35
Student	12	17	15	12	10
Not Employed	25	6	8	10	10

Source: FIND/SVP, cited in Clemente (1998)

According to the *Computer Industry Almanac*, there were more than 147 million world Internet users at the end of 1998, up from 61 million at the end of 1996. However, the world Internet population is distributed unevenly. The top 15 nations account for almost 90 percent of the world Internet users, with the United States alone accounting 52 percent (Table 7.2). The geographical distribution of the world's Internet population is consistent with the locational patterns of the Internet infrastructure (Figure7.1). Internet domains are located predominantly in the United States (58.1%) and Europe (24.3%). Both Asia and Latin America account less than 10 percent of the world's Internet domains. Africa is the least wired continent, accounting for less than 0.5 percent of world's Internet domains (Figure 7.2). America's dominance is reflected also in the distribution of the world's 100 most visited Web sites, among which 94 are U.S. Web sites (Figure 7.2). While the rest of the world claims only 6 percent of these 100 most visited sites, California alone accounts 40 percent and New York 16 percent.

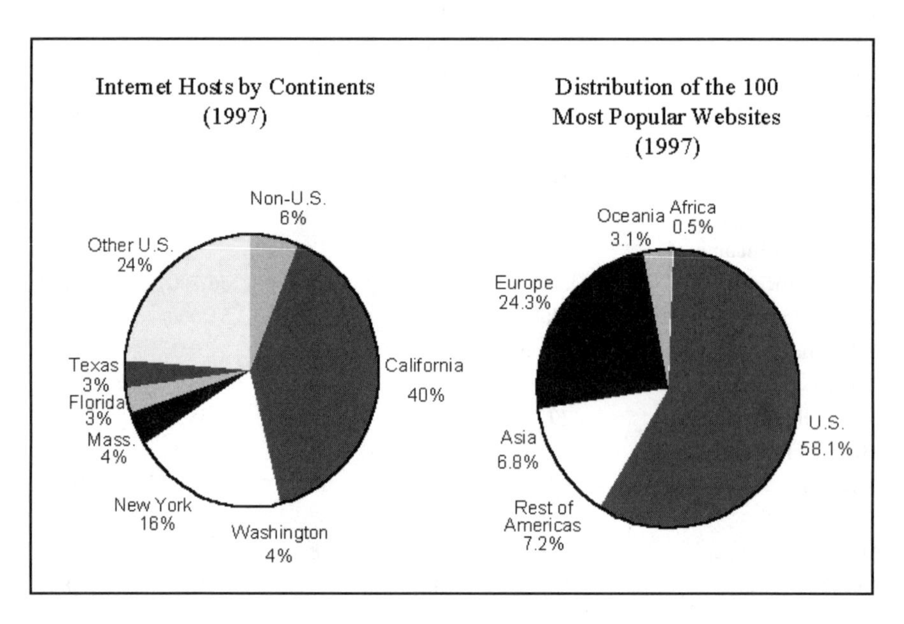

Figure 7.2. Internet hosts by continents and distribution of the 1,000 most popular Web sites in 1997 (Data source: TeleGeography, Inc.)

According to a survey conducted by INECO, by the end of December 1998, approximately 108 million adults – 55 percent of the U.S. adult population had accessed the Internet at least once in the previous 30 days. Several national surveys have also confirmed that for the first time there are more people (36 million) accessing the Net from home than from work (26 million). Approximately 24 million users have access to the Internet from schools, libraries, or community centers. The average age of Internet users increased by 3 years as more senior citizens are moving on-line (Table 7.3). Women, the less educated, the less affluent, and blue-collar workers are increasingly accessing the Internet (Table 7.3). It is estimated that by 2003, Internet users in the United States will reach 207 million. On average in 1998, Internet users spent 8.2 hours per week on the Net (Table 7.4), with the majority logging on in the evenings daily between 6 p.m. and 10 p.m. Since, the Internet will impact significantly on the temporal rhythms of social life, we will need to pay more attention to 'the urbanization of time' (Janelle 1993, see Chapter 9 by Harvey and Macnab).

Table 7.4. Online trends among Internet user households

Internet Usage Characteristics	Percent of Internet User Households
Access Method:	
Internet Service Provider (ISP)	60
Commercial Online Service	48
America Online	38
Microsoft Network	12
CompuServe	7
Access Speed:	
14.4kbs or slower	25
28.8-36.6kbs	51
Other	10
Average Hours/Week Use Internet from Home:	
Under 5 Hours	45
5-9 Hours	24
10 or more Hours	29
Average Number of Hours	8.2 hours
Average Session Length:	
< 15 Minutes	2
15-44 Minutes	20
45-74 Minutes	40
75-119 Minutes	8
2 Hours or more	30
Average Number of Minutes	90 minutes
Primary Use by Time of Day:	
Before 8 a.m.	3
8 a.m. –Noon	13
Noon-6 p.m.	14
6 p.m.-10 p.m.	41
After 10 p.m.	12
Varies too much to say	16

Source: FIND/SVP, cited in Clemente (1998).

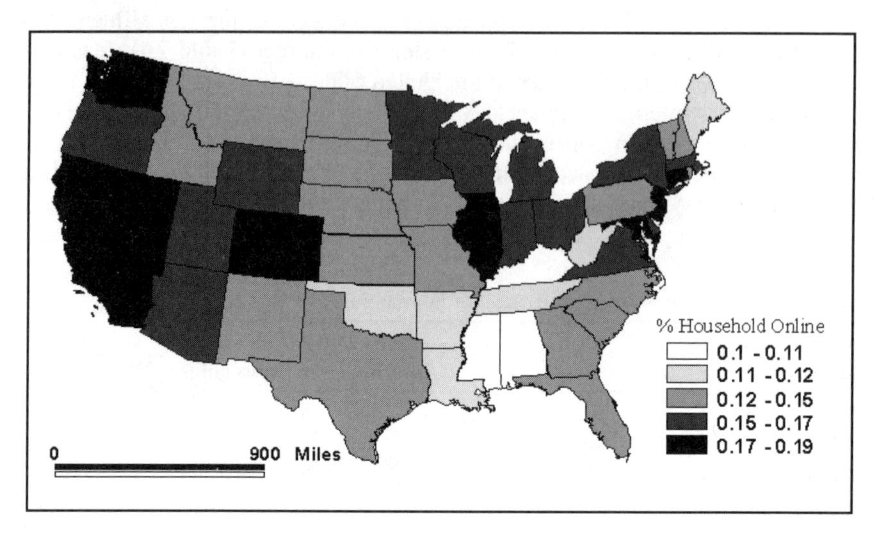

Figure 7.3. Location of Internet households in the United States, 1997 (Data source: CyberAtlas, Inc.)

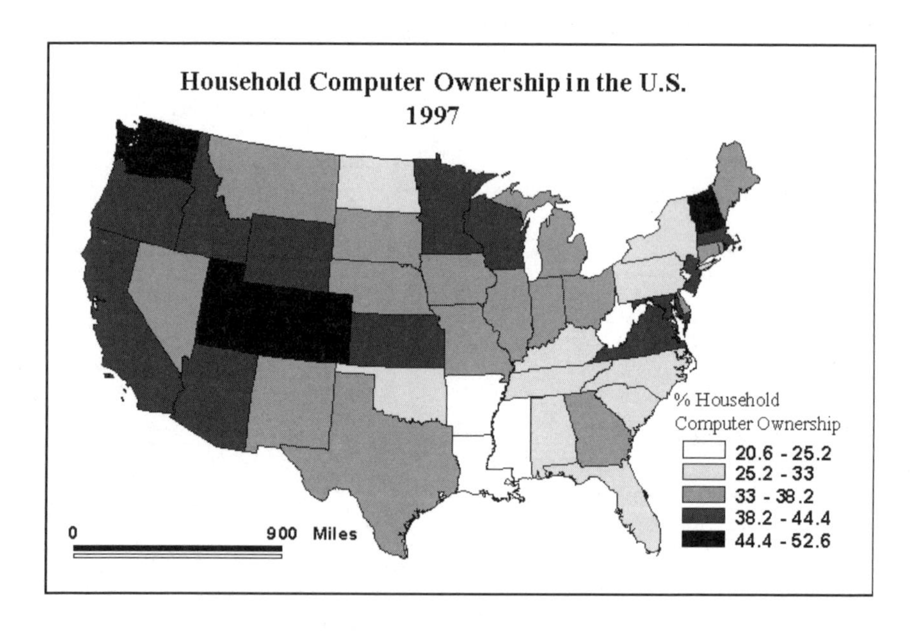

Figure 7.4. Household computer ownership in the United States, 1997. (Data source: CyberAtlas, Inc.)

At the household level, about 20 percent of U.S. households had access to the Internet by the end of 1997. But it is estimated that this Internet penetration will reach 70 percent by 2003. With the dramatic drop of prices of personal computers, 40 percent of U.S. households own PCs. However, the geographical distribution of the Internet households and PC ownership in the United States is uneven (Figures. 7.3 and 7.4), with strong concentrations in states where high-tech industry and R&D activities are located, such as California, Massachusetts, Washington, and Colorado. Many rural southern states, such as Mississippi, Arkansas, Alabama, and Kentucky, lag. The disparity in information infrastructure is reflected clearly in the distribution of Internet hosts at the state level (Figure 7.5).

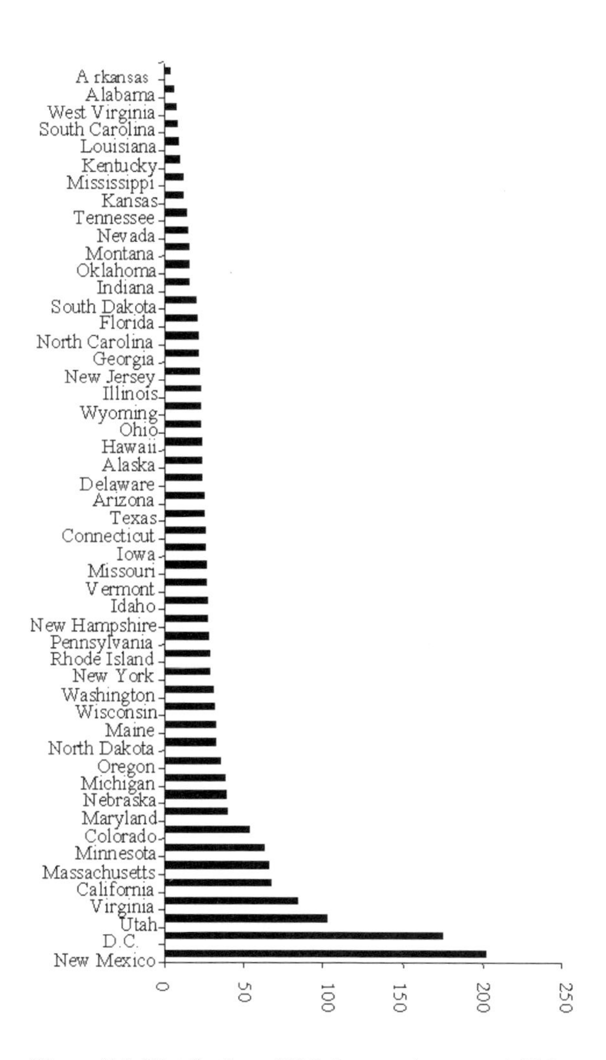

Figure 7.5. Distribution of U.S. Internet hosts per 1,000 people by state, 1997

Table 7.5. America's most wired cities: Top U.S. cities ranked by percent of population online

Metropolitan Statistical Area	Percent Online
San Francisco Bay Area CA	72
Miami FL	67
Houston TX	65
Seattle/Tacoma WA	65
Washington DC	64
San Diego CA	64
Cleveland/Akron OH	62
Atlanta GA	61
Dallas TX	60
Philadelphia PA	60
Sacramento CA	59
Los Angeles CA	59
Chicago IL	58
New York NY	58
Phoenix AZ	57
Boston MA	57
Denver CO	55
U.S. as a nation	55
Detroit MI	52
Minneapolis/St. Paul MN	52
Pittsburgh PA	49

Source: CyberAtlas, Inc.

Table 7.5 shows America's most wired cities. Not surprisingly, 72 percent of adults in the San Francisco Bay area access the Internet regularly, making it the most wired community in the nation. This relates obviously to the nearby presence of Silicon Valley and its concentrations of high-tech companies and highly skilled people. Other most wired cities in the United States tend to be the traditional economic, financial, and political centers, or the new booming Sunbelt cities, where new corporate headquarters tend to be located or relocated. Figure 7.6 shows the distribution of Internet domains at the zip code level in Texas. There exist, obviously, heavy concentrations in the four largest Texas cities – Houston, Dallas/Forth Worth, Austin, and San Antonio. Several zip code areas in Austin claim the heaviest concentrations of the Internet domains. This map also shows the growing dispersal to suburban areas of these four big cities, and even to small towns and less populated areas in rural Texas.

Table 7.6. Possible methods and techniques to handle surprises in modeling

Surprise-generating Mechanisms	Methods/Techniques
Instability	Catastrophe/Bifurcation Theory
Unpredictability	Non-linear Dynamics
Irreducibility	Holistic Approach, Q-analysis
Uncomputablity	Neural computing/Genetic Algorithms
Emergence:	Self-organizing, Cellular Automata

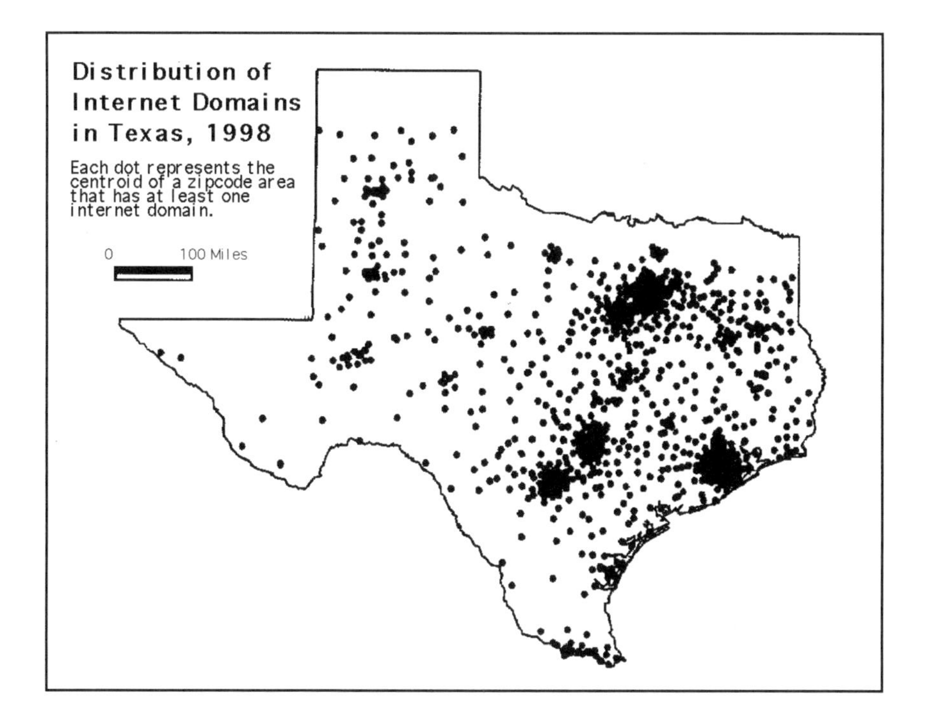

Figure 7.6. Distribution of Internet domains in Texas, 1998 (Data source: MIDS, Inc.)

Although the results are far from conclusive, preliminary findings do provide hints to the E-merging geography of the information society. Contrary to the simplistic, utopian expectations that telecommunications have sentenced distance to death and render geography meaningless, the E-merging geography of the information age is producing new rounds of unevenness and geographical clustering. The ubiquitous availability of information has not created (and will not create) a more egalitarian even distribution. Despite businesses that increasingly conduct commerce without geographical propinquity, the development of the information society is coalescing into an increasingly clustered spatial form at global and national scales, with a rapid dispersal at local scales. The Internet-led telecommunication revolution simultaneously reflects and transforms the topologies of capitalism, creating and rapidly recreating nested hierarchies of spaces technically articulated in the global information infrastructure led by the architecture of computer networks. Indeed, as Warf (1997, 9) observed,

> far from eliminating variations among places, such systems permit the exploitation of differences between areas with renewed ferocity. The new geography engendered by the telecommunication revolution dos not entail the obliteration of local uniqueness, only its transformation and reconfiguration.

Findings of this research are consistent with Feldman and Florida's (1994) and Gertler's (1995) work on spatial diffusion of innovations and on conclusions reached from a more conventional political economy approach (Harrison, Gant, and Kelly 1996, Scott 1996, Storper 1997, Cooke and Morgan 1998). Such uneven development reveals the tendency of path dependence to lock-in already-established patterns of regional development (Rosenberg 1982). History has a strong vote (although not a veto) in determining what geographic patterns emerge. Exciting theoretical groundbreaking work has been conducted to explain path dependence through the mechanism of increasing returns and positive feedback in situations of multiple equilibrium (Arthur 1994, Krugman 1997; 1998).

7.5 Spinning the Web: Bring Life back to Geographic Research

From the perspective of evolutionary economics, the current Internet-led telematic revolution is obviously a major genetic mutation in societal evolution. The digital economy is emerging as a new species in the dense economic jungle (Tapscott, Lowy, and Ticoll 1998, Kelly 1998) and may very likely become a predator in the new business ecosystem (Moore 1993, 1996). Those who can best adapt at the technical, behavioral, and cognitive levels will succeed (Gates 1999). This is an enormously complex process, which is subject to many random perturbations. Access and accessibility are only a necessary condition, but insufficient to guarantee success. The shift in focus from accessibility to adaptability challenges us to rethink research questions and methodologies.

Marshall (1961, viii) noticed that 'the Mecca of the economist lies in economic biology rather than economic mechanics,' but he was also fully aware that biological conceptions are more complex than those of mechanics are. Biologically motivated thinking defies some of the conventional mathematical languages to describe reality.

Future research on the geography of the information society should be conducted at the individual (ontogenetic), organizational (phylogenetic), or societal (ecosystem) levels to better understand how individuals, organizations, and society as a whole adapt to new technological innovations. We need to seek better theoretical links of the adaptive perspective to theories in evolutionary economics and in the new economic geography. The adaptive perspective also necessitates studies of a person's access to technological media (either electronic or printing press) and to various social networks via personal face-to-face contacts. Recent studies have indicated that simply having access to information will not be sufficient for a person to succeed in the socio-economic arena if he/she is not connected to the appropriate social network (van der Poel 1993, Powell and Smith-Doerr 1994). Rogerson's (1996) recent work on estimating social networks may help us to understand how people are adapting to the new technical innovations. One of the rallying concepts in the adaptability perspective is learning and learning capabilities (Hanssen-Bauer and Snow 1996). Regions are evolving from production systems developed in the industrial age into massive learning systems facilitated by both technical and social networks in the knowledge economy (Petchell 1993, Lundvall and Johnson 1994). Understanding how a region or a locale *learns* will be one of the most important challenges faced by geographers in the information age (Jin and Stough 1998). Preliminary results indicate that the E-merging geography of the information age will be differentiated increasingly by a region's learning (adaptive) capabilities (Florida 1995, Asheim 1996).

These research issues require new research techniques to model the complexity and adaptability of the information society. Although linear and deterministic techniques are still applicable in certain situations, we need to expand efforts to apply concepts and theories developed in non-linear dynamics. Recent developments in complexity and chaos theory seem to provide the most appropriate language to describe reality from the adaptive perspective (Capra 1996). Incorporating insights gained from non-linear dynamics is the best way to handle surprises in future modeling efforts. Table 7.6 summarizes major techniques to handle each of the five surprise-generating mechanisms. Although there are overlaps among the five possible solutions listed in Table 7.6, they are good starting points for developing a unified framework. The goal would be to integrate the fragmented modeling based on non-linear dynamics to unpack the geographic patterns of the information society.

Chaos and complexity theory will play a central role in modeling the geography of the information society. Chaos theory offers a possibility of elegantly reconciling the simultaneous presence of complexity/irregularity and simplicity/regularity in a complex system. Chaos theory implies apparent randomness out of order and order out of randomness. According to chaos theory, complex non-linear systems are inherently unpredictable, no matter how sophisticated or detailed the model may be. However, it is generally quite possible, even easier, to model the overall behavior of a system. The way to express such an unpredictable system lies not in exact equations, but in representations of the behavior of the system – in plots of strange attractors or in fractals. Pioneering works have already revealed that urban forms are essentially fractals in nature, and that urban processes can be simulated as self-organizing cellular automata and neural networks (Batty 1995; Couclelis 1997, Fischer and Gopal 1998). We can expect that new developments in non-linear dynamics will play an increasingly more important role as we switch our modeling focus on industrial cities as bounded regions and networks to information cities as space of flows and quantum states. Evolutionary economics has reached similar conclusions as those based upon chaos theory and non-linear dynamics (Leydesdorff and van den Besselaar 1994, Arthur 1994).

These alternative techniques are consistent with major paradigmatic changes in computing, as evidenced by various biologically motivated computing innovations, such as genetic algorithms, evolutionary programming, and neural computing (Paton 1994). John Holland's (1998) groundbreaking work in learning algorithms is particularly pertinent to studies of how regions learn in the information age.

7.6 Untangling the Web: Beyond Accessibility and Adaptability

Like it or not, the information age has arrived (at least in the United States, if not in the world) and it is here to stay. The U.S. economy has undergone a major genetic mutation from the production and shipment of tangible goods to the generation and circulation of intangible information and knowledge. To better understand the geography of the E-merging information society, we must liberate ourselves from the tyranny of mechanistic conceptualizations of reality in the Cartesian mode. The information society is obviously a complex adaptive nonlinear network. Conventional conceptualizations of reality, according to deterministic, reductionistic, ahistorical, and teleological assumptions, may be sufficient for the industrial age, but are incapable of capturing the unpredictable, interconnected, and constantly evolving new information society.

The central arguments of this chapter are that the importance of access and accessibility declines as new innovations mature and that information overload is as harmful as information scarcity. What really matters in the information age is adaptability, which entails how individuals, organizations, or regions learn from what they have access to and translate knowledge into productive use. By shifting focus from accessibility to adaptability, we will not only enhance understanding of the complexity and evolution of the information age better, but also bring life back into geographic research by ridding it of the mechanistic metaphor.

The geographical understanding of the information age demands more than just another twist of accessibility measurements by modifying the various versions of the Newtonian gravity model. The adaptability perspective proposed in this chapter demands a fundamental paradigm shift in geographic research at the semantic and syntactic, as well as pragmatic, levels. At the semantic level, we are dealing with a fundamentally different kind of society as a result of the telematic revolution. Many old concepts and theories are no longer applicable and new theories for understanding this new reality have yet to be developed. At the syntactic level, developments in science and technology during the later half of 20th century provided new languages to describe and model various facets of society. These new theories and concepts, reflected in chaos theory, cellular automata, fractal geography, and self-organizing theory, are coalescing rapidly into a non-linear science that challenges deterministic and linear thinking of the Newtonian tradition. Although we may never be able to eliminate surprises from our models, we can still hold out the possibility of creating something approximating what Casti (1994) called a science of the surprising.

As for policy implications, the adaptability perspective emphasizes the cognitive and learning capabilities of individuals, organizations, and geographic regions (Senge 1990) instead of the single-minded pursuit to satisfy the insatiable desire to have unlimited access to resources and information. The adaptive paradigm seeks to balance economic efficiency, social equity, and environmental sustainability so that society and environment can co-evolve in harmony and avoid further tragedies for the commons. Philosophically, adaptability shifts the focus from *accessing without* to *learning within*. From the adaptability perspective, it is far more important for an individual, an organization, or a geographic region to be *inspired* than to be *wired*. Studies on the geography of information society will eventually gravitate toward the three fundamental questions that Immanuel Kant (1998) raised in *Critique of Pure Reason*. Who are we? What can we do? What should we do? Research on the information society may fail miserably if we do not step back to ponder on these more fundamental questions. I strongly believe that a research focus on adaptability may provide better answers to Kant's questions than would a focus on accessibility. Adaptability implies a change of human consciousness that is essential if the world is ever to be transformed to a sustainable state.

Acknowledgements

The author would like to thank Don Janelle for substantive and editorial comments that significantly improved the quality of this chapter; Shawn Dawson and Kristi Rudy at MIDS, Inc. for helping me compile the Internet domain data at the zip code level for the state of Texas; Paul Adams for referring me to TeleGeography'1999; and Fayu Lai for research assistance in data processing.

References

Abler, R.F. 1991. Hardware, software, and brainware: mapping and understanding telecommunications technologies. In Brunn, S.D. and Leinbach, T.R. (eds.) *Collapsing Space and Time: Geographic aspects of communications and information*. New York: Harpercollins Academic, 31-48.

Arrow, K.J. 1962. The economic implications of learning by doing. *Review of Economic Studies* 29:155-73.

Arthur, W.B. 1994. *Increasing Returns and Path Dependence in the Economy*. Ann Arbor MI: The University of Michigan Press.

Asheim, B. 1996. Industrial districts as 'learning regions': a condition for prosperity? *European Planning Studies* 4:379-400.

Barnes, T.J. 1997. Theories of accumulation and regulation: bring life back into economic geography. In Lee, R. and Wills, J. (ed.) *Geographies of Economies*. New York: Arnold, 231-47.

Batty, M. 1995. Cities and complexity: implications for modeling sustainability. In Brotchie, J., Batty, M., Blakely, E., Hall, P., and Newton, P. (eds.) *Cities in Competition: Productive and Sustainable Cities for the 21st Century*. Melbourne: Longman Australia 469-86.

Bednarz, R.S. 1998. Editor as information filter. *Journal of Geography* 98:40-1.

Brin, D. 1990. *Earth.* New York: Bantam.

Brynjolfsson, E. and Yang, S. 1996. Information Technology and Productivity: a review of the literature. *Advances in Computers* 43:179-214.

Cairncross, F. 1997. *The Death of Distance: How the Communications Revolution will Change Our Lives*. Cambridge MA: Harvard Business School Press.

Capra, F. 1996. *The Web of Life: A New Scientific Understanding of Living Systems*. New York: Anchor Books.

Casti, J.L. 1994. *Complexification: Cxplaining A Paradoxical World Through the Science of Surprise.* New York: HarperCollins.

Clemente, P. 1998. *The State of the Net: The New Frontier*. New York: McGraw-Hill.

Cooke P. and Morgan K. 1998. *The Assocational Economy: Firms, Regions, And Innovation.* Oxford: Oxford University Press.

Couclelis, H. (ed.) 1996. *Spatial Technologies, Geographic Information, and the City.* Santa Barbara, CA: University of California, Santa Barbara, NCGIA Technical Report 96-10.

Couclelis, H. 1997. From cellular automata to urban models: new principles for model development and implementation. *Environment and Planning B* 24:165-74.

Davenport, T.H. 1997. *Information Ecology: Mastering the Information and Knowledge Environment.* New York: Oxford University Press.

Dawkins, R. 1976. *The Selfish Gene*. Oxford: Oxford University Press.

England, R.W. (ed.) 1994. *Evolutionary Concepts in Contemporary Economics*. Ann Arbor, MI: The University of Michigan Press.

Feldman, M.P. and Florida R. 1994. The geographic sources of innovation: technological infrastructure and product innovation in the United States. *Annals of the Association of American Geographer* 84:210-29.

Fischer, M. and Gopal, S. 1998. Artificial neural networks: a new approach to modeling interregional telecommunication flows. In Haynes, K.E., Button, K.J., and Nijkamp, P. (eds.) *Regional Dynamics.* Cheltenham: Edward Elgar, 167-89.

Fleck, J. 1993. Learning by trying: the implementation of configuration technology. *Research Policy,* 23 637-52.

Florida, R. 1995. Toward the learning region. *Futures,* 27:527-36.

Gates, B. 1999. *Business @ Speed of Thought: Using a Digital Nervous System*. New York: Warner Books.

Gertler, M. 1995. Being there: Proximity, organization and culture in the development and adoption of advanced manufacturing technologies. *Economic Geography* 71:1-26.

Hägerstrand, T. 1975. Space, time and human condition. In Karlkvist, A., Lundqvist, L., and Snickars, F. (eds.), *Dynamic Allocation of Urban Space,* Farnborough: Saxon House, 3-12.

Handy, S.L. and Niemeier, D.A. 1997. Measuring accessibility: an exploration of issues and alternatives. *Environment and Planning A* 29:1175-94.

Hanson, S. 1998. Off the road? reflections on the transportation geography in the information age. *Journal of Transport Geography* 6:241-9.

Hanssen-Bauer, J. and Snow, C. 1996. Responding to hypercompetition: the structure and processes of a regional learning network organization. *Organization Science* 7:413-27.

Harrison, B., Gant, J., and Kelly, M. 1996. Innovative firm behavior and local milieu: exploring the intersection of agglomeration, industrial organization, and technological change. *Economic Geography* 72:233-58.

Harvey, D. 1996. *Justice, Nature and the Geography of Difference.* Cambridge MA: Blackwell Publishers.

Hodge, D.C. 1997. Accessibility-related issues. *Journal of Transport Geography* 5:33-4.

Hodgson, G.M. 1993. *Economics and Evolution: Bringing Life Back into Economics.* Ann Arbor MI: University of Michigan Press.

Holland, J.H. 1998. *Emergence : From Chaos to Order.* Reading, MA: Addison-Wesley.

Hughes, T.P. 1983. *Networks of Power: Electrification in Western Society, 1880-1930.* Baltimore MD: John Hopkins University Press.

Janelle, D.G. 1993. Urban social behaviour in time and space. In Bourne, L.S, and Ley, D.F. (eds.) *The Changing Social Geography of Canadian Cities* Montreal and Kingston: McGill-Queens University Press, 103-18.

Jin, D.J. and Stough, R.R. 1998. Learning and learning capability in the Fordist and Post-Fordist age: an integrative framework. *Environmental and Planning A* 30:1255-78.

Kant, I. (Guyer, P., and Wood, A.W., trans.) 1998. *Critique of Pure Reason.* New York: Cambridge University Press.

Kelly, K. 1998. *New Rules for the New Economy: Ten radical strategies for a connected world.* New York: Viking.

Krugman, P. 1997. How the economy organizes itself in space: A survey of the new economic geography. In Arthur, W.B., Durlauf, S.N. and Lane, D.A. (eds.) *The Economy as an Evolving Complex Systems II.* Reading MA: Addison-Wesley, 239-62.

Krugman, P. 1998. What's new about the new economic geography? *Oxford Review of Economic Policy* 14:7-17.

Kwan, M.P. 1998. Space-time and integral measures of individual accessibility: A comparative analysis using a point-based framework. *Geographical Analysis* 30:191-216.

Leebaert, D. (ed.) 1998. *The Future of the Electronic Marketplace.* Cambridge MA: The MIT Press.

Leydesdorff, L.A. and van den Besselaar, P. 1994. *Evolutionary Economics and Chaos Theory: New Directions in Technology Studies.* New York: St. Martin's.

Litan, R.E. and Niskanen, W.A. 1998. *Going Digital: A guide to Policy in the Digital Age.* Washington DC: Brookings Institutions.

Lundvall, B. and Johnson, B. 1994. The learning economy. *Journal of Industry Studies* 1:23-41.

Lynch, A. 1999. *Thought Contagion : How Belief Spreads Through Society.* New York: Basic Books.

Margherio, L. 1997. *The Emerging Digital Economy*. Project report prepared for the Department of Commerce. Available on-line at http://www.ecommerce.gov

Marshall, A. 1961. *Principles of Economics* (9th ed., with annotations by C. W. Guillebaud). London: Macmillan for The Royal Economic Society.

Miller, H.J. 1999. Measuring space-time accessibility benefits within transportation networks: Basic theory and computational procedures. *Geographical Analysis*, 31:1-26.

Mirowski, P. (ed.).1994. *Natural Image in Economic Thought: 'Markets Red in Tooth and Claw.'* Cambridge: Cambridge University Press.

Mirowski, P. 1989. *More Heat Than Light: Economics as Social Physics and Physics as Nature's Economics*. New York: Cambridge University Press.

Mitchell, W.J. 1995. *City of Bits: Space, Place, and the Infobahn*. Cambridge MA: MIT Press.

Mody, A. 1993. Learning through alliances. *Journal of Economic Behavior and Organization* 20:151-70.

Mokyr, J. 1997. Are we living in the middle of an industrial revolution? *Economic Review* 82:31-44.

Moore, J.F. 1993. Predators and prey: A new ecology of competition. *Harvard Business Review* May/June, 75-86.

Moore, J.F. 1996. *The Death of Competition: Leadership and Strategy in the Age of Business Ecosystems*. New York: Harper Business.

Nelson, R.R. 1995. Recent evolutionary theorizing about economic change. *Journal of Economic Literature*, 53:48-90.

NUA. 1998. *How Many Online?*
(http://www.nua.ie/surveys/how_many_online/index.html).

Ohmae, K. 1990. *The Boderless World*. New York: Harper Business.

Ohmae, K. 1995. *The End of the Nation Sate: The Rise of Regional Economics*. New York: Free Press.

Paton, R. 1994. *Computing with Biological Metaphors*. New York: Chapman & Hall.

Petchell, J. 1993. From production systems to learning systems. *Environment and Planning* A 15:797-815.

Powell, W. and Smith-Doerr, L. 1994. Networks and economic life. In Smelser, N. and Swedberg, R. (eds.) *The Handbook of Economic Sociology*. Princeton NJ: Princeton University Press, 368-402.

Rogers, J.D. 1998. Executive summary of GVU's WWW User Survey. Available at http://www.cc.gatech.edu/gvu/user_surveys/survey-1998-04/

Rogerson, P.A. 1996. Estimating the size of social networks. *Geographical Analysis* 29:50-63.

Rosenberg, N. 1982. *Inside the Black Box: Technology and Economics*. Cambridge: Cambridge University Press.

Schindler, P.E. 1998. Home Internet use breeds depression.
http://www.techweb.com:3030/internet/story/NIN19980831S0001

Scott, A. 1996. Regional motors of the global economy. *Futures*, 28:391-411.

Senge, P.M. 1990. *The Fifth Discipline: The Art of Practice of the Learning Organization*. New York: Doubleday.

Shen, Q. 1998. Location characteristics of inner-city neighborhoods and employment accessibility of low-wage workers. *Environment and Planning B* 25:345-65.

Simon, H.A. 1997. Designing organizations for an information-rich world. In Lamberton, D.M. (ed.) *The Economics of Communication and Information*. Cheltenham UK: Edward Elgar Publishing, 187-202.

Storper, M. 1997. *The Regional World: Territorial Development in a Global Economy*. New York: Guilford.

Sui, D.Z. 1998. Environmental impacts of the E-merging digital economy: Toward the greening of electronic commerce via information ecology. Working paper, Department of Geography, Texas A&M University, College Station, Texas (available from the author upon request).

Sui, D.Z. 2000. Geography without vision? Talking about geography in the new millenium. *Annals of the Association of American Geographers* (in press).

Tapscott, D., Lowy, A., and Ticoll, D. (eds.) 1998. *Blueprint to the Digital Economy: Wealth Creation in the Rra of E-Business*. New York: McGraw-Hill.

van der Poel, M.G.M. 1993. Delineating personal support networks. *Social Networks* 15:49-70.

Warf, B. 1997. Telecommunications and the changing geographies of knowledge transmission in the late 20th century. In Alberts, D.S. and Papp, D.S. (eds.) *The Information Age: An Anthology on its Impact and Consequences.* Available from http://www.ndu.edu/ndu/inss/books/anthology1/index.html.

Wilson, E.O. 1975. *Sociobiology: The New Synthesis*. Cambridge MA: Harvard University Press.

Part II

Visualization and Representation

8 Representing and Visualizing Physical, Virtual and Hybrid Information Spaces

Michael Batty[1], **and Harvey J. Miller**[2]

[1] Centre for Advanced Spatial Analysis, University College London, 1-19 Torrington Place, London WC1E 6BT, UK. Email: m.batty@ucl.ac.uk

[2] Department of Geography, University of Utah, 260 S. Central Campus Drive Room 270, Salt Lake City UT 84112-9155, USA. Email: harvey.miller@geog.utah.edu

8.1 Introduction: Varieties of Space

The strongest convention in contemporary geographic thought is the notion that geographic space is rooted in a Euclidean geometry that defines the physical world. Although geographers have long sought to escape this paradigm through a rich array of perceptions based on ways in which we might imagine space, physical distance or its economic surrogates still provide the basic logic used by geographers to order their world and to make sense of the way activities locate in time and space. There is however a sea change in the making. As the world moves from one organized around energy to one based on information, the role of physical distance is changing as it is complemented by near instantaneous transactions that dramatically distort the effect of distance, thus changing the traditional bonds that have led to the current geographical organization of cities, regions, and nation states (Cairncross 1997).

This transition is from a society dominated by the movement and manipulation of materials to one dominated by the movement and manipulation of information. In Negroponte's (1995) terms, it is a transition from a world based on *atoms* to one based on *bits*, from a material world to an ethereal one where the convergence of computers and communications – the devices used to manipulate and transport the *bits* – has evolved new varieties of space, collectively referred to, in the popular lexicon, as *cyberspace*.

Despite the rapid emergence of cyberspace and the many attempts to chart and measure its morphology and spread, a more complete and focused conception is in the idea of the *information space*. The new information spaces that are emerging are rooted in both the material and ethereal worlds of commodities and flows, and cannot be understood without each other (Castells 1996). If we are to explore the continued relevance of ideas based on the measurement of accessibility or propinquity defined traditionally in relation to physical locations and interactions, then we need to examine the ways in which information and energy are combining to create new spaces and new patterns of human behavior. In short, new definitions and conceptualizations of accessibility can only be defined by mapping physical or material space onto virtual or ethereal space, thus defining a nexus of hybrid space which, we will argue here, represents the appropriate focus for a new geography of the information age.

A popular example serves to make our point. The current fascination with the online bookshop Amazon.com, which is mentioned many times in this book, is based on the notion we can substitute making a physical visit to the bookshop with a virtual visit, even engaging in price comparison, reducing the need for several visits to different places. However although much of our behavior in browsing and purchasing is removed from the physical realm, ultimately the book needs to be delivered in its material form and this depends on where it is warehoused and how it is shipped. In the case of Amazon.com, there are 6 warehouses in the United States and 2 in Europe, strategically located to minimize physical transfer costs and to maximize access to population centers, thus reinforcing long established ideas that location ultimately depends upon physical accessibility (Dodge 1999). Of course bookshops are probably not the best example as the product itself can easily be made virtual – it does not need to be material – although food shopping and other popular kinds of e-commerce reinforce the point.

We can visualize the interpenetration of these two kinds of space, although in themselves they are much variegated, as an intersection of two worlds. It is even possible that there is simply one world, for one cannot exist without the other, although there is an assumption that the physical world existed prior to the virtual. In the sense of the virtual world being exclusively cyberspace, this is indeed the case. However, it is convenient to consider these domains as intersecting but not being coincident, for this serves to show how geographers (and all other social scientists, of course) often abstract them as one or the other, thus providing perspectives from one viewpoint or the other.

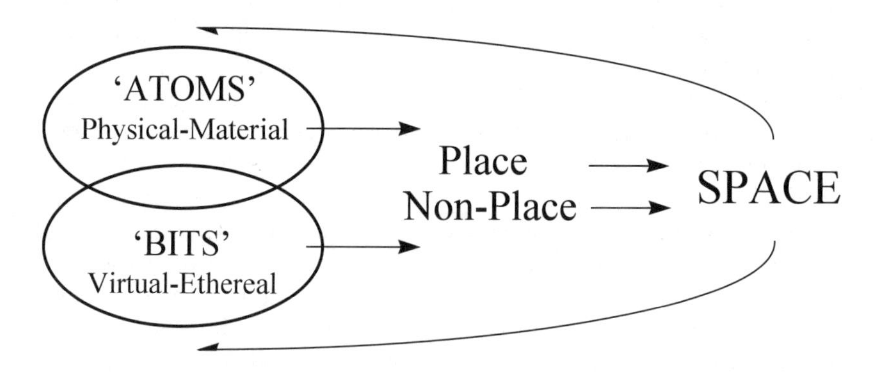

Figure 8.1. Geographic abstraction of physical, virtual and hybrid worlds.

The first level of geographic abstraction is in terms of the place or non-place urban realm (Webber 1964) but spatial analysis usually lies beyond this to a level of abstraction where there is rarely any distinction between the aspatial and the spatial in geographic terms. For our current focus on the role of accessibility in articulating and understanding the morphologies and morphogenesis of these new information spaces, Figure 8.1 shows that traditional accessibility and related spatial analysis might be applied to one or the other but the message of this introduction is that neither are any longer complete as information space is necessarily a product of both. There is still a place for analyses of one or the other, which emphasize certain characteristics to the exclusion of others, but an integrated understanding must be based on the analysis of hybrid information spaces.

So far there are few meaningful classifications of physical and virtual space, and there have been even fewer attempts at exploring how concepts and models developed for one can transfer to the other. In short, the extent to which we can generalize geographic theory from the predominantly physical domain to the virtual or rather adapt such theory to the hybrid domain is unclear. To an extent, this is what all the contributors to this book are attempting from different perspectives. There are many ideas, but most are preliminary and somewhat rudimentary. It appears that the economic geography of the production and consumption of these new technologies bears an uncanny resemblance to the old order, with the role of historical accident and agglomeration economies still being significant in where such activity is located. For the ways cities are restructuring, then the effect of distance is being distorted as physical ties on single locations are loosened – witness the edge-city effect, particularly in North America, and the growth of the global city – but, as yet, we have little idea of how our collective access to facilities is changing. Clearly the transition to an information age is increasing opportunities for different kinds of physical and virtual interaction dramatically for some groups, while locking out others, although even the contemporary geography of disadvantage seems to follow established social, ethnic, and gender lines.

In our introduction to this part of the book, we will consider how the traditional notion of accessibility is relevant to an understanding of the way these new information spaces are being structured and how old spaces are being restructured. We will first explore these traditions and then we will speculate on how existing spatial analysis and geographic information systems and science might be used to detect the new morphologies of information (see Batty, *et al.* 1998, Clarke 1998). We will then suggest a wider research agenda, concluding with some more specific ideas as to where research might be immediately targeted.

8.2 Prevailing Themes

The Inadequacy of Traditional Definitions

The traditional definition of accessibility focuses on physical proximity. For example, three major approaches to measuring accessibility are constraint-based measures, attraction-accessibility measures, and benefit measures (see Miller 1999). Constraint-based measures demarcate the activity locations that are available to an individual, typically from a space-time perspective. Attraction-accessibility measures measure the trade-off between the attractiveness of destinations versus their interaction costs. Benefit measures involve a similar trade-off but attempt to measure explicitly the benefits accruing to the individual from a choice set. In all three approaches, there is an explicit assumption that physical distance is a major structuring factor that influences spatial choices and therefore accessibility.

In contrast, the virtual world ignores (or at least greatly discounts) physical space. The *cost* of virtual interaction has little to do with relative location. Instead, virtual interaction cost relates to factors such as network capacity, server capacity, and current load; these translate into the latency (delay time) experienced by the user. Another cost is the difficulty in navigating the space and extracting useful information (see Dodge, Chapter 11). This suggests the need for a major reconceptualization and expansion of our definitions of accessibility.

It is also not obvious that traditional mapping techniques can yield significant insights into virtual spaces. The terms 'virtual space' and 'cyberspace' may in fact be oxymorons since there is little that is 'spatial' in these realms, at least in the traditional sense. For representational and analytical tractability, formal and computational models of physical space invoke certain restrictions on its topological and geometric properties. These typically include metric-space properties for distance measures such as non-negativity, identity, symmetry and triangular inequality (see Beguin and Thisse 1979, Smith 1989). A cyber-spatial analysis by Murion (Chapter 12) suggests that latency, the most obvious distance measure in virtual space, does not obey the properties of metric space. We can handle the relaxation of metric-space properties only if they are carefully controlled (see Muller 1982, Smith 1989). Moreover, current GIS software does not treat non-Euclidean space in an appropriate way.

An alternative to direct mapping of virtual space, implied by virtual interaction, is to map locations of the physical and logical components of virtual space within physical space. An example is mapping the locations of Internet hubs, host computers, domain names, or backbone networks within physical space as a measure of accessibility to cyber-space. Empirical analysis by Moss and Townsend (Chapter 10) suggests caution. The spatial/geographical metaphor may not be appropriate, particularly since information flow in most networks apparently does not correlate with geographical space (see Mitchell 1995).

Interactions between Virtual Space and Physical Space

Although virtual space is aspatial and does not correlate well with geographic space, it is clear that virtual space and physical space influence each other. Activity in virtual space can affect activity in physical space and vice-versa. For example, virtual interaction can be both a substitute and a complement to physical interaction. An example of the former relationship is when an individual shops online rather than visiting a retail establishment. An example of the latter case is when an individual uses the Web to find a new restaurant or plan a vacation.

To date, there is little research on the interactions between activities in virtual space and physical space (but see, however, Salomon (1986); and Shen (1998)). There is a tradition within the geographic literature on measuring the interaction between physical interaction and perceived/experienced geographic space. For example, Abler (1975) explores the impact of space-adjusting technologies on human activities in geographic space while Janelle (1968, 1969, 1991) has pioneered the concept of time-space convergence to describe the radical impacts of transportation on spatial relationships. Reginald Golledge and Waldo Tobler have developed analytical techniques for transforming geographic space based on the perceived distances and observed interaction patterns (see, e.g., Golledge and Spector 1978, Tobler and Wineburg 1971, Tobler 1976, 1978), but the usefulness of these techniques for visualizing hybrid space has not yet been explored.

Identifying significant centers and locations in both the virtual and material worlds is also an important task for future research. Of equal importance to measuring flows is research into the content of such flows and into the processes that mediate these flows. Markets are increasingly structured in real time across electronic networks. This poses a level of complexity on the real world that makes traditional market analytical techniques untenable. Also important is developing clear notions of the demand for and supply of information, particularly with respect to highly diverse networks where there are already very clear distinctions in terms of usage. What is available and what is required for what purposes are very different notions that must be identified, not only in relation to new information spaces but also to how these spaces map onto existing physical spaces. These may be articulated at various scales from social networks to global markets.

The Quality of Interaction in Physical, Virtual and Hybrid Worlds

Questions as to the quality of interactions in and hybrid worlds are also central themes. Despite progress in immersive and virtual reality technologies, it is likely that qualitative differences between interaction using physical and virtual modes will persist for some time. Previous research on consumer search behavior clearly demonstrates that individuals use different modes depending on the type of information sought (for a review, see Miller 1993). As virtual and hybrid worlds expand to encompass more aspects of daily life, it is likely that a complex partitioning of activities among these modes will occur. The nature of this partitioning is far from clear.

A related theme concerns the quality of information within physical, virtual, and hybrid spaces. Vagueness and fluidity characterize virtual spaces. Apparently good and bad information spaces can be contrasted with good and bad information within these spaces with no one-to-one correspondence between each. The transitory nature of the digital world contrasts with the material world where information spaces are usually structured in terms of built form that has a life span with some permanence. This focus on temporality is an issue that serves to test the limits of our debate, reinforcing the long standing idea that time geography and accessibility in time, as well as or rather than space, is of much more significance here than we had hitherto thought.

The cliché of the digital world – that networks enable people to interact with *anyone*, *anywhere*, at *any time* and in *any place* – illustrates our crude vision of the emerging digital world. Here our focus is much more considered with an emphasis on how humans interact with one another in space and time, adapting to access the *right amount* and the *right* information in the *right time* and in the *right place*. Harvey and Macnab (Chapter 9) clearly illustrate these issues with their discussion of interpersonal temporal accessibility at the global scale. The ability to interact in real-time may be the critical factor that distinguishes among world cultures rather than traditional notions of geographic separation and determinism.

Navigation Tools for Information Space

Another theme that weaves its way through the debate involves the development of tools and protocols to enable efficient navigation through information space. This instrumental viewpoint suggests that good geographical metaphors, grounded in good theory about the information society, should be at the basis of navigation tools that link behavior to purpose. Much visualization work is concerned with the development of better tools. These tools are being developed and tempered by the various institutional structures that require them. At the same time, we are beginning to learn more about new information spaces using the very tools designed for using these spaces in a routine fashion. As we develop tools to explore these new spaces, these tools are being used in routine ways to navigate them. At the same time, the existence of these tools modifies these spaces. This is a kind of relativism that is rarely highlighted in the material world.

8.3 The Morphology of Cyberspace

Although there are few indicators of the kinds of structure and behavior, form and process, that are determining the shape of these new information spaces, there are already some hints as to what we might expect. As in traditional studies of geographic space, these are dominated by explanations and models of the morphology of cyberspace in its many forms, rather then the processes through which it is evolving. This emphasis on form rather than process is one of the most problematic features of social research. It is not only difficult to observe social and economic processes at work, it is often impossible to infer the *modus operandi* of human decision-making that is determined by multiple causes and contextual circumstances. In the case of cyberspace, this is doubly difficult in that getting access to observe processes that take place invisibly across networks requires very special analytical skills and even then, the completeness of any survey is forever in doubt.

Cyberspace like other spaces has a form that is being mapped, and a natural starting point is to see whether or not the frictionless world that has emerged has any parallel in traditional geographic spaces. The macro-properties of traditional physical space has largely been charted and explained using ideas from social physics. Interaction patterns and accessibility measures were originally developed in analogy with the laws of classical physics, with gravitational force and potential energy being concepts of great relevance in *explaining* or at least summarizing how space becomes structured. In terms of interaction, Murnion (Chapter 12) has made several studies of interaction over the Internet using traditional models but with distance being replaced by new measures of *latency*, which have shifted the focus away from Euclidean distance and its surrogates – time or cost – to network measures, which depend upon switching and relays and the capacity of telecommunications.

Even more specific results that appeal to social physics are being produced. Bernado Huberman and his colleagues at Xerox PARC, in a series of studies of the size and shape of the Web have produced quite conclusive evidence that the net is scaling in that servers and server capacity are distributed according to the rank-size rule (Markoff 1999). This has implications for the structure of the net as a hierarchical system – remember that Christaller's central place system generates rank-size or at least scaling laws of center size – while work grounded in the distribution of Web hosts geographically bears out similar scaling. Moss and Townsend (Chapter 10) reveal that same kind of pattern for the New York region, while Shiode and Dodge (1998) provide a dramatic graphic of this kind of spatial organization in their picture of Web hosts in South East England, shown in Figure 8.2. Contrast this with Stewart and Warntz's (1958) maps of population potential of North America and Britain, which is at the basis of the physical measurement of accessibility across many scales.

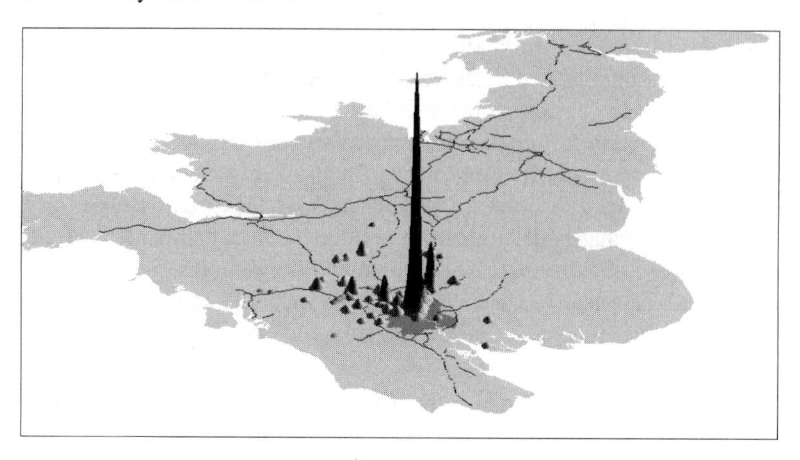

Figure 8.2. The new geography of economic potential: accessibility to Internet hubs (from Shiode and Dodge 1998).

Casual evidence also reveals the importance of the local and the global in the morphology of cyberspace. Any casual examination of your email log will reveal distance decay around your local site that is obvious enough in that most email deals with human activity through a virtual medium in a local physical space. However for academics and increasingly for the public at large logged into the Web, such local interaction is being supplemented with global interaction, which binds our social networks together in ways that serve to strengthen the global economy. New network studies of the small-world problem (Watts and Strogatz 1998) suggest that such occasional global ties increase interaction much more significantly than the number of such ties might suggest and, thus, studies of the net and nets using such ideas appear promising.

For the shape of the net, there is enough evidence to show that this too is scaling – fractal – in that links (through Web sites for example) are scaling in importance, and follow the classic dendritic pattern that we find in many areas of morphology from the growth of crystals to the growth of cities and other organisms and organizations. In Figure 8.3(a), we show a piece of the network diagram of the net produced by Bill Cheswick and Hal Burch at Bell Labs, which has clear fractal structure. The particular segment – its nodes and links – is not important *per se* but its fractal structure is. In Figure 8.3(b) we show the morphology of a virtual community in cyberspace – Alphaworld – reported on by Dodge (Chapter 11), which has also evolved as a fractal around the point that members enter this virtual world – at ground zero. There is no friction of distance in Alphaworld but its physics is dictated by who came first and the town has evolved around ground zero and along easily recognizable radial routes.

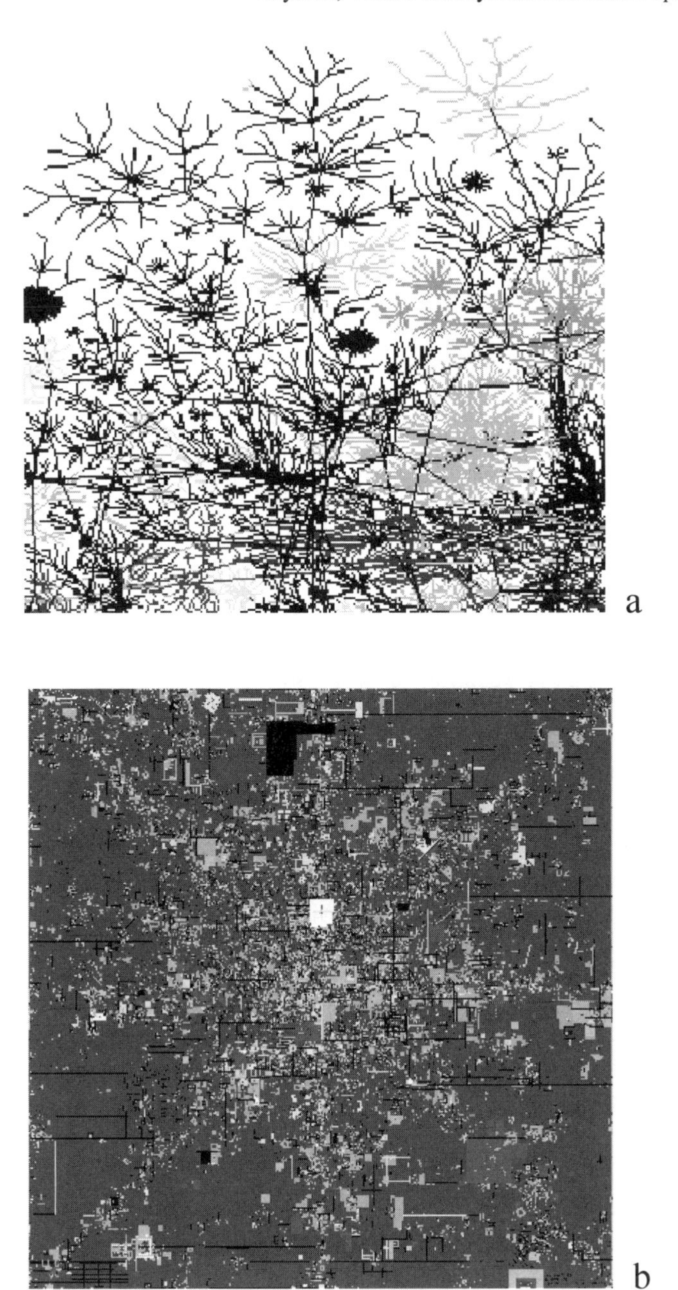

Figure 8.3. Morphologies of virtual space: cyberspace. (a) the fractal structure of the Internet (http://cm.bell-labs.com/cm/cs/who/ches/map/index.html); (b) the spatial structure of a virtual world – Alphaworld (see Dodge, Chapter 11).

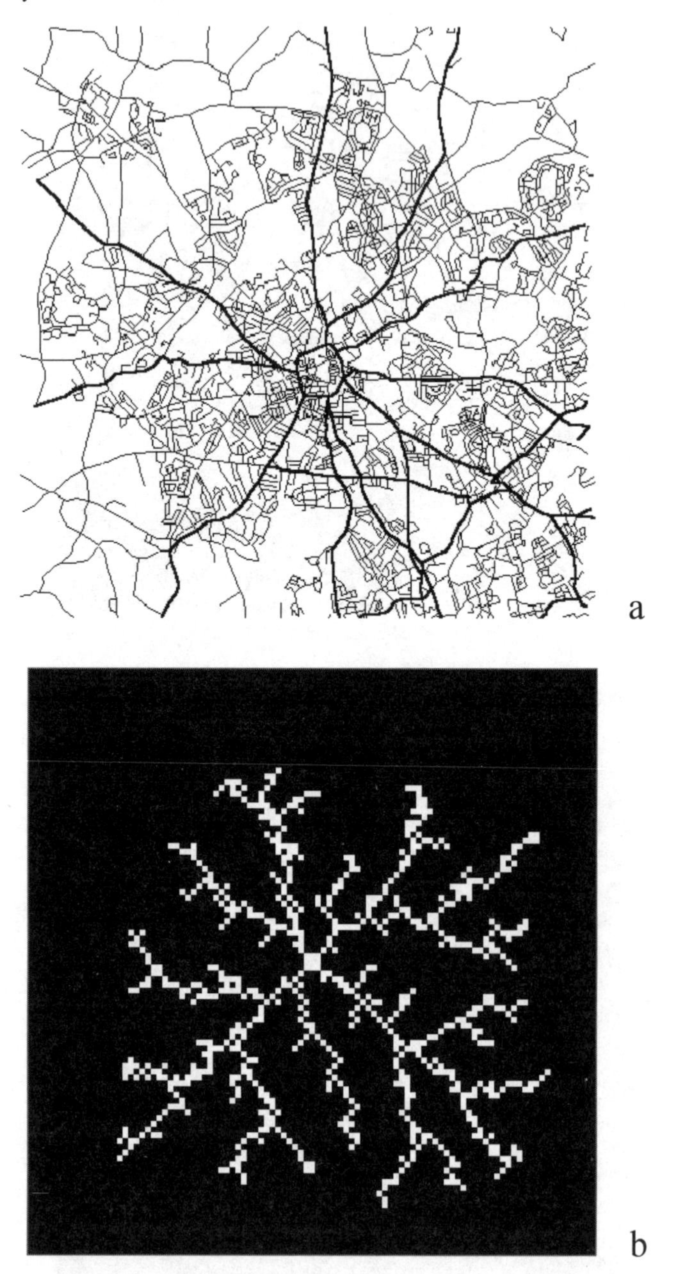

Figure 8.4. Morphologies of real space: Euclidean geographical space. (a) route structure of a medium-sized English town; and (d) model of urban growth based on diffusion-limited aggregation.

We can compare these to more traditional ways of measuring urban space. In Figure 8.4(a), there is a picture of the road network in a medium-sized English town (Wolverhampton: population circa 250,000) that is clearly fractal, although this might be at any scale, and finally in Figure 8.4(b) we show crystal growth using the process of diffusion-limited aggregation, which has been used to measure and model several real towns. All these examples show that the physical and the virtual worlds have much in common, and this suggests that models of physical space, such as accessibility, may have more to offer in the study of new information spaces than we have assumed hitherto.

These are but brief forays into a speculative realm that forms a much wider research agenda set by all the contributions in this book. Here we will conclude our introduction to this section by sketching this agenda for measuring and modeling hybrid information spaces.

8.4 A Research Agenda

We will now attempt to cull all these deliberations into a plan for future research based on (i) what do we know? and what do we have? (ii) what are appropriate future research directions? and; (iii) what are appropriate research questions? We will deal with these in turn. Although the various debates that follow in the chapters within this section cover a wide range of issues, we will map out our research agenda in terms of the initial themes introduced here, stating these as questions that frame research directions.

We begin by considering 'What Do We Know about Visualizing and Representing Information Space?' A thorough review of these questions is required. This might be accomplished through research projects but it is more likely to come from the current generation of researchers, such as those of us writing here, coming to conclusions similar to our own and spontaneously developing such reviews and statements. There are several themes that might spin-off from such reviews and we will list these:

- **Representing Networks**. This includes reviewing different ways of coding and identifying networks based on extensions of graph theoretic measures, methods of sampling and so on that can account for their virtual as well as physical/logical nature.
- **Conceptualizing Activity Spaces and Accessibility Measures.** These are relevant to the virtual world but have developed to date largely for spatial issues in the real world.
- **Cataloguing Market Data**. This includes reviewing methods for counting and observing network flows and new concentrations of information in real and virtual space.

- **Exploring the Role of Geographic Information Science in the Analysis of Virtual and Hybrid Information Spaces**. This focuses on assessing how far existing methods of GIS in particular and spatial analysis in general are useful for mapping new information spaces.
- **Exploring the Role of Scientific Visualization in Measuring and Mapping**. This requires reviewing how new methods of visualization for spatial and non-spatial data in spaces with many dimensions might be used to chart new information spaces.

These reviews begin to merge into major research questions to define the research frontier, and there are some obvious areas that require research programs: again we state four of these to provide some sense of where we consider the focus should be:

- **Researching the Flow and Cost of Information**. How flows can be identified and linked to the emergence of new spaces, which, in turn, map onto existing market, social, and institutional processes.
- **Tools of Cybernavigation**. The development of new tools for both exploring and moving through information spaces that are based on insights into the emergence of such spaces, the interface between activity in real and virtual worlds, and developments in human-computer interaction.
- **Mapping Activity Spaces**. Exploring ways in which existing approaches within time geography can be informed and extended by network paradigms, network flow data, and scientific visualization.
- **Visualization of Connections between Virtual and Real / Material Geographies**. Providing insights into how information spaces are connected to real spaces through augmenting existing measures of accessibility and the development of new ones.

Much more specific research issues can also be identified, which might drive forward the research agenda. There is an urgent need for a major initiative in the collection of network data and its subsequent analysis with respect to the search for new information spaces. These initiatives could take many forms and we list four here:

- an Internet census: a data archive for the Internet
- the definition of private networks
- the collection of behavioral data associated with many varieties of network
- the role of time sampling in the use of networks

We also need to evaluate the role of existing tools in spatial analysis and to develop new tools relevant to the issues we have identified pertaining to the analysis of information spaces within an information society. This should focus in particular on principles and tools in contemporary cartography and scientific visualization. We also need to develop new theories that generalize the concept of distance from physical to the virtual domains, and from this would flow models and visualizations of accessibility in real, virtual, and hybrid spaces based on generalizations of geographic distance in formal, logical, and computational terms.

This research agenda is wide and deep but although there are many straws in the wind, it would appear that much geographic and spatial knowledge developed in the last 100 years, although broadly relevant in a philosophic sense to new inquiries into information spaces, needs to be redefined, reworked, and restructured in ways that meet the challenges we have identified here, and the ways these are elaborated in the articles that follow.

References

Abler, R. 1974. Effects of space-adjusting technologies on the human geography of the future. In Abler, R., Janelle, D., Philbrick, A., and Sommer, J. (eds.) *Human Geography in a Shrinking World*. North Sciutuate MA: Duxbury Press, 35-52.

Batty, M., Dodge, M., Doyle, S. and Smith, A. 1998. Modelling virtual environments. In Longley, P.A., Brooks, S.M., McDonnell, R., and Macmillan, B. (eds.) *Geocomputation: A Primer*. Chichester UK: John Wiley, 139-61.

Beguin, H. and Thisse, J.-F. 1979. An axiomatic approach to geographical space. *Geographical Analysis*, 11:325-41.

Cairncross, F. 1997. *The Death of Distance*. Cambridge MA: Harvard Business Press.

Castells, M. 1996. *The Rise of the Network Society*. Oxford UK: Blackwells.

Clarke, K. 1998. Visualising different geofutures. In P.A. Longley, Brooks, S.M., McDonnell, R., and Macmillan, B. (eds.) *Geocomputation: A Primer*. Chichester, UK: John Wiley, 119-37.

Dodge, M. 1999. Finding the source of the Amazon.com: Examining the truth behind the hype of the 'earth's biggest bookshop'. A paper presented to the *5th E*Space Conference*, Cape Town, South Africa, July 1999.

Golledge, R.G., and Spector, A. N. 1978. Comprehending the urban environment: Theory and practice. *Geographical Analysis* 10:403-26.

Janelle, D.G. 1968. Central place development in a time-space framework. *Professional Geographer*, 20:5-10.

Janelle, D.G. 1969. Spatial organization: A model and concept. *Annals of the Association of American Geographers* 59:348-64.

Janelle, D.G. 1991. Global interdependence and its consequences. In Brunn, S.D., and Leinbach, T.R. (eds.) *Collapsing Space and Time*. London: Harper-Collins, 49-81.

Markoff, J. 1999. Not a great equalizer after all ? On the Web, as elsewhere, popularity is self-reinforcing. *The New York Times* (Monday June 21, C4).

Miller, H.J. 1993. Consumer search and retail analysis. *Journal of Retailing*, 69:160-92.

Miller, H.J. 1999. Measuring space-time accessibility benefits within transportation networks: Basic theory and computational procedures. *Geographical Analysis*, 31:187-212.

Mitchell, W.J. 1995. *City of Bits: Space, Place and the Infobahn*. Cambridge MA: MIT Press.

Muller, J-C. 1982. Non-Euclidean geographic spaces: Mapping functional distances. *Geographical Analysis*, 14:189-203.

Negroponte, N. 1995. *Being Digital.* New York: A. A. Knopf.

Salomon, I. 1986. Telecommunications and travel relationships: A review. *Transportation Research A*, 20:223-38.

Shen, Q. 1998. Spatial technologies, accessibility and the social construction of urban space. *Computers, Environment and Urban Systems*, 22:447-64.

Shiode, N., and Dodge, M. 1998. Using GIS to analyse the spatial pattern of the Internet in the United Kingdom. A paper presented at the GIS Research UK 6th National Conference, 31st March - 2nd April 1998, Edinburgh UK.

Smith, T.E. 1989. Shortest path distances: An axiomatic approach. *Geographical Analysis*, 21:1-31.

Stewart, J.Q., and Warntz, W. 1958. Physics of population distribution. *Journal of Regional Science*, 1:99-113.

Tobler, W.R. 1976. Spatial interaction patterns. *Journal of Environmental Systems*, 6:271-301.

Tobler, W.R., 1978. Migration fields. In Clark, W.A.V. and Moore, E.G. (eds.) *Population Mobility and Residential Change.* Northwestern University Studies in Geography Number 25:215-32.

Tobler, W.R., and Wineburg, S. 1971. A Cappadocian speculation. *Nature* 231:39-41.

Watts, D.J., and Strogatz, S.H. 1998. Collective dynamics of 'small world' networks. *Nature* 393:440-42.

Webber, M.M. 1964. The urban place and the non-place urban realm, in Webber, M.M., Dyckman, J.W., Foley, D.L., Guttenberg, A.Z., Wheaton, W.L.C., and Wurster, C.B. (eds.) *Explorations into Urban Structure.* Philadelphia PA: University of Pennsylvania Press, 79-153.

9 Who's Up? Global Interpersonal Temporal Accessibility

Andrew S. Harvey[1] **and Paul A. Macnab**[2]

[1] Time Use Research Program, Department of Economics
Halifax, Nova Scotia, Canada, B3H 3C3. Email: Andrew.Harvey@stmarys.ca

[2] Department of Geography, Saint Mary's University
Halifax, Nova Scotia, Canada, B3H 3C3. Email: Macnab@ns.sympatico.ca

9.1 Introduction

A petroleum executive sitting at her desk in Houston knows when a satellite phone call might find her chief geologist hard at work in Uzbekistan, but this is nothing new. Since the end of the World War II, many a British veteran has maintained scheduled weekly contact with ham radio comrades half-a-day away in Australasia. And long before the Tokyo Stock Exchange operated on a 24-hour cycle, brokers in New York City knew when to reach colleagues at the office in Japan; if they weren't certain, all they had to do was glance up at one of the half dozen or so clocks mounted on the wall. Even new initiates to the World Wide Web learn quickly that file transfers can be expedited by downloading from sites in countries where most of the population lies dormant. Computer-mediated communication across time zones surely must reach its zenith in emergency tele-medicine: interactive specialists in Toronto, Paris, and Auckland confer while viewing digital injuries sent from backpack transmitters by doctors on the ground in war-torn Bosnia. Fine and well, but when might a Japanese child log on in the classroom to chat with a virtual pen-pal in Canada? Chances are, never. Even in the near future of affordable webcams, speech recognition, and simultaneous translation, it will be impossible for students in Halifax to converse with sister-city pupils in Hakodate. Why? When Japanese students start school at 9 o'clock in the morning, it is 10 o'clock at night on Canada's east coast. Simply put, one city or the other is always going to be asleep. So obvious, perhaps, are such points that they have largely escaped the gaze of academic researchers in the social sciences. The ability of individuals to interact in real-time around the planet by way of the Internet forms the basis of this paper. Beginning with the assumption that the time of day will always be one of the few things truly separating the world's cultures, we explore interpersonal accessibility with a particular emphasis on the role of daily activity patterns. At this early stage in our investigation, we conceptualize a 'personal real-time accessibility index' to measure the potential for direct, immediate, and responsive face-to-face global remote interaction. Following some discussion of research in accessibility and of studies in time use, a case study of of research in accessibility and of studies in time use, a case study of Canada is presented to illustrate diurnal measures of accessibility across six time zones.

9.2 Interaction and Personal Real-time Accessibility

There has been increasing concern over an impending decline in social interaction. Several factors continue to stimulate this concern. First is the declining role of the workplace, at one time, the main nucleus for networking and social mobility in modern society. Next is the escalating complexity related to the synchronization of individuals' time budgets. This *colonization of time*, as it has come to be known, is expected to deter social interaction considerably. A future society might be envisioned in which individual units remain connected through cyberspace yet isolated in time and geographic space. This can be seen as a dangerous trend since social interaction has been a strategy of individual survival both within and away from the workplace. In spite of spatial separation, emerging communication technologies enhance the ability of individuals to share experiences and information. Computers, for example, '... are being used more and more ... for all types of communication, which traditionally have taken, place face to face' (Batty 1997). The most timely and meaningful exchanges still require personal accessibility and direct interaction in real-time. While phone systems and conference calls most often enable these exchanges, various Internet tools such as 'instant messaging' are gaining in popularity.[1] Significant barriers do, however, still exist to such real-time sharing of experiences and information at the global level.

Accessibility and Potential

As numerous authors note, the concept of accessibility is easy to use and hard to define (Pirie 1979). We will not enter that debate in this paper except to indicate that by accessibility we are interested primarily in access to people (e.g., telephoning someone) and not in access to places (e.g., lunch at an exclusive country club). Two particular aspects of accessibility that have been identified in the literature are drawn upon for our explorations: first is the notion of connectivity, and second the notion of potential (Fellman, Getis, and Getis 1990). The Internet provides connectivity. What needs to be determined is the potential for individual interaction.

Interpersonal temporal accessibility (ITA) or interpersonal reachability means an individual is available to interact reciprocally with other individuals in the normal course of daily activity. In its simplest form, the *personal real-time acces-*

[1] In a discussion of the proliferation of IRC sites (Internet Relay Chat), Rheingold (1993) characterizes the assembled interlocutors as *real-time tribes*.

sibility index is the population reach (i.e., the effective number of persons reachable) from a given location at any given time. Extensions could be the role group (e.g., worker, student, other) and possibly the nature of the contact. The index varies both spatially and temporally. The potential for ITA in some locations greatly exceeds the potential in other locations. For any given location, potential will vary over the diurnal cycle as distant locations and peoples become more or less accessible.

In its simplest form, the ITA index at any hour of the day can be given as:

$$IA = \sum_{j=1}^{24} a_j * u_j * P_j \tag{9.1}$$

Where:

IA = index of interpersonal accessibility measured in persons
P = population
u = capability factor (proportion of persons using the Internet)
a = coupling factor (proportion of the population awake and alone)
j = time zone (j=1..24)

Hence, the total potential population reach from a given time zone is the sum of the number of persons available for interaction in all of the time zones. This measure shows only how many individuals can be contacted with no indication of the duration of time they are available for communication. The total time availability can be calculated as the duration (d) of time persons are alone and awake:

$$IAT_j = \sum a_j * u_j * P_j * d_j \tag{9.2}$$

Where:

IAT = index of interpersonal accessibility measured in terms of time
d = average duration of time alone per person

Personal Accessibility

Activities are undertaken within an objective and subjective context. Objectively they occur at a location, both in time and over time and in the presence or absence of other individuals. These spheres are manifest in activity settings (Harvey 1982). Both space (location) and time are themselves multi-faceted. On one hand, fixed coordinates immutably define space. On the other hand, built space is defined generically as structures or activities (e.g., home and work).

Giddens (1990) identifies two types of social interaction: (1) face-to-face and, (2) remote. Both transport and communications mediate the latter, identified as time-space distanciation. Since the rapid diffusion of information technology has considerably diminished the friction of distance as an impediment to interaction (Graham and Marvin 1996, Cairncross 1995), the present investigation will focus on the role played by communication. In Janelle's (1995) view, communication can require temporal coincidence, or not, and spatial coincidence, or not. Various possibilities are shown in Table 9.1.

Table 9.1. Spatial and temporal constraints on communication systems

<table>
<tr><th colspan="2" rowspan="2"></th><th colspan="2">Spatial coincidence of communicating parties required</th></tr>
<tr><th>Yes</th><th>No</th></tr>
<tr><td rowspan="2">Temporal coincidence of communicating parties required</td><td>Yes</td><td>Face-to-face meeting A</td><td>Picture phone B
Phone –
(wire/cell/satellite)
Teleconference
(audio or audio-visual)
Radio - CB/HAM/VHF
Net phone
Instant messaging
Cuseeme</td></tr>
<tr><td>No</td><td>Refrigerator notes C
Hospital charts</td><td>Answering and D
recording machines
Mail/E-mail
Telegrams, telex, fax
Printed publications
Computer conferencing</td></tr>
</table>

Source: Adapted from Janelle 1995

Combining the perspectives of Giddens and Janelle provides a useful departure point for our exploration. Quadrant A in Janelle's typology corresponds with Gidden's category of *face-to-face* communication while quadrants B, C and D parallel Gidden's notion of *distanciated contact*. Note, however, that the modes of communication listed in quadrant B are significantly different from those in C and D. Real-time distanciation (B) must be distinguished from delayed distanciation (C and D). Real-time distanciation indicates that the means of communication enables an immediate exchange of information. Such an exchange might be thought of as mediated face-to-face contact.[2] Delayed distanciation involves a separation in both time and space. We have three meaningful categories with respect to communication:

(1) Face-to-face contact (A) – spatial and temporal coincidence
(2) Mediated face-to-face contact (B) – temporal coincidence
(3) Mediated distanciation (C & D) – no temporal or spatial coincidence

The remainder of this investigation is concerned with the second category, mediated face-to-face interaction. Our focus is communication, rather than transportation, and temporal constraints, rather than spatial constraints. Temporal constraints take two forms and the significance of each must be recognized. First, temporal location determines when activities occur (e.g., travel to work in the morning). Second, duration determines the amount of time needed (e.g., viewing a movie requires about two hours). In order for mediated face-to-face contact to take place, individuals must be free of both constraints; that is, they must be temporally accessible at the required moment and have sufficient time available to interact meaningfully with others.

9.3 Determinants of Accessibility

Hägerstrand (1970) posits three sets of constraints that will be drawn upon for our discussion of accessibility: (1) capability, (2) coupling, and (3) authority. *Capability* constraints are imposed by physiology, abilities and access to tools. *Coupling* constraints define when, where, and for how long one needs to join with other individuals. An individual or group imposes *authority* constraints through control of a time-space entity. Each constraint plays a role in determining global ITA. The increasing availability and affordability of computers are enhancing the capability for global ITA by way of the Internet. Although barriers have been greatly diminished, they are far from having been eliminated (Schuler 1996, Salomon 1988). There is still a need for equipment and software (a *vehicle*) as well as an access point and time (an *on-ramp* and *fuel*).

[2] In a sense, this is analogous to an encounter between a deaf person and a hearing person where a third party interprets with sign language to facilitate an exchange of information in the present.

Capability Constraints

The ability to communicate randomly via the Internet demands *personal* and *technological* capabilities. Personal capabilities comprise adequate computer skills and language abilities – possibly foreign languages – while technological capabilities entail the availability of a computer, communications software, and some kind of network interface (e.g., modem or Ethernet card). Table 9.2 illustrates these requirements and others with a corresponding measure of constraint for three modes of communication: text, voice, and full audio-visual. Successive levels of interaction require more elaborate and costly devices (i.e., for voice a microphone is required, and for video, a camera).[3] The degree of constraint varies by device and mode; for example, the exchange of text is minimally affected by computer power, but moderately affected by the need for compatible communications software.

Capability constraints, measured by access to appropriate technology, represent the initial constraint on accessibility. This is true for two reasons. First, the technology must be in place for communication to take place. Second, the development of personal skills will follow the availability of technology since computing skills are primarily learned through practice. Hence, the speed with which populations become potentially accessible depends on the three factors: (1) the speed of diffusion of the technology, (2) the speed of adoption by individuals, and (3) the length of the learning curve to bring one online in an active fashion. Understanding each of these factors is in itself a study that we will leave for future explorations.

Language differences remain a significant barrier to real-time interpersonal communication. Individuals must share a common spoken or written form of language if they are to take full advantage of existing communication technologies. The barrier posed by language will diminish as voice recognition and translation software improves, but for the present discussion, language abilities are taken to be a prerequisite for functional communication.[4] Even without considering the built-in bias towards Roman characters, the dominance of English language on the Internet, trailed at some distance by Japanese, German, and Spanish (see Table 9.3), must be regarded as a major constraint to widespread communication. While there is a strong possibility that the numbers presented in Table 9.3 fail to account for the use of English by non-native speakers, there can be little doubt that the use of English by non-anglophones far exceeds the use of foreign languages by native speakers of English.

---

[3] Combinations of levels exist as well, e.g., in the burgeoning online adult entertainment trade, a performer often responds to customers' typed requests with a one-way video feed.

[4] Consider the manner in which various technologies have reduced barriers for persons with hearing, visual and motor impairments. For example, optical character recognition enables scanned text to be converted and enlarged on the screen or embossed in Braille; voice generation software reads from computer files; and voice recognition tools capture lectures in print and word prediction programs to reduce keyboarding.

Table 9.2. Constraints on random personal real-time Internet accessibility

Constraint	Text	Voice	Video
Capability			
Personal			
Computer skills	medium	medium	medium
Language	medium	high	medium
Technological			
Computer power	low	medium	high
Communications software	medium	medium	high
Network interface (modem)	low	medium	high
Camera	--	--	high
Network connectivity (ISP)			
Speed	low	high	high
Reliability	medium	medium	medium
Bandwidth	low	medium	high
Coupling			
Temporal coincidence	medium	high	high
Authority			
Access rights	low	low	low

Table 9.3. Internet use by language

Language	Count (millions)
English	55.0
Japanese	9.0
German	6.9
Spanish	5.3
French	3.7
Korean	2.1
Italian	1.55
Portuguese	1.2
Mandarin (Simplified Chinese)	0.589
Mandarin (Traditional Chinese)	0.576
Cantonese	0.346
Hebrew	0.300

Coupling Constraints

Assuming appropriate technology is in place, ITA requires temporal availability. Accordingly, coupling constraints center on the temporal coincidence of the communicating parties. For interactive communication to take place there must be temporal coincidence of individuals who are also free to communicate. Such coincidence will be a function of the personal and social times of individuals and will be manifest in their daily activity patterns. Such patterns can best be explored as activity settings that encompass one's location, when, for how long and with whom (Harvey 1982, Harvey 1997). These characteristics represent the major objective context of an activity and provide a useful framework for examining behavior within a time-use structure.

The space where one is situated can be characterized not only by geographic coordinates, but also land use (e.g., home, transport, work, shopping, and entertainment). Each venue imposes conditions and/or provides opportunities, thus promoting or coercing certain behavior (Barker 1951). Hence one can anticipate individuals will be available for direct interchange at some places and not at others.[5] The differential nature of social contact across differing venues (i.e., social spaces) has been noted (Harvey *et al.* 1997). Hence it is necessary to identify when and where people are likely to be reachable. The timing and location of social contact is integrally related to the significance of the contact and can best be understood in that light.

Drawing on Maslow's (1954) ratings of social welfare, Allardt (1990) identifies two significant aspects of contact. One is *loving*, or the individual need for attachment to other people in the immediate neighborhood. *Being*, the relationships between the individual, the community at large and the social system, relates to the need for self-realization in contrast to alienation (Gronmo 1982). Emerging technology presents individuals with opportunities and threats as they attempt to achieve these needs.

According to Lewin (1951), behavior will vary depending on who one is with (i.e., social circle), where they are (i.e., social space) and when they are there (i.e., time). To properly understand and measure the impact of social interaction it is necessary to incorporate all facets of the social environment. Employing time-use data, Schneider (1972) identifies three major circles of social interaction: (1) the family, (2) the workplace, and (3) the social or recreational circle. Each group fills all the conditions set by Lewin to identify valid social settings. The social environment incorporates Lewin's spheres: people, life space, and time, all of which are encompassed in activity settings. Time-use studies provide a solid basis for implementing such a paradigm as noted in Table 9.4.

---

[5] Portable phone systems in all of their forms continue to have an undeniable effect on when and where people can be reached.

Table 9.4. Spheres of the social environment

Spheres	Lewin	Time-use studies
1	Social circles	With whom
2	Life space	Location
3	Time	Time

Source: Harvey et al. 1997

Ideally, one would divide social circles into at least three categories: (1) family, (2) friends, and (3) other acquaintances. However, existing data and survey limitations require that the social circles be generalized down to a pair. First is the family circle, the principal unit of social interaction in society. This circle includes spouse or partner, children, and any other family member, whether they live in the household or not. Second is the circle of other acquaintances; any person with whom individuals interact who is not related by blood or marriage.

Social space, physical surroundings or location can be meaningfully identified by three main venues. First is home, the household or place of residence. The second encompasses work or education sites, professional associations, and office or school clubs. Third is community space, composed of social or recreational clubs (not related to work or school), church, community organizations, restaurants, cafes, and houses of friends and relatives.

Preliminary research has shown that the segregation of social circles is not easily accomplished. In reality they intermingle, at times becoming the single entity generally referred to as society (Schneider, 1972; Harvey *et al.* 1997). Still, with time-use data it is possible to formulate an operational measure of social contact for analysis purposes. Three types of social interaction are integrated with social space in Table 9.5 to create a social environment matrix (Harvey *et al.* 1997). Since an individual's availability is a function of both his or her social space and social circle, the social environment provides a framework for evaluating the availability of individuals.

In our view time spent awake and alone is the significant factor in individual accessibility. As Hägerstrand argues,

> when an activity has started or a room has become occupied these cells become closed for some duration of time ... projects looking for the same locations in space and time have to await their turn or go elsewhere (Hägerstrand 1973, 80).

Individuals will occasionally interrupt conversations or activities to interact with others, but this is generally viewed as unacceptable; therefore, a primary requirement of ITA is being alone. Location must also be considered since the mere act of being alone will not necessarily guarantee accessibility. Assuming contact over the Internet, it is reasonable to assume further that individuals will normally have access primarily at home or work. We readily acknowledge airplane modems, portable e-mail addressing, Internet cafes, public terminals, and community-oriented networks, but for the present discussion, we shall regard such forms of access as minimal.[6] Hence, for communication to take place, individuals must be at a location with sufficient technology to satisfy the capability conditions identified earlier in Section 9.3.

Table 9.5. Social environment

Social	Social space			
	Household	Workplace	Community	Transit
Alone & awake	Alone at home	Alone at workplace	Alone in community	Alone in transit
With family	With family at home	With family at workplace	With family in community	With family in transit
With others & multiple	With others/ multiple at home	With others/ multiple at workplace	With others/ multiple in community	With others/ multiple in transit

Source: Harvey et al. 1997

Authority Constraints

Authority constraints are less obvious and less amenable to quantification than those related to capability and coupling. Employer's banning of Internet usage for personal communication during working hours is a growing example of this type of authority. Other constraints emanate from religious prohibition and ritual observance, both of which usurp time available for accessing others. The V-chip, encryption technology and World Wide Web content filters have all been developed to limit access to sites and material by specific groups. This rapidly evolving area requires further consideration and analysis in the context of ITA.[7]

---

[6] In a survey conducted more than two years ago, Schuler (1996) uncovered nearly 300 computer systems designed exclusively for community use and hundreds more in the planning stages. While most community networks offer Internet access through community-based computing centers, many are also interested in *rearguard* technologies such as universal voice-mail, local news provision, community-based radio stations, and local cable television. Several federal governments, including Canada's, continue to embrace *cutting edge* communication technologies before less advanced forms reach their potential (e.g., see Information Highway Advisory Council 1995).

[7] For an extended treatment of authority-related issues in computerization, see the collection edited by Kling (1996).

9.4 Measurement and the Role of Time-use Data

Capability Measurement

Detailed capability measurement will require considerable data on the availability, adequacy, and adoption of technology as well as the skill levels needed for successful use. Currently there is extensive interest in quantifying many factors, but for the moment, we shall focus on revealed capability as measured in actual Internet usage. A survey conducted in September of 1998 by Neilsen Canada shows that about twice as many people access the Internet from home as access it from the office (26% versus 14% respectively; see Table 9.6 and Figure 9.1). A similar ratio is apparent in work and community settings: twice as many access the Internet from their place of employment as access it through schools and other locations.

Table 9.6. Location of Internet access in Canada, September 1998

Location	Percent using Internet
Home	26
Work	14
School	6
Other	7

Source: Neilsen Canada 1998

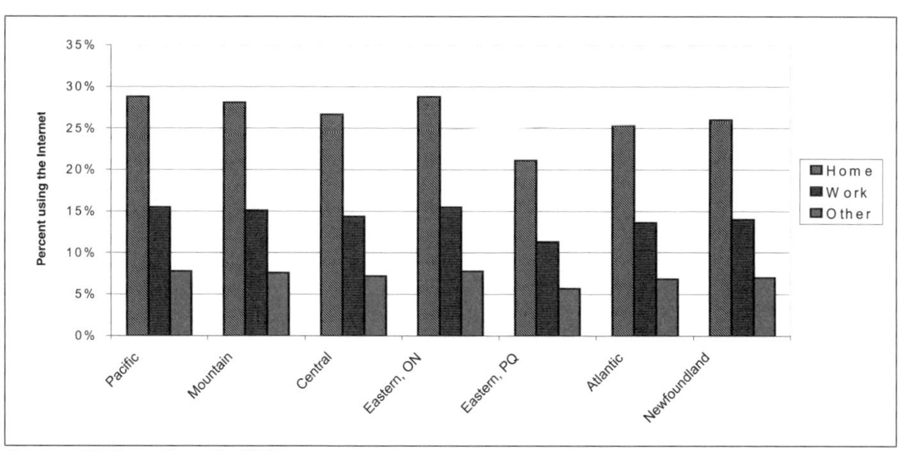

Source: Neilsen Canada 1998

Figure 9.1. Location of Canadian Internet use by region, September 1998.

Time-use Data

Time-use studies provide ample evidence of temporal variation in the city's social geography due to activity patterns and settings (Goodchild and Janelle 1984, Harvey *et al.* 1997, Janelle, Klinkenberg, and Goodchild 1998). While there is an extreme dearth of time-space data, time-diary data do exist for a large and growing number of countries. Studies typically capture extensive information about what each participant is doing, where they are, and who they are with. People account for every minute throughout a given recording period, usually 24 or 48 hours. While most of the existing databases lack detailed geographical coordinates, individuals can usually be identified by region, state, or province. Time zones, the geographical characteristic of most relevance for the present undertaking, can be determined readily from these fields.

Most time-use studies provide some spatial detail of activity and generic location (e.g., home, workplace, other place, or traveling). Generic locations are a major indicator for the type of contact likely to be made with respect to connections at home, at work, or in the community. Data drawn from time-use studies, conducted by central statistical offices in Canada, Norway, and Sweden, provide numbers describing involvement in various social environments. People spend approximately 950 to 1,000 minutes of the day awake (see Table 9.7). Slightly over 500 minutes of this time is spent in the household and between 170 and 250 minutes are spent alone in the workplace. Of the remaining time, 146 to 186 minutes is allotted to activities in the community, while 70 to 80 minutes are spent in transit.

Table 9.7. Dimensions of the social environment

		Allocation of time in minutes				
		Social space				
Country	**Social circle**	**Home**	**Workplace**	**Community**	**Transit**	**Total**
Canada	Alone	294.9	29.7	32.4	33.0	390.0
(1992)	Family	187.9	3.5	49.1	23.4	263.9
	Others & multiple	35.9	137.5	104.7	13.7	291.8
	Total awake	518.7	170.7	186.2	70.0	945.6
Norway	Alone	211.3	71.6	31.8	30.6	345.4
(1990)	Family	252.1	6.1	60.2	24.6	343.0
	Others & multiple	39.8	128.8	90.4	17.1	276.0
	Total awake	503.2	206.5	182.5	72.3	964.5
Sweden	Alone	217.3	53.0	27.2	37.7	335.1
(1991)	Family	271.3	4.0	22.1	19.9	317.3
	Others & multiple	59.8	192.0	96.8	23.3	371.9
	Total awake	548.3	249.1	146.1	80.9	1024.4

Individuals in modern societies of the Northern Hemisphere appear to spend a significant amount of time alone. The Canadian, Norwegian, and Swedish data presented in Table 9.7 suggest that 335 to 390 minutes of the waking day are spent in isolation, with the majority of that time (between 211 and 295 minutes) spent in the household. In other social spaces, people tend not to spend comparable lengths of time alone. In contrast, they spend little time alone at the workplace.

9.5 Canada Wakes Up

The extent to which ITA is affected by the diurnal cycle is amply illustrated with a case study of Canada, one of the few countries to span six time zones. The remainder of this investigation draws primarily on time-use data collected by Statistics Canada in 1992 during Cycle 7 of the General Social Survey Program. Figure 9.2 depicts the provincial boundaries and time zones of the country.

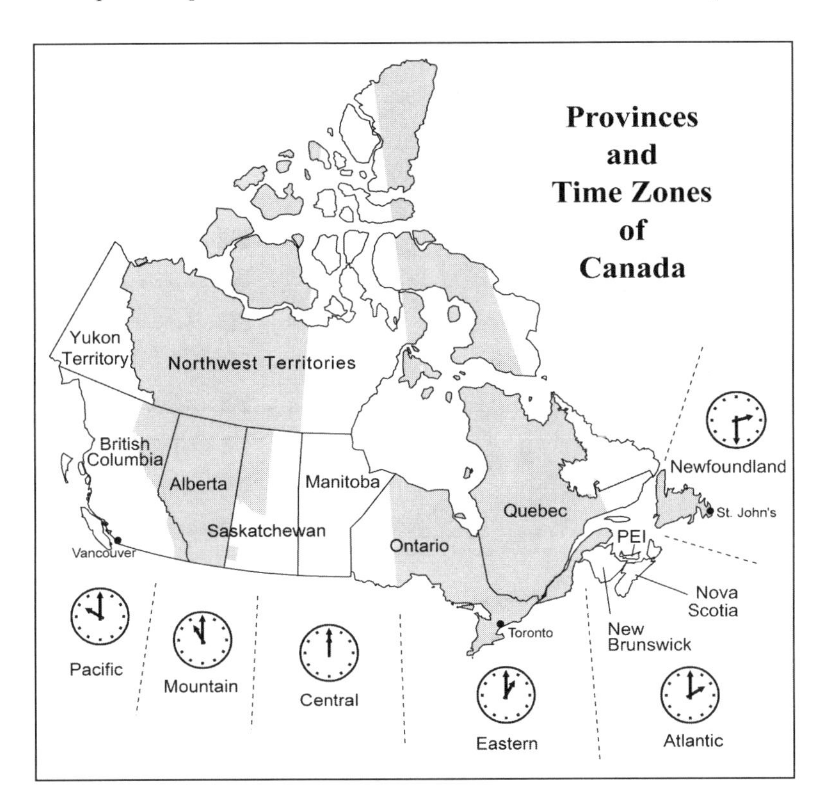

Figure 9.2. Provinces and Time Zones of Canada

When it is 10:00 a.m. in Newfoundland, the easternmost province, it is 9:00 a.m. in the neighboring provinces of Nova Scotia, New Brunswick, and Prince Edward Island.[8] At the same moment in British Columbia, the westernmost province, the clocks will read 5:00 a.m. Reconciling the time of day with time-use data from Statistics Canada reveals some interesting patterns (see Figure 9.3). For example, at 10:00 a.m. in Newfoundland, 85 percent of the population is awake while at the same moment in British Columbia, 95 percent of the population is asleep. Meanwhile, in Ontario and Quebec, where it is 8:00 a.m., close to half the population is still in bed.[9]

An entirely different pattern is revealed when it is 3:00 p.m. in Newfoundland. Figure 9.4 shows the social environment of Canadians at that time of day. Only about five percent of the population is still asleep in British Columbia where it is 10:00 a.m. and approximately 25 to 30 percent of Canadians are accessible either at home or work. Table 9.8 presents the numbers from Figure 9.4 in tabular form.

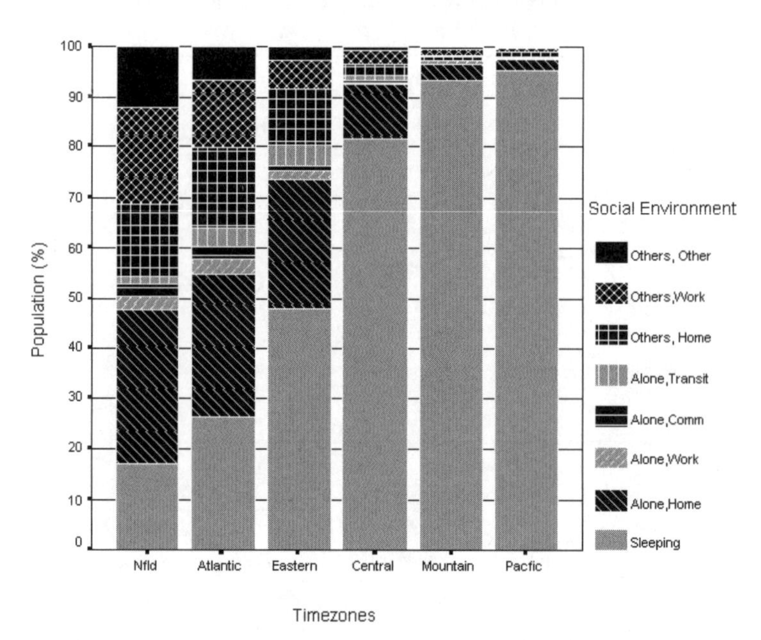

Figure 9.3. Social environment across Canada at 10:00 a.m. Newfoundland time.

---

[8] Liberty has been taken in the example by treating Newfoundland as if there was a full hour difference from the Atlantic time zone when the difference is only a half-hour.

[9] The totals depicted here are raw percentages. In actual numbers of people, 50 percent of the eight million Ontarians represents a far greater population than 95 percent of the one million Newfoundlanders.

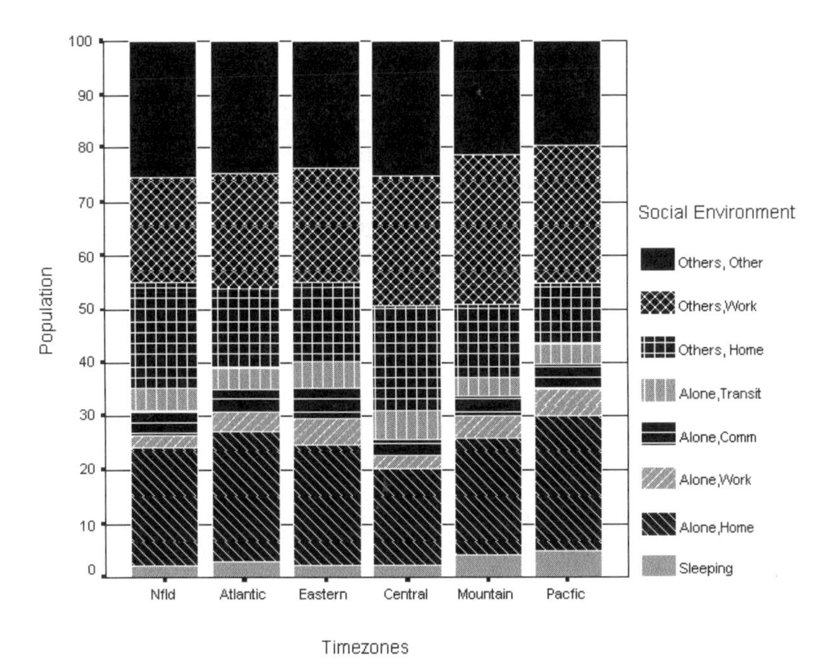

Figure 9.4. Social environment across Canada at 3:00 p.m. Newfoundland time.

As the day progresses, there is a constantly changing pattern of accessibility for personal interaction. When accessibility decreases in one direction, it opens in another. For example, by 3:00 p.m. in Toronto, workers to the east, in St. John's, are inaccessible at the office, as most have finished their working day. To the west, however, Torontonians will reach a growing number of workers in Vancouver where it is only 12:00 p.m. This gives rise to an accessibility topography shaped by the location of the population and the attendant behavior patterns. From Table 9.9, access to Canadians from Newfoundland is calculated for three different times. Expressed as persons accessible, the Canadian ITA for Newfoundland ranges from 1.14 million at 8:30 a.m. to 1.48 million at 2:30 p.m., and to 1.21 million at 7:30 p.m. These numbers are then adjusted to the Internet-usage statistics from Section 9.4 and presented in Figure 9.5.

Many have argued that spatial data are best represented with maps. Graphs, tables, and matrices like those demonstrated here provide a familiar if not cumbersome way of representing spatial-temporal patterns. Time is less amenable to mapping than space, particularly given present methods and software tools, but the potential has attracted the attention of geographers, albeit with a sizable emphasis on such tangibles as weather modeling, ozone depletion, disease spread, shore line erosion, and deforestation (e.g., see Egenhofer and Golledge 1998, Vasiliev 1996, MacEachren 1995, Langran 1992). Urban settlements and activity patterns (e.g., Janelle, Klinkenberg, and Goodchild 1998, Batty 1996, Janelle 1995) and Internet traffic monitoring (e.g., Batty 1997, Jiang and Ormeling 1997, Dodge 1996) are two areas in which the mapping of time has been applied extensively.

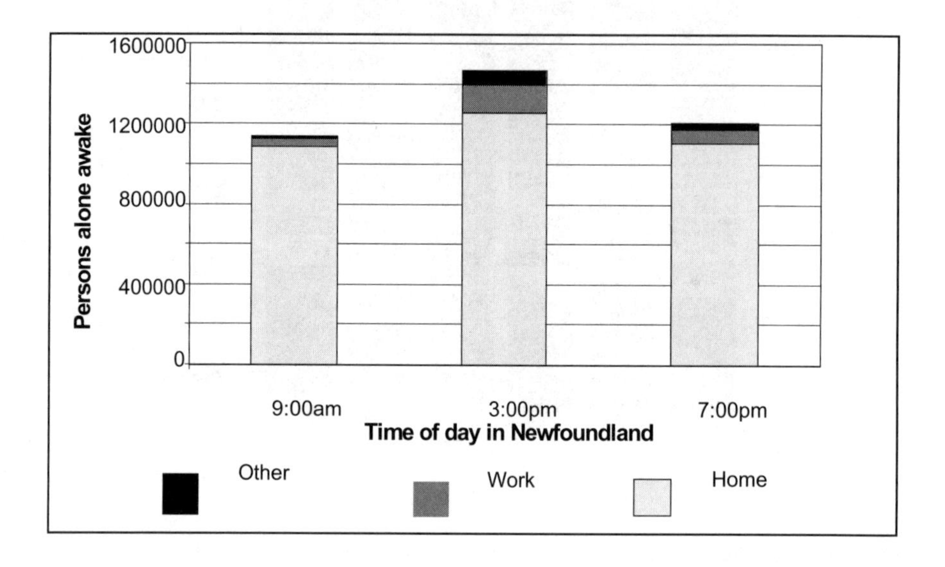

Figure 9.5. Canadian ITA adjusted to Internet usage showing Internet-accessible Canadians at selected times for Newfoundland.

Spatial changes over time are well represented on maps, but what about time changes over space? How often has a phone book been opened to check international time zones on the long-distance pages?[10] Classic representations, typified by static maps and textual descriptors, have been augmented in recent years by a range of software and online conversion tools (e.g., CLOX World Time Zone

---

[10] Say I wish to call a cousin in New Zealand and the phone book tells me that her country is Atlantic Standard Time plus 14 hours. That means if it's 8 o'clock in the evening here, it's 10 o'clock in the morning there -- darn, she's already left for work!

Table 9.8. Social environment across Canada, 3:00 p.m. Newfoundland time

		TIME ZONES						
		Nfld. 3:00	**Atlantic 2:00**	**Eastern 1:00**	**Central 12:00**	**Mountain 11:00**	**Pacific 10:00**	**Total**
Sleeping	Count	10218	43656	297772	19788	107016	130320	608770
	Row %	1.7	7.2	48.9	3.3	17.6	21.4	100.0
	Col %	2.3	3.2	2.2	2.4	4.0	5.0	2.9
	% Total	.0	.2	1.4	.1	.5	.6	2.9
Alone, Home	Count	95865	329778	2989528	151224	584226	651540	4802161
	Row %	2.0	6.9	62.3	3.1	12.2	13.6	100.0
	Col %	21.7	23.9	22.4	18.0	21.8	25.1	22.6
	% Total	.5	1.5	14.0	.7	2.7	3.1	22.6
Alone, Work	Count	10868	50770	668205	17781	114694	125613	987931
	Row %	1.1	5.1	67.6	1.8	11.6	12.7	100.0
	Col %	2.5	3.7	5.0	2.1	4.3	4.8	4.6
	% Total	.1	.2	3.1	.1	.5	.6	4.6
Alone, Comm.	Count	19902	56558	733251	27372	96565	124243	1057891
	Row %	1.9	5.3	69.3	2.6	9.1	11.7	100.0
	Col %	4.5	4.1	5.5	3.3	3.6	4.8	5.0
	% Total	.1	.3	3.4	.1	.5	.6	5.0
Alone, Transit	Count	18133	55526	692708	43836	94853	94766	999822
	Row %	1.8	5.6	69.3	4.4	9.5	9.5	100.0
	Col %	4.1	4.0	5.2	5.2	3.5	3.6	4.7
	% Total	.1	.3	3.3	.2	.4	.4	4.7
Others, Home	Count	88539	204690	1974478	165648	370461	295884	3099700
	Row %	2.9	6.6	63.7	5.3	12.0	9.5	100.0%
	Col %	20.1	14.9	14.8	19.7	13.8	11.4	14.6%
	% Total	.4	1.0	9.3	.8	1.7	1.4	14.6%
Others, Work	Count	86194	293240	2813188	203551	729583	660674	4786430
	Row %	1.8	6.1	58.8	4.3	15.2	13.8	100.0
	Col %	19.5	21.3	21.1	24.2	27.3	25.4	22.5
	% Total	.4	1.4	13.2	1.0	3.4	3.1	22.5
Others, Other	Count	111550	343366	3193094	210922	578856	513819	4951607
	Row %	2.3	6.9	64.5	4.3	11.7	10.4	100.0
	Col %	25.3	24.9	23.9	25.1	21.6	19.8	23.3
	% Total	.5	1.6	15.0	1.0	2.7	2.4	23.3
Total	Count	441269	1377584	13362224	840122	2676254	2596859	21294312
	Row %	2.1	6.5	62.8	3.9	12.6	12.2	100.0
	Col %	100	100	100	100	100	100	100
	% Total	2.1	6.5	62.8	3.9	12.6	12.2	100

Source: Calculated from Statistics Canada, General Social Survey, Cycle 7.

Table 9.9. Canadian accessibility at selected Newfoundland times

Time Zone	Time	Home	Work	Other	Total
Nfld	8:30 a.m.	34500	1722	713	36935
Atlantic	8:00 a.m.	99545	5482	2181	107208
Eastern	7:00 a.m.	876444	29421	8413	914278
Central	6:00 a.m.	25085	344	271	25700
Mountain	5:00 a.m.	26405	362	285	27053
Pacific	4:00 a.m.	27134	1497	0	28631
All Canada		1089113	38827	11864	1139804
Nfld	2:30 p.m.	24251	1480	1355	27087
Atlantic	2:00 p.m.	83425	6916	3852	94193
Eastern	1:00 p.m.	755750	91350	48808	895908
Central	12:00 a.m.	40381	2557	1968	44905
Mountain	11:00 a.m.	164215	17359	7308	188882
Pacific	10:00 a.m.	187714	19487	9637	216838
All Canada		1255736	139149	72928	1467813
Nfld	7:30 p.m.	27514	610	445	28569
Atlantic	7:00 p.m.	85620	1901	1552	89074
Eastern	6:00 p.m.	492622	21622	15067	529311
Central	5:00 p.m.	163336	12446	5586	181368
Mountain	4:00 p.m.	171932	13102	5880	190914
Pacific	3:00 p.m.	168127	15091	8612	191830
All Canada		1109151	64772	37142	1211065

Clock, ID LOGIC World Time Zone Clock for Windows). Interactive time zone converters vary in their utility and mode of global visualization, but generally, most of the packages appear to permit queries such as: if it is 6:00 p.m. in country A, what time is it in country B? Some converters are text-based while others permit graphic database queries, a useful function likely inspired by GIS.[11] Figures 9.6 and 9.7 offer preliminary cartographic depictions of ITA in Canada at 8:30 a.m. and 7:30 p.m. Newfoundland time.

[11] At least one software package provides a visual indication of who's awake. The *Moon and Earth Viewer* offers a map or image of the world shaded to indicate which countries are experiencing daylight and which fall under the shadow of the moon.

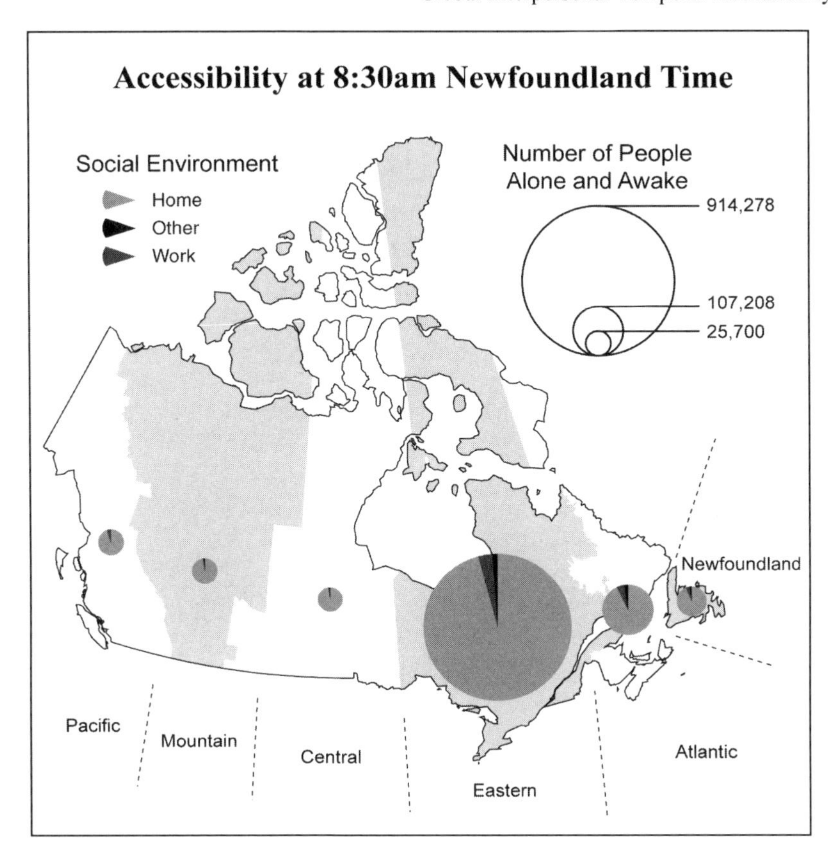

Figure 9.6. Preliminary cartographic depiction of ITA in Canada, 8:30 a.m., Newfoundland time

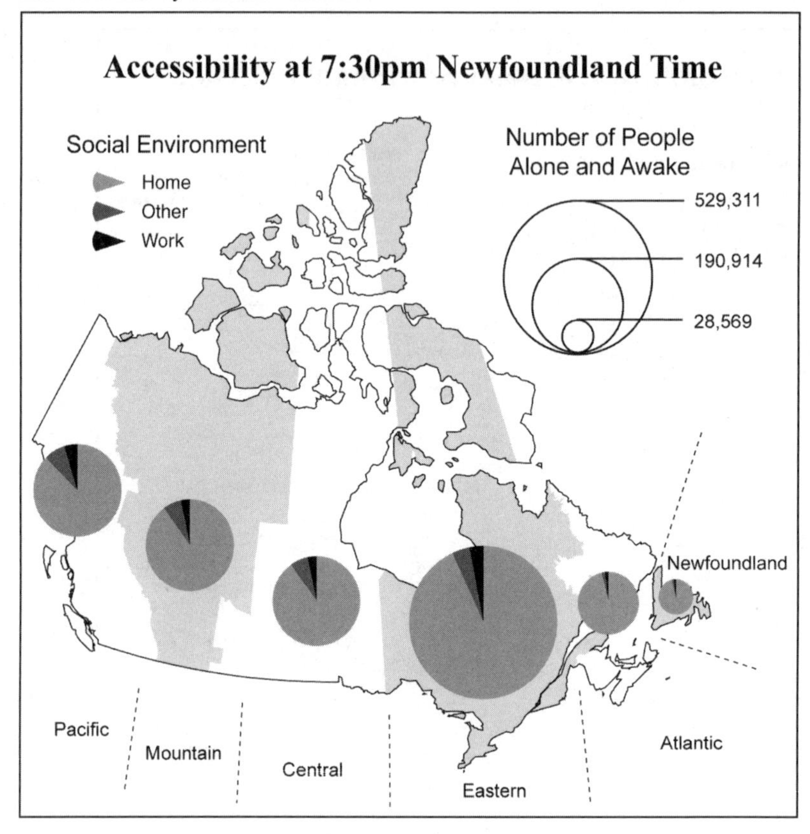

Figure 9.7. Preliminary cartographic depiction of ITA in Canada, 7:30 p.m., Newfoundland time

9.6 Research Directions

Our preliminary inquiries suggest research needs in five principal areas: (1) interactive geographic visualizations, (2) diminishing accessibility as a result of increasing accessibility, (3) Internet-traffic monitoring to compare ITA potential with real-time communication volumes, (4) expansion of time-use and Internet surveys to collect detail on the nature and duration of mediated face-to-face interaction, and (5) continued exploration around the notion of temporal *regions* within and between the planet's geographic *regions*.

(1) For geographic visualization, future research could be directed to a number of enhancements. With some additional experimenting, it would be possible to construct an animated sequence using 10-20 minute intervals per frame that

would show Canada *waking up*. The potential for further visualization includes several interactive GIS *add-ons* that would enhance the options described above for time-zone converters. What if you wished to determine where and when the most people might be reached? By extending a time-zone database to include diurnal activity patterns at a larger scale (e.g., telephone area codes or Internet-service providers), one could query the statistically optimal time of day for accessibility between two cities. Conversely, one could determine the location where the maximum number of people could be reached at a given time. The World Time Zones data set available from ESRI contains 35 zones at a scale of 1:3 million. Coded to indicate the hours of difference from Greenwich Mean Time, this coverage might be a useful starting point. We anticipate and welcome any additional suggestions for GIS treatment of these connections across time and space.

(2) There is a paradox in the trend towards real-time electronic communication: as people's communication reach expands, their real-time accessibility to other people decreases. Put another way, the more accessible an individual, the greater the probability that he or she will be occupied. Consider the net decrease in accessibility when multiple modes of communication (e.g., phone, voice mail, pager, e-mail) occupy an individual's time. The probability of catching a person unoccupied diminishes at some functional rate with respect to the increase in potential contacts. This contrasts sharply with the normal notion of accessibility, typically framed as public goods (e.g., shopping centers, stadiums) that can be enjoyed concurrently by many users.

(3) The exploratory calculations presented here offer one possible index of interpersonal-temporal accessibility. What other measures should be factored in? The numbers derived for potential accessibility on the Internet across Canada bear comparison with actual counts and duration volumes of mediated real-time connections. At present, it is difficult to extract these measures from electronic-traffic reports; but, in time, such monitoring is expected to become commonplace. Are the numbers embedded in time-use datasets useful and accurate for predicting accessibility? If so, might call centers, community groups, fundraisers, NGOs, entertainers, retailers, educators, business interests, and media outlets around the world be keen to make use of the numbers?

(4) Time-use statistics are time-consuming and expensive to collect. Furthermore, it may take several cycles before survey designs and sampling strategies shift to accommodate emerging trends. As recently as the 1996 Canadian Census, very little was queried as to the online dallying of respondents. Internet surveyors abound, no doubt spurred on by the dynamic nature of cyberspace, but numbers to date have proven limited for penetrating analyses of real-time, direct interpersonal connectivity. Clearly there is a need to expand time-use and Internet surveys to collect detail on the nature and duration of mediated face-

to-face interaction. Some effort in these directions might also be extended data collection and statistical measures related to authority constraints.

(5) In the realm of real-time communications, the fundamental geographical notion of the *region* is in need of a temporal overhaul. In the past, regions have been defined by physical measures such as climate, landscape, vegetation, and wildlife, alongside human determinants such as politics, language, religion, economics, and ethnicity. In the increasingly homogenous global village, where culture is being commodified along reflective glass tendrils, perhaps the 'region' needs to be recast in a temporal framework. To what degree will traditional east-west channels, like those between the French in Québec, Vietnam, and France give way to north-south alignments more in keeping with the time of day? Who will be awake and available for face-to-face commentary when soccer's next World Cup begins, whenever and wherever that should be?

9.7 Summary

In this paper we discuss preliminary measures of interpersonal-temporal accessibility with a particular emphasis on face-to-face contact mediated by emerging technologies. Our investigation centers on the real-time communications potential associated with Internet connectivity. Although the need for spatial coincidence is diminishing, temporal coincidence remains a prerequisite for meaningful exchange between individuals. Constraints, defined in terms of capability, coupling, and authority, help to frame our analysis. Capability relates to the availability and quality of the technology needed to access the Internet. Coupling relates to the temporal availability of communicating parties. We introduce data on time zones, time-use, and Internet use in an effort to determine when people are awake, alone, and in the vicinity of a computer – at home, work, or school. Calculations based on activity patterns and corresponding social environments across Canada enable us to chart potential accessibility at three points in the day according to Newfoundland time. A second set of calculations incorporates the percentage of Internet use for preliminary measures of Internet accessibility. With some extension, these analyses could be expanded to encompass time zones around the world. If it remains true that trust and understanding are most strongly built through real-time communication, we can expect future alliances to be forged between individuals living in regions that permit interaction in real-time. Students in Australia and New Zealand will continue to be the best *live* English connection for Japanese students, at least for the foreseeable future.

Acknowledgements

Staff of the Time Use Research Center at Saint Mary's University assisted at several stages during the course of our investigation. We are grateful for the contributions of Jennifer, Barbara and Wendy.

References

Allardt, E. 1990. Challenges for comparative social research. *Acta Sociologica* 33(3):183.

Barker, R.G. 1951. *One Boy's Day: A Specimen Record of Behaviour*. New York: Harper.

Batty, M. 1996. Visualizing Urban Dynamics. In Longley, P. and Batty, M. (eds.) *Spatial Modelling in a GIS Environment*. Cambridge UK: Geoinformation International, 297-320.

Batty, M. 1997. Virtual geography. *Future* 29(4/5):337-52.

Cairncross, F. 1995. The death of distance: a survey of telecommunications. *The Economist* 336(7934):5-28.

Dodge, M. 1996. Mapping the World Wide Web. *GIS Europe* 5(9):22-4.

Egenhofer, M.J. and Golledge, R.G. (eds.). 1998. *Spatial and Temporal Reasoning in Geographic Information Systems*. New York: Oxford University Press.

Fellmann, J., Getis, A. and Getis J. 1990. *Human Geography*, Second Edition. Dubuque IA: Wm. C. Brown.

Giddens, A. 1990. *The Consequences of Modernity*. California: Stanford University Press.

Goodchild, M.F. and Janelle, D.G. 1984. The city around the clock: space-time patterns of urban ecological structure. *Environment and Planning A* 16:807-20.

Graham, S. and Marvin, S. 1996. *Telecommunications and the City: Electronic Spaces, Urban Places*. London: Routledge.

Hägerstrand, T. 1970. What about people in regional science? *Papers and Proceedings of the Regional Science Association*, 24:7-24.

Hägerstrand, T. 1973. The domain of human geography. In Chorley, R.J. (ed.) *Directions in Geography*. London: ethuen & Co. Ltd, 67-87.

Harvey, A.S. 1997. From activities to activity settings. In Ettema, D. and Timmermans, H. (eds.) *Activity-based Approaches to Travel Analysis*. Tarrytown NY: Pergamon.

Harvey, A.S. 1982. Role and context: Shapers of behaviour. *Studies of Broadcasting* 18:70-92.

Harvey, A.S., Taylor, M.E., Ellis, S. and Aas, D. 1997. *24-Hour Society and Passenger Travel*. Report cmmissioned by Transport Research Centre, Ministry of Transport, Public Works and Water Management. Rotterdam.

Information Highway Advisory Council. 1995. *Access, Affordability and Universal Service on the Canadian Information Highway*. Ottawa: Supply and Services Canada.

Janelle, D.G. 1995. Metropolitan Expansion, Telecommuting, and Transportation. In S. Hanson, S. (ed.) *The Geography of Urban Transportation*, Second Edition. New York: Guilford Press, 407-34.

Janelle, D.G., Klinkenberg, B. and Goodchild, M.F. 1998. The temporal ordering of urban space and daily activity patterns for population role groups. *Geographical Systems* 5:117-37.

Jiang, B. and Ormeling, F.J. 1997. Cybermap: The map for cyberspace. *The Cartographic Journal* 34(2):111-16.

Kling, R. (ed.). 1996. *Computerization and Controversy: Value Conflicts and Social Choices*. Second Edition. San Diego: Academic Press.

Langran, G. 1992. *Time in Geographic Information Systems*. Bristol PA: Taylor and Francis.

Lewin, K. 1951. *Field Theory in Social Science: Selected Theoretical Papers*. New York: Harper.

Neilsen Canada Ltd. Personal communication, 1998.

MacEachren, A.M. 1995. *How Maps Work: Representation, Visualization and Design*. New York: The Guilford Press.

Maslow, A.H. 1954. Motivation and personality. *Harper's Psychological Series*. New York: Harper.

Pirie, G.H. 1979. Measuring accessibility: A review and proposal. *Environment and Planning A*:11:299-312.

Rheingold, H. 1993. *The Virtual Community: Homesteading on the Electronic Frontier*. Reading MA: Addison-Wesley.

Salomon, I. 1988. Geographical variations in telecommunications systems: The implications for location of activities. *Transportation* 14:311-27.

Schneider, A. 1972. *Patterns of Social Interaction. The Use of Time: Daily Activities of Urban and Suburban Populations in Twelve Countries*. The Netherlands: Mouton & Co.

Schuler, D. 1996. *New Community Networks: Wired for Change*. Reading MA: Addison-Wesley.

Statistics Canada, *General Social Survey*, Cycle 7, 1992.

Vasiliev, I. 1996. Design issues to be considered when mapping time. In Wood, C.H. and Keller, C.P. (eds.) *Cartographic Design: Theoretical and Practical Perspectives*. New York: John Wiley and Sons, 137-46.

10 The Role of the Real City in Cyberspace: Understanding Regional Variations in Internet Accessibility

Mitchell L. Moss[1] and Anthony M. Townsend[2]

[1] Taub Research Center, New York University, 4 Washington Square North, New York NY 10003, USA. Email: mitchell.moss@nyu.edu

[2] Department of Urban Studies and Planning, Massachusetts Institute of Technology, 77 Massachusetts Avenue, Cambridge MA 02139, USA. Email: amt@mit.edu

10.1 Introduction

Since 1993, when the first graphical web browser, Mosaic, was released into the public domain, the Internet has evolved from an obscure academic and military research network into an international agglomeration of public and private, local and global telecommunications systems. Much of the academic and popular literature has emphasized the *distance-shrinking* implications and *placelessness* inherent in these rapidly developing networks. However, the relationship between the physical and political geography of cities and regions and the virtual (or logical) geography of the Internet lacks a strong body of empirical evidence upon which to base such speculation.

This chapter presents the results of a series of studies conducted from June 1996 to August 1998. Our research suggests there is a metropolitan dominance of Internet development by a handful of cities and regions. We identify and describe an emerging structure of 'virtual' hubs and pathways which are linking a set of major cities in the United States, suggesting that there is a complex emerging inter-urban communications network that goes far beyond Castells' (1989) *informational mode of development*.

More importantly, we analyze the utility and relevance of three measurements of Internet development for urban planning and policy research. The first measure is the number of computers connected to the Internet on a full-time basis, based on data collected by Matrix Information and Demography Services of Austin, Texas. We find this measure to be of limited usefulness, as it only measures location-specific hardware installations with little reference to their purpose or function. The second measurement system we explore is based on the Internet's addressing scheme, known as the domain name system. Unique to individual organizations (business, education, non-profit, government), domain names are registered with InterNIC, an administrative clearinghouse contracted by the National Science Foundation. Associated with each domain name is a unique billing address, which permits the localization of the organization using that name. This is the most informative measure for assessing variations in Internet use across a variety of geographic units, from states to individual ZIP codes. Finally, we examine the capacity and topography of nationwide Internet backbone networks that transport data between metropolitan areas.

Based on these three measurements, we find that a limited number of cities and metropolitan areas dominate the rapidly emerging telecommunications landscape of the United States, leading in the development of increasingly sophisticated applications and technologies. Accessibility to the most highly developed real and virtual Internet infrastructure is a metropolitan phenomenon, and highly stratified among regions and cities. Furthermore, we describe each of the three measurements used – host counts, domain counts, and backbone network capacity – and their unique advantages and disadvantages for research. Successfully applying them to urban analysis requires not only an understanding of urban and regional development processes, but also the purpose, design, and function of these complex technical systems.

10.2 Cities, Regions, and Telecommunications

The Internet and other telecommunications advances pose a serious challenge to the study of urban life. Electronic commerce and the decline of distance-sensitivity in telecommunications pricing have encouraged speculation that the dispersion of human settlement is imminent. However, as Peter Hall (1997, 316) states, 'the urban world of the 1990's… is a world in which cities deconcentrate and spread to become complex systems of cities linked together by flows of people and information.' Thus, while advanced telecommunications permits the evolution of an increasingly complex urban system with multiple linkages and hierarchies, place and centrality remain extremely important in the information economy. Gaspar and Glaeser (1996) present strong evidence that telecommunications and travel are synergistic, suggesting the intrinsic value of face-to-face interaction beyond what can be communicated at a distance. The Internet has the potential to subsume other communications media such as telephony, print, television and radio, as well routine activities like shopping, learning, entertaining, and socializing.

We have sought to quantify and localize various measurements of Internet development in an effort to understand the relationship of the Internet to urban and regional development. Hall (1997, 318) asserts that cities' competitiveness in the global economy 'depends on their capacity to generate, process, and exchange information'. To address those concerns, our measurements of Internet development seek to describe the emerging structure of systems by which cities exchange information through electronically mediated communications. Just as the Interstate

Highway System transformed urban development in 20th century America, the Internet will help shape urban activity patterns in the 21st century.

In the past, researchers have studied the flow of information in colonial American cities through newspapers (Pred 1973), by telephone in the urban complex of the Northeastern United States (Gottmann 1961), and in the information economy by means of office buildings and overnight letter delivery (Sui and Wheeler 1993; Mitchelson and Wheeler 1994). More recently, others have tried to quantify the informational capacity of cities and regions, identifying clusters of both human (Nunn and Warren 1997) and physical (Greenstein, Lizardo, and Spiller 1997) *information age* capital. However, with the notable exception of Dodge and Shiode's (2000) recent work on Internet *real estate* in the United Kingdom and Murnion's work (see Chapter 12), little research has yet to directly address the geography of the evolving Internet.

There are several reasons why the Internet has eluded urban scholars, at least in the United States. One is that the telecommunications industry in general, and particularly the Internet sector, is extremely competitive. Because the Internet services sector emerged during a period of rapid deregulation in the American telecommunications industry, it has grown unsupervised by federal regulators, and there is a lack of systematically gathered data on its operations. While individual companies may possess substantial information that could be used to advance scholarly research or public understanding of these new networks, there are few available sources of geographic data. Second, the Internet operates primarily over pre-existing telephone network infrastructure. With the exception of Qwest, Inc.'s new national fiber optic network, there is little physical construction activity solely associated with the deployment of new Internet infrastructure. Rather, the flip of a switch to light a dark strand of glass fiber is all that is required to deploy new capacity. Compare this to the spread of the cellular telephone network, which can be measured either visually by counting towers or by consulting the U.S. Federal Communications Commission's (FCC) extensive public database of transmission antennas.

This chapter describes three basic measurements used over a two-year period to attempt to assess Internet accessibility among cities and metropolitan areas in the United States. The difficulty of obtaining reliable data at comparable levels of geographic aggregation limit the usefulness of presenting the results together, further highlighting the need for new empirical tools and methods to be developed. However, we seek to identify methods of measuring Internet development and diffusion to guide further empirical research in this area, and stimulate debate about the implications of these observations and methods for urban theory.

10.3 Measuring Regional Variations in Internet Development

The primary question this series of studies seeks to address is which cities and metropolitan areas exhibit a rapid buildup of Internet-related telecommunications infrastructure, and secondly, how can we effectively measure the geographic distribution of the Internet? Measuring this phenomenon proved very difficult. First, as noted earlier, telecommunications providers have strong incentives to keep data on their operations closely guarded. Second, from a very early stage in its development the Internet was designed to subvert geography by using a packet-switched message routing system, which operates more like the postal service than the telephone network. As a result, messages can be rerouted around faulty switches or stations. Thus, the few aspects of the Internet that can be localized and measured have only limited relevance to physical geography. Third, each of the three data sources measured aspects of the Internet that did not always have overlapping geography. Finally, we needed to consider what these measurable criteria could indicate about the level of human activity on the Internet in different cities and regions and the socioeconomic consequences of these information flows. This section describes the three measurements we used and the strengths and limitations of each. The final section presents a composite picture of the North American system of cities we have derived from these observations.

Computers Connected to the Internet on a Full-time Basis

Matrix Information and Demography Services of Austin, Texas has measured the geography of the Internet since the early 1990's, and is considered a leader in analyzing what it calls *the matrix* of inter-connected global computer networks. MIDS' methods permit the localization of Internet hosts (computers connected to the Internet on a full-time basis) to geographic areas as specific as a street address. Our study was based on data provided by MIDS for January 1996, and indicated the number of Internet hosts per county for all 50 states. The data included a substantial number of manual corrections to the automated survey that generates the data, based on MIDS' proprietary knowledge of the known geographic location of large clusters of host computers in corporate research and development centers, for example.

This measurement was a useful first step towards visualizing the emerging telecommunications landscape of the United States. As anticipated, Santa Clara County, California (*Silicon Valley*) and Middlesex County, Massachusetts (*Route 128*) were the largest clusters of permanently connected Internet hosts. Table 1 shows the top 25 counties by number of host computers. However, we did not anticipate the large number of central cities that appear on this list. Furthermore, 12 of the 25 top counties were located within just four metropolitan areas: San Francisco, Los Angeles, Washington DC, and New York. An asterisk indicates these twelve counties. While these results confirmed our suspicions regarding the concentration of high levels of telecommunications activity in a select group of metropolitan areas, the overwhelming presence of Silicon Valley, Route 128, and several academic clusters in Michigan forced us to re-evaluate the relevance of these findings. We believed these high figures to be artifacts of these regions' role as the birthplace of the Internet in academic and industrial settings. Our interest thus turned to identifying locations where a broad-based adoption of these technologies had rapidly occurred across a wide range of industries and population groups.

Table 10.1. Top 25 counties by number of Internet hosts, January 1996

County, State	Description	Hosts
**Santa Clara, CA	Silicon Valley	554,967
Middlesex, MA	Route 128	243,765
*Los Angeles, CA	Central city	159,944
*New York, NY	Central city	146,371
*Fairfax, VA	Edge city	131,874
*Orange, CA	Edge city	123,685
*San Diego, CA	Central city	111,981
Cook, IL (Chicago)	Central city	110,726
Hennepin, MN (Minneapolis)	Central city	109,047
*San Mateo, CA	Edge city	92,781
Salt Lake, UT	Central city	90,693
*Alameda, CA	Edge city	89,851
Washtenaw, MI	Universities	82,790
King, WA (Seattle)	Central city	79,142
Allegheny, PA (Pittsburgh)	Central city	64,616
Philadelphia, PA	Central city	62,387
Travis, TX (Austin)	Central city	62,371
Dallas, TX	Central city	61,811
Bay, MI	Universities	57,726
*District of Columbia, DC	Central city	55,755
*San Francisco, CA	Central city	53,183
*Prince George's, MD	Edge city	46,450
*Montgomery, MD	Edge city	42,156
Mecklenburg, NC (Charlotte)	Central city	37,369
Fulton, GA (Atlanta)	Central city	34,103

Source: Moss and Townsend 1996

* indicates that county is located within the metropolitan area of New York, Los Angeles, San Francisco, or Washington DC

At this point, we encountered the first of many instances in which the technological realities of the Internet drastically affected the interpretation of our results. The measurement of host counts was extremely coarse because it failed to differentiate between various types of computer equipment. For example, if a financial services company in Manhattan uses a highly centralized computer system based on a mainframe and *dumb* terminals, this method might only count a single host – the mainframe being the only machine directly connected to the Internet. Conversely, a small software company in Silicon Valley that uses a Local Area Network to connect its dozen microcomputers to the Internet individually would have a much higher host count. In some cases, networked printers might even be counted by this method. Finally, the increased use of firewalls, computers designed to mediate external Internet connections and shield institutional networks from intruders, excludes a significant number of hosts from detection, especially those of large corporations. These factors generate significant variations in the effectiveness of comparing host counts across regions and industrial sectors.

It also seemed counter-intuitive to use a measurement based purely on the technical organization of the Internet to infer some understanding of the rate of adoption of these technologies across regions. The nature of the modern American economy and the production systems of the Internet services industry further complicates this approach. Many organizations do not physically house their Internet-accessible information at their physical location, preferring to hire contractors who provide expertise and equipment. While the information-producing jobs and economic activity associated with a website may take place at a centralized office in a dense urban area, it is just as likely that the fruits of this labor are electronically disseminated from a remote location, which could conceivably be located anywhere. These limitations in the host count measurement led us to seek other indicators of Internet use, which proved more useful in understanding the spread of the Internet among cities.

The Location of Organizations Using the Internet

Domain names are one of the basic forms of Internet addressing, which map groups of numeric Internet addresses to intuitive names like *nyu.edu* or *att.com*. Each domain name is registered with Network Solutions, an organization chartered by the National Science Foundation to administer the domain name system. As Figure 10.1 shows, each name is registered to an individual or organization, and the publicly available registration record contains a billing address for that entity. From this information, it is possible to localize the location of the entity that owns that domain name as specifically as the postal code (ZIP) level. This geographical specificity of the domain name makes it a highly attractive measure for Internet activity. Network Solutions has enjoyed a monopoly over domain registrations for the most popular commercial, non-profit, educational, and government domains in the United States, leaving only a small portion of the American Internet beyond the scope of these data.

The strengths of this measurement for urban research stem from its representation of a social phenomenon, rather than a technical one. Because each domain name roughly corresponds to a corporate, government, or educational entity, this measurement indicates spatial variations in the adoption and use of Internet-based communications by organizations. Since nearly 90 percent of Internet growth over the period between 1994 and 1997 was from the addition of commercial domain names, these results primarily measure the extent to which businesses deployed these new technologies. Domain registrations also indicate the date each domain was first registered, permitting us to identify those regions that had the most rapid growth in Internet use.

Registrant:

Five Points Internet Solutions (FIVEPOINTS3-DOM)

45 Havemeyer St. #2R

Brooklyn, NY 11211

US

Domain Name: FIVEPOINTS.NET

Figure 10.1. A typical domain name registration record

However, the domain measurement is seriously handicapped as well. First, it does not take into account an organization's size or its dependence upon or capacity to generate flows of information over the Internet. There is no practical method for sorting through the hundreds of thousands of records and assigning weights to entities of differing sizes, revenues, or information processing and production capacity. As a result, this technique weights Microsoft's immense Internet presence little more than the small website maintained by Redmond, Washington's municipal government. Also, the geographic data associated with domain registrations do not always correspond to the true physical location of a domain's primary users (who may be dispersed over multiple *continents)*, but rather to an administrative or MIS headquarters location. For example, while AT&T is headquartered in New York City, its domain registration for *att.com* is in Florida. Furthermore, most of its data networking operations are controlled from a center in the St. Louis, Missouri area. Finally, companies are increasingly registering multiple domain names:

those of the company's products, or variations upon the company's name. This practice may be responsible for some distortion in the overall results.

The primary source of data for our research on the geographic distribution of domain names was Imperative! of Pittsburgh, Pennsylvania. Imperative! sells mailing lists to direct marketers who offer Internet-related products to the owners of domain names and maintains a database of currently registered domain names that can be aggregated by nearly any geographical unit. We analyzed regional variations in domain name registration at three geographic levels; (1) U.S. Census regions, (2) 85 major U.S. cities, and (3) postal code areas in New York City (Moss and Townsend 1997a, 1997b, 1998). Among major U.S. cities, the largest clusters (Manhattan and San Francisco), were also the most densely networked. Furthermore, in the 15 cities with the largest number of Internet domains, which accounted for 12.6 percent of all U.S. domain registrations in April 1994, new domains were registered faster than other areas. By 1997, these 15 cities accounted for nearly one-fifth (19.7 %) of all U.S. domain registrations. Clearly, Internet technologies were being more rapidly deployed in major urban areas. Outside the dense urban areas of the Atlantic, Pacific, and Gulf coasts, only Phoenix and Chicago had a significant number of domains.

We also computed the density of domains with respect to population, or domains per 1000 persons. Domain density was typically highest among cities whose primary function was as a resort, government, or education center. Nodal cities also showed high concentrations (Moss and Townsend 1998).

Comparing these results to a limited set of domain counts for approximately a dozen cities in April 1994 permitted us to track the rate of domain growth over a 3-year period. These cities registered domains far faster than the national average, adjusted for population. The growth rate of domains between 1994 and 1997 was linked to a city's relative position in the national urban system, with the advanced service centers growing most rapidly. This indicates a strong relationship between information-intensive economic activity and early adoption of the Internet among businesses. The cities primarily fall into four broad categories of growth rates, summarized in Table 10.2.

Table 10.2. Trends in growth of domain name registrations, 1994-1997

Description	Examples	Domain growth rate (Multiple of national average)
Global information centers	Manhattan (NYC), San Francisco	6+
Mid-sized *information* cities	Atlanta, Boston, Miami, Seattle	4-5+
Regional centers	Denver, Dallas, Phoenix	2-3+
World cities	New York, Los Angeles, Chicago	1-2+

Based on Moss and Townsend 1998.

The fact that the set of cities most commonly referred to as *world cities* or *global cities* recorded the slowest growth rates among large Internet clusters is disturbing, for it indicates an averaging function. As our detailed analysis of domain registrations in Manhattan indicates, adoption of the Internet is not a widespread phenomenon across urban populations. Rather, it is almost entirely limited to the central business districts, with moderate adoption rates in the more successful immigrant communities (Moss and Townsend 1997a). The fact that smaller cities such as Austin or Boston exhibit higher growth rates is most likely due to a more even spread of technological opportunities among their more homogeneous populations. On the other hand, world cities appear to be characterized by a *digital elite* co-existing with a vast, largely disconnected *information ghetto*. As one example of this disturbing trend, a preliminary survey of domain registrations in the Los Angeles area indicates a much slower diffusion of Internet technologies among Spanish-speaking and immigrant communities.

Jed Kolko, a doctoral student at Harvard University, is currently conducting research that will extend the analysis of domain name registrations to all 285 U.S. Metropolitan Statistical Areas (MSAs) over a four-year period from 1995-1998. This work addresses a shortcoming of our research, in that we neglected to explore Internet adoption and use in suburban areas surrounding the major cities. In fact, the bulk of new office space in recent decades has emerged not in the central cities that we focused upon, but rather in the *edge cities* that surround them. (Garreau 1991) Some preliminary results, in Table 10.3, show that even where including the surrounding metropolitan areas there is strong evidence of a select group of cities that dominate Internet activity.

Table 10.3. Domain name registrations by metropolitan area: January 1998

Consolidated Metropolitan Statistical Area	Domains Jan-1998	Percent of all U.S. domains
New York	112,524	8.6
Los Angeles	109,917	8.4
San Francisco/Silicon Valley	89,584	6.8
Washington, DC	43,766	3.3
Boston/Route 128	41,736	3.2
Chicago	38,447	2.9
Philadelphia	28,693	2.2
Miami	27,993	2.1
Dallas	26,520	2.0
Seattle	25,238	1.9
These 10 Metropolitan Areas	544,418	41.4
Rest of U.S.	771,393	58.6
Entire U.S.	1,315,811	100.0

Source: Kolko 1998

Once again, the four metropolitan areas that contained the largest clusters of Internet hosts, and the densest concentrations of domains, also account for the largest nodes of Internet activity by this measure.

Internet Backbone Networks

While it was important to identify centers of Internet activity and variations in the concentration of Internet indicators among American cities, we also need to quantify the flows of information *between* cities to understand how these networks are developing within the American urban system. Communications on the Internet is primarily carried over fiber optic networks, portions of which have been adapted from their original use (the transmission of voice telephone calls), although the first dedicated networks optimized for the TCP/IP protocol are now under construction. Traversing the country along traditional rights-of way, such as railroad tracks, interstate highways, and even abandoned canals, these networks are the physical manifestation of the 'information superhighway'. Like the host and domain-name measurement, our analysis of the conglomeration of national data networks collectively known as the Internet 'backbone' indicates a high degree of centralization in the deployment and use of Internet technologies.

Again, the technology in question dictated careful interpretation of the results. While the host count described a physical, real phenomenon (the connection of computers to the Internet), and the domain count a thoroughly virtual, logical one (the organization of Internet Protocol addresses into convenient hierarchies), the measurement of backbone capacity is a hybrid. Although some networks operate on isolated fiber optic cables, many are merely *virtual networks* operated over lines leased from national and regional telephone companies. Often, a backbone provider's only capital equipment are the powerful routing computers that manage the flow of data packets at network junctions (Rickard 1997).

While the geography and topography of these networks has received more attention from scholars than the identification of nodes of Internet activity, no other studies have analyzed the aggregate topology and capacity of the major backbone networks (Moss and Townsend 1998). Using maps and data from Boardwatch Magazine's *Quarterly Directory of Internet Service Providers*, we compiled a list of the capacity and endpoints of every major backbone link for 29 major backbone operators in the United States. We estimate that these 29 wholesale providers supply at least 95 percent of long-haul Internet data transport services in the United States. Based on the assumption that barriers to backbone accessibility were likely to be found at the inter-metropolitan, rather than intra-metropolitan level, we further aggregated these links by metropolitan area. The existence of networks, such as Metropolitan Fiber System's Metropolitan Area Ethernets, and the rapid proliferation of the baby Bells' high-speed regional fiber networks underlies this assumption. This aggregation allowed us to focus on the largest capacity fiber optic networks, constructed of DS-3 (45 MBps), OC-3 (155 MBps) and OC-12 (622

MBps) technology. One megabit per second (MBps) indicates a data transfer rate equal to approximately 128 pages of text per second.

The analysis is also confined to direct network connections. Theoretically, any city on a given network has access to all locations served by all networks. However, there is significant traffic congestion of data packets at inter-network gateways (bridges), and there are strong indications that providers have established direct links on the most highly trafficked inter-metropolitan routes. For example, several providers have established direct links between New York and Washington, D.C., even though they already operate a route connecting these two metropolitan areas through intermediate cities, such as Baltimore, Philadelphia or Wilmington, Delaware. The San Francisco Bay Area is directly linked to almost every metropolitan area in the United States, although many could presumably have been served indirectly through another node. These patterns of investment indicate the superiority of direct connections and their importance to a metropolitan area's ability of to import and export information via the Internet.

The results of the backbone analysis were the most striking and conclusive of the three measurements we used. Summarized in Table 10.4, the data show that a group of seven metropolitan areas (San Francisco/Silicon Valley, Washington, D.C., Chicago, New York, Dallas, Los Angeles, and Atlanta) form a core group of urban areas that dominate the Internet in the United States.

Table 10.4. Top 10 metropolitan areas by backbone capacity

Metropolitan area	Percent of total national backbone capacity	Total inter-metropolitan backbone capacity (in MBps)
San Francisco/Silicon Valley	11.6	7,506
Washington DC	10.4	7,826
Chicago	9.8	7,663
New York	9.7	6,766
Dallas	7.1	5,646
Los Angeles	6.7	5,056
Atlanta	6.6	5,196
Denver CO	3.7	2,901
Seattle WA	2.5	1,972
Houston TX	2.4	1,890

Source: Moss and Townsend 1998

These seven metropolitan areas each have the capacity for over 5,000 megabits per second (MBps) of Internet data throughput, sufficient to transfer text across the Internet at a rate of over 640,000 pages per second. No other metropolitan areas approach this level of capacity. Denver, ranked eighth, has only 60 percent of the backbone capacity of seventh-ranked Atlanta. As a result, these seven metropolitan areas share 62.0 percent of the nation's backbone capacity. The next 14 metropolitan areas together account for an additional 25.5 percent of the nation's backbone capacity, while the remainder of the United States houses the remaining 12.5 percent. Outside the major metropolitan backbone hubs, communities are linked to the Internet backbone via less robust data pipelines such as DS-1 lines (1.5 MBps, also known as T-1), frame relay (0.05 to 0.25 MBps), ISDN (0.125 Mbps), and modem (0.025 to 0.5 MBps) lines, which substantially limits their ability to move large amounts of information quickly. Many of these technologies, especially the popular DS-1 lines, have notoriously distance-sensitive price structures that have severely restricted their proliferation in non-metropolitan areas.

These seven metropolitan Internet hubs also connect directly to a great variety of other metropolitan areas and cities. The San Francisco Bay Area has the largest number of external connections, with 153 links to dozens of other metropolitan areas. Washington, Chicago, and New York have over 100 links maintained by various network companies to other cities. Dallas, Los Angeles and Atlanta each have 75 or more direct external links. By contrast, the next 14 metropolitan areas each have 42 or fewer external linkages. The variety of linkages associated with the top seven metropolitan areas is yet further evidence of their key role as central switching centers for flows of information on the Internet.

The most striking finding of this analysis is that among the top seven metropolitan areas, the majority of backbone capacity was used to link these regions to each other or to provide service within the region, rather than to connect less important cities and outlying areas. This aspect of the emerging backbone network merits further research as it suggests that these metropolitan areas do not serve as conduits for a national system of data distribution, but instead have coalesced into a separate, highly networked urban system containing both the major producers and most consumers of Internet services. This configuration is very similar to observations about so-called 'global cities' like New York, London, and Tokyo, which have become largely disconnected from their national economies while being increasingly integrated with each other.

The backbone analysis was the most complete and accurate of these three analyses because the data set involved was more limited and manageable with existing analytic techniques. However, like the domain and host count measures, it should serve only as a general indicator of relative differences in magnitude of Internet accessibility and adoption among cities, regions, and metropolitan areas. Notwithstanding, the stark difference in backbone capacity between the top seven metropolitan areas and the rest of the nation was the clearest conclusion drawn from this series of studies.

10.4 Conclusions

The research on the geography of the Internet summarized in this paper was conducted over a two-year period from 1996 to 1998. Despite substantial difficulty in obtaining accurate, timely, and comprehensive data sets, our results indicate several important trends, as well as questions for future research. When compared, the three sets of data indicate a consistent metropolitan dominance of the Internet in the United States. This section summarizes our collective findings and observations.

Most significantly, the results of this research demonstrate conclusively that a select group of metropolitan areas, and in particular their central cities, overwhelmingly dominate the Internet in the United States. Above and beyond their status as centers of population and employment, these regions consistently lead the nation in the magnitude, density, and growth of Internet clusters. This fact calls into question many assumptions regarding the spatial effects of information technology and telecommunications, as many influential authors have promulgated a deterministic view that new technologies will lead to a radical *decentralization* of population and economic activity (Gilder 1995, Negroponte 1995, Toffler 1980). This view is not borne out by our research. The New York, San Francisco, Washington, and Los Angeles metropolitan areas that are the largest clusters of Internet activity appear to be employing these technologies to reassert their economic importance in the American urban system. Furthermore, the presence of a set of smaller, Internet-savvy metropolitan areas, including Austin, Boston, Miami, and Seattle, suggest that rather than causing decentralization, the Internet may permit the development of a more complex, networked urban system.

These studies also indicate the lingering importance geographical location in determining accessibility to advanced telecommunications services. Cities like Atlanta, Chicago, and Dallas are extremely important hubs for national Internet backbone networks out of proportion to their share of other measures of Internet activity. We infer that their geographic centrality is an important factor in this allocation of network capacity. As a result, these cities have access to more, faster, and less expensive Internet infrastructure than cities of comparable size that are less ideally located. However, the absence of comparable levels of Internet development in other centrally located cities, such as Detroit and Philadelphia, lead us to believe that a certain set of socioeconomic prerequisites influence the allocation of these technologies.

In our effort to measure and describe the emerging telecommunications landscape of the United States, we anticipated being able to measure flows of information across the Internet to gain a greater understanding of which regions were net information producers, and which were net information consumers. Abler (1970) and Mitchelson and Wheeler (1994) used this methodology in the past to determine the relative position of cities within an urban hierarchy. While we were unable to gather such data, future research may benefit from a proposal to integrate geographic information in the Internet's domain name system. The Request For

Comment 1976 proposal (RFC 1976), *Putting Locations in the Domain Name System,* would theoretically permit researchers to measure the flow of information across monitoring stations along the Internet in real time and to more accurately determine the location of host computers than is currently possible (Davis 1998).

Perhaps the most important conclusion we have made is that researchers studying the Internet need to be very sensitive to the technical intricacies of the systems they investigate. The relevance of the host count measurement was seriously undermined by several factors discussed earlier, such as the use of firewalls to mask corporate networks. Without a clear understanding of exactly how the Internet is constructed, we probably would have accepted those results beyond their actual significance. Furthermore, because much of this *infrastructure* is actually an array of intangible data and logical constructs (domains, virtual backbones) or easily reconfigured electronic equipment (host computers and fiber optic networks), means it can be reallocated almost instantly in response to market shifts, natural disasters, etc. By definition, the Internet is highly volatile and in constant flux. Research runs the risk of being an anachronism before it is ever published. For cities, this means that unlike the televised urban riots that launched Lyndon Johnson's *War on Poverty* or the graphic images of Rust Belt cities in decay that helped put Ronald Reagan in the White House, Internet communities can disappear with the flip of a switch.

Finally, there is a strong need for systematic data gathering on the geography of Internet development and information flows. The RFC 1976 proposal is a good candidate, and we hope its key points will be incorporated in the new version of the basic Internet Protocol, IPv6, now being designed.

One of the many interesting ideas to emerge from the NCGIA Project Varenius specialist meeting, from which the chapters in this volume are drawn, was the concept of a federally funded Internet Census. Historically, Congress has often charged the Census Bureau to conduct exhaustive surveys on the emerging telecommunications industries. A full twelve pages of the 1852 economic Census were dedicated to the telegraph (Standage 1998), and two special studies of the telephone industry in 1902 and 1907 each contained over 500 pages of statistics about the structure and geography of the fledgling industry. These important documents were very influential in highlighting the uneven geographic distribution of telecommunications capabilities and paved the way for the universal service provisions that governed most national telecommunications policies for the remainder of the 20th century.

As this chapter has shown, the Internet in the United States is spreading in a highly uneven fashion, strongly favoring profitable markets and punishing economically distressed cities and remote areas. If the Internet will be the primary vehicle for commerce and communications over the next 50 years, as many in the Clinton-Gore Administration believe, accurate data about the location of Internet infrastructure and users will be essential for making sound policy decisions. Public officials have chosen to defer to the market in regulating the development of the Internet, but this research shows that a laissez-faire approach has not led to an equitable distribution of access and technology among and within cities and metropolitan areas. It follows that further systematic collection of data is necessary to more precisely determine the cause of these disparities.

References

Abler, R. 1970. What makes cities important? *Bell Telephone Magazine.*

Castells, M. 1989. *The Information City: Information Technology, Economic Restructuring, and the Urban-Regional Process.* Cambridge MA: Blackwell.

Davis, C. 1998 *RFC 1976 Resources: Putting Locations in the Domain Name System.* Electronic document [http://www.kei.com/homepages/ckd/dns-loc/]. Viewed 28 July 1998.

Dodge, M. and Shiode, N. 2000. Where on the Earth is the Internet?: An empirical investigation of the spatial patterns of Internet 'real estate' in realtion to geospace in the United Kingdom. In Wheeler, J.O., Aoyama, Y. and Warf, B. (eds.). *Cities in the Communications Age: The Fracturing of Geographies.* New York: Routledge.

Garreau, J. 1991. *Edge City: Life on the New Frontier.* New York: Doubleday.

Gaspar, J. and Glaeser, E. 1996. Information technology and the future of cities. *National Bureau of Economic Reseach Working Paper,* No. 5562. Cambridge MA.

Gilder, G. 1995. *Forbes ASAP*, 27 February, p.56.

Gottmann, J. 1961 *Megalopolis.* Cambridge MA: MIT Press.

Greenstein, S., Lizardo, M. and Spiller, P. 1997.The evolution of the distribution of advanced large scale information infrastructure the United States. *National Bureau of Economic Research Working Paper,* No. 5929. Cambridge MA.

Hall, P. 1997. Modeling the post-industrial city. *Futures* 29: No. 4/5

Kolko, J. 1998. Personal communication. J. Kolko can be reached at email: kolko@nber.org

Matrix Information and Directory Services, Inc. 1996. *Matrix Maps Quarterly, No.* 303 (January). Austin TX.

Mitchelson, R.L., and Wheeler, J.O. 1994. The flow of information in a global economy: the role of the American urban system in 1990. *Annals of the Association of American Geographers*, 84:87-107.

Moss, M.L. and Townsend, A.M. 1996. Leaders and losers on the Internet. Taub Urban Research Center, New York University, September [http://urban.nyu.edu/research/l-and-l/report.html]

Moss, M.L. and Townsend, A.M. 1997a. Manhattan leads the 'net nation. Taub Urban Research Center, New York University, August. [http://urban.nyu.edu/research/domains/domains.html]

Moss, M.L. and Townsend, A.M. 1997b. Tracking the 'net: Using domain names to measure the growth of the Internet in U.S. cities. *Journal of Urban Technology.* 4: No. 3.

Moss, M.L., and Townsend, A.M. 1998. Spatial analysis of the Internet in U.S. cities and states. Paper presented at the *Urban Futures – Technological Futures Conference*, Durham UK, 23-25 April. [http://urban.nyu.edu/research/newcastle/newcastle.html]

Negroponte, N. 1995. *Being Digital.* New York: Knopf.

Nunn, S. and Warren, R. 1997. Metropolitan telematics infrastructure and capacity for economic development in the information society. Center for Urban Policy and the Environment, Indiana University.

Pred, A.R. 1973. *Urban Growth and the Circulation of Information.* Cambridge MA: Harvard University Press.

Rickard, J. 1997. The Internet – what is it?. *Boardwatch Magazine – Internet Service Providers Quarterly Directory*. Littleton CO, Fall.

Standage, T. 1998. *The Victorian Internet*. New York: Walker and Co.

Sui, D.Z., and Wheeler, J. 1993. The location of office space in the metropolitan service economy of the United States, 1985-1990, *The Professional Geographer* 45:33-43.

Toffler, A. 1980. *The Third Wave*. New York: William Morrow & Co.

11 Accessibility to Information within the Internet: How can it be Measured and Mapped?

Martin Dodge

Centre for Advanced Spatial Analysis, University College London, 1-19 Torrington Place, Gower Street, London, WC1E 6BT, UK. Email: M.Dodge@ucl.ac.uk

11.1 Introduction

One definition of the Internet is '. . . a collection of resources that can be reached from those networks' (Krol and Hoffman 1993, 1). This definition provides the starting point for my conceptualization of accessibility in the Information Age. I will examine how one can begin to measure and visualize the aspects of accessibility to information resources *within* the Internet. My discussion starts with the assumption that a person has *physical* access to the Internet, via a networked computer[1]. Once this 'physical' connectivity has been overcome, how accessible are the information resources, people, and electronic places available online? What are the future accessibility issues that need to be considered to realize the full potential of the Internet beyond basic connectivity? As the Microsoft mantra says, *where do you want to go today?*, so what resources are accessible within the Internet and how do you reach them? This is very much a concern with individual accessibility and with developing a behavioral geography for Cyberspace (Brunn 1998, Kwan 1998).

It is also important to consider how these dimensions of accessibility can be represented, particularly given the abstract nature of information spaces, which often do not have natural spatial structures. The online activities of searching, browsing and communicating in Cyberspace are ethereal and non-tangible to conventional mapping techniques. Future representations need to be dynamic and interactive, as well as being readily available while navigating the Internet (December 1995, Dodge 1999b).

I will argue that the scope of geographical accessibility needs to be expanded to encompass notions of *information accessibility*. The growing importance of the Internet, and its layered services, for receiving and distributing all manner of information and for personal interaction, will require us to consider how concepts of

---

[1] This is, of course, a big assumption and one that requires serious investigation; see, for example, Wresch (1996) and Holderness (1998).

accessibility are played out within these electronic spaces. Gaining access in a timely fashion to the right information resource, be it a Web page, an email, a video clip, a chat room, or a particular place in a virtual world, is problematic for a number of human and technical reasons. The Web may well facilitate easy access to vast arrays of information from servers around the world, but this does not mean one can find useful, current, reliable, and affordable information at the right time. As Pirolli, Pitkow, and Rao (1996, 1) comment, 'The apparent ease with which users can click from page to page on the World-Wide Web belies the real difficulty of understanding the what and where of available information'. The Web is a vast array of information, but the ratio of noise to useful information can be very high. The problems of information retrieval through searching and browsing this massive space are becoming important for conceptualizing accessibility in the Information Age. There is an increasing awareness of the problem of *information overload*, with 'a tsunami of data crashing onto the beaches of the civilized world' (Wurman 1997, 15). Accessibility to too much information is potentially as significant an issue as accessibility to too little information. Excessive information impedes its assimilation and therefore does little to improve knowledge and understanding (Shenk 1997). A great deal of effort is being directed by researchers in a range of disciplines to cope with the problem of information retrieval and information overload through filtering, structuring, analyzing and visualizing information to aid the limited human capacity to search for, absorb, and comprehend information (Berghel 1997, ASIS 1998). Much of this research is relevant to broadening the scope of geographic accessibility to encompass information spaces of the global Internet.

In this chapter I discuss how we can develop the theme of *information accessibility*, examining the following topics (1) nature of the different Internet information space, (2) issues of network performance and tools to try and diagnose problems, (3) the importance of search engines and the problem of *searchability* of resources, and (4) accessibility problems caused by the design of information spaces. The issues of measurement and representation of information accessibility are highlighted in detail for two particular Internet information spaces; an empirical investigation of accessibility between Web sites using the structural information contained in hyperlinks and an examination of accessibility in a 3d virtual world on the Internet.

11.2 Information Spaces of the Internet

It is important to be aware that the Internet provides users with a number of distinct services, which are often thought of as different *spaces*, with differing virtual landscapes. In particular, these different spaces support different types of information exchange, degrees of synchronicity, and levels of social interaction. Therefore, they are likely to require different measures of accessibility and forms of graphic representation to appropriately model their true nature. At a fundamental level, the different information spaces are caused by the different network protocols used by software applications to communicate over the Internet, which give rise to the different forms and functions apparent to the end-user. Figure 11.1 shows a sketch map produced by John December, showing the principal information spaces and some of the connections between them (December 1995).

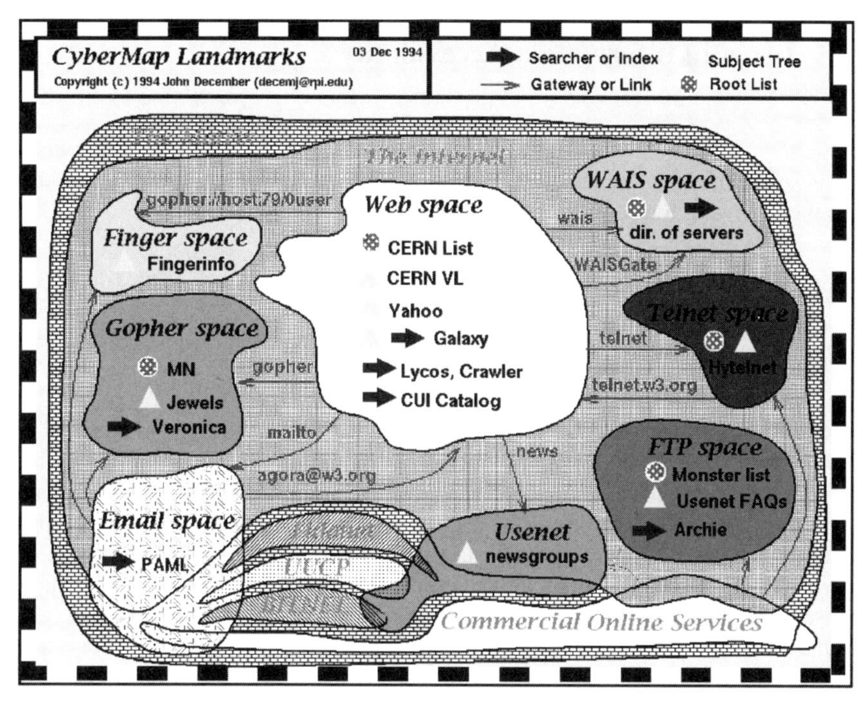

Figure 11.1. Information spaces of the Internet – circa 1994. Reproduced with permission from John December (1995).

December's map provides a good way of conceptualizing the different information spaces of the Internet as distinct and self-contained domains, but with fluid, complex boundaries and many interconnections and overlaps among them. The map was drawn at the end of 1994 and the nature of the Internet has changed markedly since then, with certain spaces dying off as they fall out of favor with users (WAIS and Gopher) and the inexorable and exponential growth of Web space. For many end-users the Web, seen through the browser interface, is the key information space, although email is still the most widely used information space (ITU 1997, Clemente 1998). Other important information spaces within the global Internet that have evolved and grown since December drew this map include multi-user chat environments and virtual worlds. Also, the rise of large private networks and Intranets are creating important information spaces, but they are largely unseen from the outside and so are difficult to quantify and map. More recent work on conceptualizing the form and structure of different information spaces includes Michael Batty's (1997) examination of *Virtual Geography*, Paul Adams's (1998) discussion on *Network Topologies and Virtual Place*, Manuel Castells's (1996) *Space of Flows,* and Brian Gaines's research on the human-factors of Internet information spaces (Gaines, Chen, and Shaw 1997).

11.3 Issues in Information Accessibility

There are a number of important issues that need to be factored into future comprehensive models of accessibility to information in networked space. I will discuss the following four issues:

- network performance,
- size of the information spaces,
- information findability and persistence, and
- information structure, design and user behavior.

A common joke is that the WWW really stands for the World-Wide *Wait* because of the poor and unpredictable performance of the Internet, particularly perceived by the Web user waiting for a page to download. It is often asserted that 'distance is dead' on the Internet because it does not matter where the information is geographically located in the world that the user is trying to access – it is only a mouse click away. Instead, what really matters is where the site is located in temporal space. Perceived response time replaces geographic distance as the key variable in access to interactive information spaces like the Web. A slow response time from a Web site that is physically nearby means that it will be perceived as being more remote and inaccessible than a fast site that is thousands of miles away on another continent. Limited human patience with interactive computer interfaces means that response time is critical and even delays of a few seconds can prove so frustrating that people simply give up and try to locate an alternative source closer in temporal space (Nielsen 2000). If network performance degrades below certain thresholds, some information spaces on the Internet become effectively infinitely *distant* and inaccessible because people will not bother to wait. The delays caused by network performance also have financial implications for those people paying by the minute for their access to the Internet. The complexity and self-organizing nature of the underlying Internet infrastructure makes it difficult to determine where performance problems that effect information accessibility are located in the network (Huberman and Lukose 1997). Problems can be at any point in the chain from the user's computer through the network links and nodes, to the target server. Examining Internet traffic-flows reveals the incredible complexity of routes that data travels through the network, often traversing fifteen or more nodes and crossing physical infrastructure owned and operated by competing Internet service providers and telecommunications companies. It is possible to explore Internet traffic routing using utilities called traceroutes (Rickard 1996) and it can be surprising just how complicated things are 'under the hood', so to speak. It is amazing that the Internet works as well as it does! To take an example, my Web site is located at a commercial hosting service that is geographically about 1.5 kilometers from my office in University College London (UCL) and yet a traceroute reveals that traffic takes twelve hops (different network nodes) to travel this distance. Whereas traffic to the mirror site that is physically located in Washington DC, around 3,700 km from UCL only takes slightly longer, in network terms, at fourteen hops. For measuring accessibility through the routing topology of the Internet, these two Web site locations are pretty much equally distant from UCL. However, there are some interesting variations in response times from these two sites. The site in London is generally equally responsive throughout the day, however the accessibility response time to the mirror site in Washington depends on the time of day. The response degrades noticeably in the afternoon when the great mass of North American Internet users wake up and log on. It is well known that for European Web surfers, the performance of the Web degrades significantly in the afternoon when America wakes up. Clearly, measuring accessibility to global information resources like the Internet and the Web will require the time dimension to be fully integrated (Harvey and Macnab explore this theme in Chapter 9).

At the network performance level, information accessibility would be measured by three key parameters, (1) *delays*, known as network latency, (2) *deliverability*, the problem of data being lost in transit and having to be resent, and (3) *availability* of the network and servers, assessed by the amount of *down-time*. Much of the research into the performance of Internet infrastructure is highly technical (e.g., Paxson 1997), however there is some research particularly relevant to the issue of geographical accessibility. This includes the work of John Quarterman who undertakes real-time monitoring of Internet performance by calculating latencies to several thousand sample nodes in the global Internet from his headquarters in Texas (Quarterman, Smoot, and Gretchen 1994, Quarterman 1997). The results are presented as animated maps that he terms Internet Weather Reports (IWR). Figure 11.2 shows one of the animation frames for California (Quarterman 1997; see also http://www.mids.org/weather/). Quarterman claims that his analysis shows that there has been a thirty- percent improvement in mean Internet latencies since he started monitoring in 1994. There is also a study by Keynote Systems / Boardwatch into Internet backbone performance in the United States. This involves large-scale testing of the performance of ten different networks from twenty-seven different sample points in different cities (Rickard 1997). Shane Murnion is also undertaking research into the geography of Internet latencies, examining information flows from UK academic Web servers to sixty-six countries (Murnion and Healey 1998; see also Chapter 12). His findings show the existence of a distance-decay effect in Web server audiences with increasing latency.

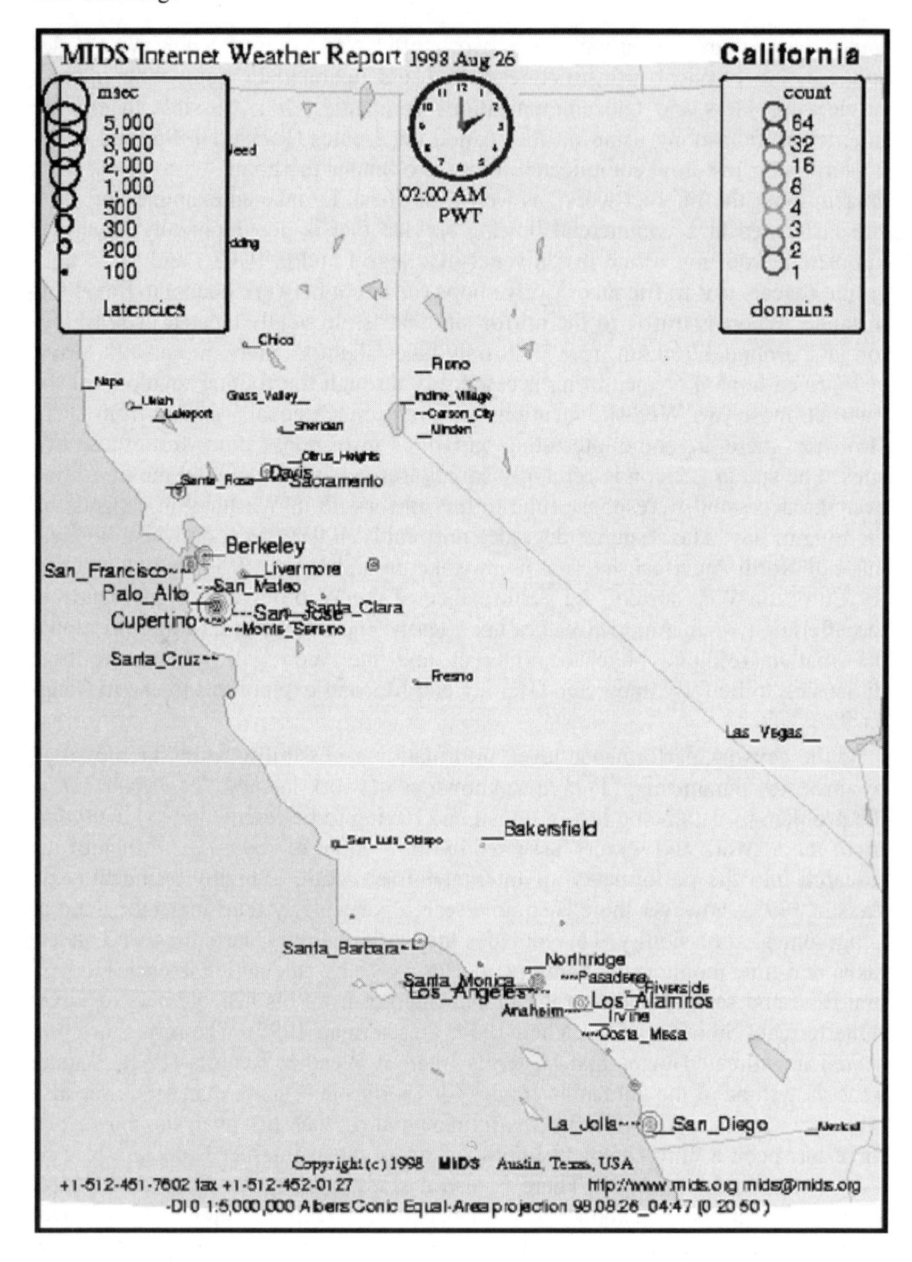

Figure 11.2. Internet Weather Report (IWR) for California. Reproduced with permission from John S. Quartermann, Matrix Information and Directory Services. http://www.mids.org/.

Another key issue for reconfiguring accessibility measures for the information age is to develop scaleable models to cope with the size, diversity, and dynamic nature of Internet information spaces. Pinning down definitive figures on the size of Internet information spaces from the cyber-hype and Net boosterism can be difficult, but it is big and, more significantly, it is the fastest growing medium of information and communication in history (ITU 1997, Clemente 1998)[2]. Recent statistics show there were an estimated 36.7 million Internet hosts (Network Wizards 1998) and approximately 153 million Internet users worldwide (NUA 1999) in January 1999. For the information resources potentially accessibility on the Web, estimates from 1998 were that 2.8 million sites contain around 300 million pages (Bharat and Broder 1998, Lawrence and Giles 1998). More importantly, these studies reveal how incomplete the databases of even the largest search engines are, only indexing at best a third of the Web. The consequence is that the Web pages people are seeking may exist, but if they are not in the search engine indexes they are invisible and, therefore, inaccessible because people will never find them. Even if they are in the index, they may come so far down the list of search results returned that most users will never see them, making them effectively inaccessible. This has been termed the *searchability* factor by Esther Dyson and is crucial in realistically modeling information accessibility (Dyson 1999).

Search engines have become the key access points – known as portals in the current terminology – to information spaces on the Internet. They are rich indexes of socially constructed information and, as such, offer a potentially useful database for geographical research. The potential is beginning to be realized, for example in research into the reproduction of the concepts of place online (Alderman and Good 1997, Jackson and Purcell 1997, Alderman 1998, Henkel 1998, Norris 1998).

Another serious issue with modeling information accessibility is the problem of persistence, or rather lack of it, on the Internet. The Internet and the Web are changing every day, with information resources, sites, and virtual places appearing and disappearing. A flick of a switch or a press of a delete key can cause whole parts of an information space to simply disappear without a trace. Information structures in Cyberspace are much less permanent than those of the real world. Can accessibility measures cope with this?

The actual design of the information space can have an impact on its accessibility (Nielsen 2000). Like real-world cities and buildings, poorly designed Web sites and pages are rendered inaccessible to certain groups of users. The virtual world of information certainly suffers from the same degree of bad architecture, as does the material world (Wurman 1997). We have all seen Web sites with poor choice of fonts, colors, or frames, for example, which make them practically unusable. The need for accessible design is especially important for visually impaired people who surf with text-only software. There was a recent high profile case reported in

---

[2] Although the Internet is a large information space, it is still small in absolute volume terms compared to other information domains, particularly broadcast television. See the paper by Lesk (1998) for a fascinating examination of the sizes of different information domains.

the United Kingdom where the newly re-vamped Web site for Number 10 Downing Street was so badly designed that it was said to be '*staggeringly inaccessible*' (Jellinek 1998). Work is ongoing to improve the *physical* design and accessibility of Web sites, coordinated through the Web Accessibility Initiative (http://www.w3.org/WAI/) of the World Wide Web Consortium. However, the problem is difficult to resolve, as many Web-site designers are keen to use the very latest technologies, which can easily make their sites inaccessible to many average users. A survey of UK Web sites in 1997 found that only 30 percent of pages were completely accessible to all users (Beckett 1997).

To develop new models of information accessibility we need to know about the content and structure of information spaces and how users behave in them. There is some useful work trying to answer these questions, but our quantitative knowledge of Cyberspace is far from complete, particularly compared to our knowledge of real-world spaces (Fagrell and Sørensen 1997). There are several reasons for this, including the sheer newness of some of the information spaces and their invisibility to the conventional monitoring and census-taking methods developed for the material world (Batty 1990). Governments have not, until very recently anyway, realized the significance of the information spaces and so have made no attempts to gather statistics on them. However, there are a number of interesting academic studies that have begun to fill in the blanks. For contents of the Web, there is the work of Bray (1996), Woodruff, et al. (1996) and Fagrell and Sørensen (1997). The structure of the information spaces, in particular the Web, has been examined; see the exemplary work of James E. Pitkow (Pirolli, Pitkow and Rao 1996, Pitkow 1998) and Bray (1996). Researchers are beginning to analyze and model how users behave in Cyberspace, as seen in the work by Huberman, et al. (1998) who have devised a *law of surfing* to describe user behavior.

11.4 Accessibility Between Web Sites

My colleague Naru Shiode and I are researching the spatial structure of the Web by analyzing the hyperlinks between sites (Dodge 1998). The data on how Web sites are linked together can be used to model accessibility for this particular information space. In a preliminary investigation, we analyzed a small, manageable subset of the Web, the site of the major universities and colleges in the United Kingdom, some 122 nodes. We used the AltaVista search engine (http://www.altavista.com/) to gather statistics on the size of each Web site (defined by the number of pages) and the number of hyperlinks between them, which took 14,884 separate queries to AltaVista. The results showed that the 122 sites contained more than one million Web pages and over 450,000 hyperlinks (although the vast majority were internal links within individual sites).

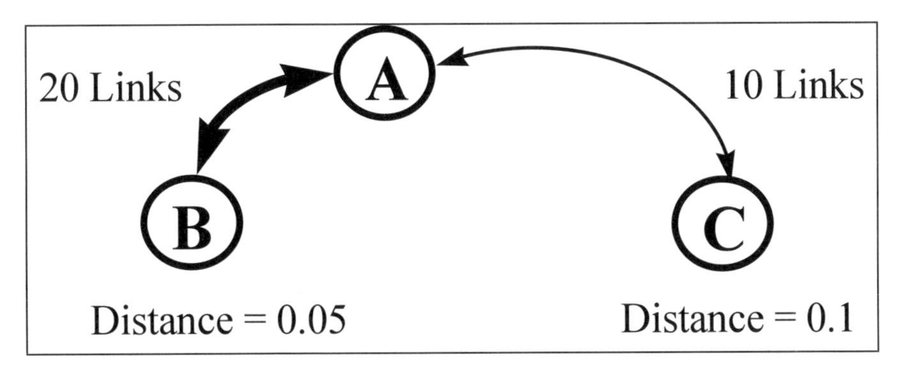

Figure 11.3. Calculating distance in virtual space.

The data on the hyperlink connectivity between sites were analyzed to determine the most accessible Web site. To do this, the distance from each university to every other one was calculated. In the Web, virtual distance was calculated as inversely proportional to the number of hyperlink connections between two points. In the example shown in Figure 11.3, site B is much closer in virtual distance to site A than is C. The Web site that had the lowest average distance, i.e., was closest to all the others, was designated the central, most accessible one.

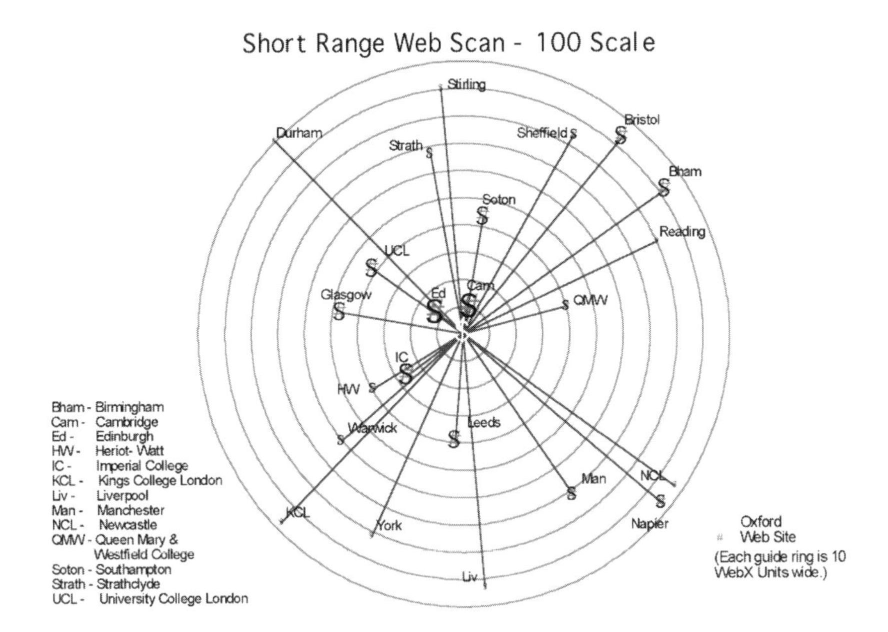

Figure 11.4. Web Scan showing the most accessible Web sites of UK universities (Source: Dodge 1998)

The virtual distances between all 122 sites were measured and stored as a large graph where the edge lengths were assigned the distance value. The graph was analyzed to find the shortest-path distance from each university to every other one using the Dijkstra algorithm. The mean of these shortest-path distances was calculated for each Web site and then normalized by dividing by the number of Web pages to take account of the influence of variations in Web-site size. The resulting ranking showed that the University of Oxford's Web site (http://www.ox.ac.uk) had the smallest, normalized, mean shortest path distance to all other site; hence, it was declared as the most accessible Web site in our experiment. The graph of virtual distance was then used to measure the shortest-path distance from Oxford to each university and this value was used as a metric of accessibility in Web space and was called the *WebX distance.*

To begin to understand the structure and differential accessibility of academic Web sites, as measured by the WebX distance, it was necessary to visualize the position of sites in relation to Oxford. To achieve this we used a radar-type map called a *Web Scan.* In the Web Scan, Oxford becomes the central point of gravity, around which planetary Web sites rotate, their orbital distances being equal to their WebX distance. Figure 11.4 shows a Web Scan for the most accessible Web sites, those closest to Oxford. What is immediately striking are the two giant sites very close to the Oxford center point. These are the University of Cambridge (http://www.cam.ac.uk/) and University of Edinburgh (http://www.ed.ac.uk/), which have large Web sites and are very closely interconnected. Cambridge has a WebX distance of 10 and Edinburgh is only slightly further out at 13. Imperial College (http://www.ic.ac.uk/) comes next with a WebX distance of 26, double that of the second-place site. Further out from the top three, there is a cluster of sites around the 40-WebX mark. These are Heriot-Watt University, the universities of Leeds, Glasgow, Southampton, Queen Mary and Westfield College, and my own institution – University College London (http://www.ucl.ac.uk/). All these universities are well connected and accessible in the academic Web. UCL has a WebX score of 42, placing it in seventh place away from Oxford, a respectable place given its historic place in the development of Cyberspace in the United Kingdom, as it was the first organization in Britain connected to ARPANET, the Internet's forerunner, back in 1973 (Salus 1995). There is, then, a slight gap until the next Web sites are encountered, including large metropolitan universities such as Birmingham, Liverpool, Bristol, Sheffield, and Manchester, and, as well, smaller provincial institutions like York and Stirling. Finally, right on the edge of this scan is the University of Durham (http://www.dur.ac.uk/) with a WebX score of exactly one hundred. Further results of our preliminary work are presented in Dodge (1998). We are extending this type of virtual-network accessibility analysis using a larger, more realistic subset of the World-Wide Web. This type of analysis of the structure of the Web, I believe, provides a potentially useful avenue for developing accessibility measures suitable for the Information Age.

11.5 Accessibility in Virtual Worlds

Virtual worlds are a form of information space on the Internet that provides a simulated environment in which multiple users can interact with each other in real-time. Crucially, the users have a bodily representation in the space as an avatar, and the simulated environment is graphically rendered in 2d, 2.5d or full 3d. A number of virtual worlds from competing companies have emerged on the Internet since mid-1995 (Rossney 1996, Damer 1998). They are used by many thousands of people who have constructed new forms of social interaction and a distinct sense of community and cultural identity to suit the unique characteristics of these spaces (Rossney 1996, Donath 1997, Schroeder 1997, Damer 1998, Rafaeli, Sudweeks and McLaughlin 1998). In many respects the users can be said to inhabit these worlds and have developed a real sense of place in Cyberspace. Virtual worlds have pre-defined and programmed geographical dimensions, architectural structures, and rules of avatar movement, they are truly information *spaces* (Anders 1998). They provide a fascinating new realm, arguably at the cutting-edge of the Information Age, in which to explore the meaning of geographical accessibility. The spatial nature of virtual reality and Internet virtual worlds has, so far, received little attention from academic geographers, notable exceptions being Hillis (1996) and Taylor (1997).

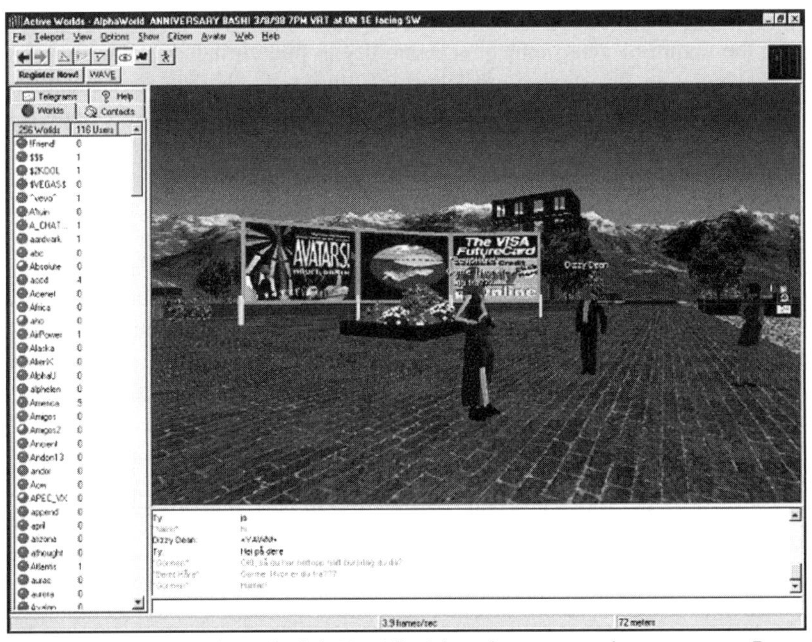

Figure 11.5. Screen-shot of AlphaWorld's 3d environment and user avatars. Reproduced with permission from Circle of Fire Studios, Inc (http://www.activeworlds.com/).

One of the most popular, technologically advanced, and geographically interesting virtual worlds is called Active Worlds, owned by Circle of Fire Studios, Inc (http://www.activeworlds.com/). Their flagship world is called AlphaWorld and it is one of the oldest (it opened in the summer of 1995) and most developed virtual worlds on the Internet. Figure 11.5 shows a screen-shot of a typical view of AlphaWorld with its realistic 3d environment and users represented by avatars. Research has begun to explore the physical and social geography of AlphaWorld (Schroeder 1997, Dodge 1999a). This has been made easier because, uniquely, it has been mapped in remarkable detail and also has a rich recorded history.

Figure 11.6 shows two satellite style maps of the city growing at the center of AlphaWorld at two snap-shots in time, December 1996 and February 1998 (Vilett 1999). A technical innovation, unique amongst Internet virtual worlds, is that registered citizens of AlphaWorld are able to claim plots of vacant land and build a homestead to their own design. This has had profound consequences on the nature of urban development in the world, facilitating spontaneous and organic growth of towns and cities. The geographical extent of AlphaWorld is huge, covering some 429,000 km^2, larger than California, so it can easily contain the virtual building boom in which around thirty thousand people built over 27 million objects in the world (Vevo 1999). Most of the development has taken place in the center of the world, around what the locals call Ground Zero, located at 0,0 in the Cartesian co-ordinate space of this world. A sprawling city has grown outwards from Ground Zero in a totally unplanned way. To give you an idea of the scale of the city, the maps of it (Figure 11.6) cover an area of four hundred square kilometers.

The morphology of urban growth, revealed by the maps, provides useful information on the nature of geographical accessibility in this virtual world and on the impact of changes to the conventional laws of physics for human movement. In AlphaWorld there are no cars, trains or planes, so people are reliant on teleportation to travel any distance. Teleportation in AlphaWorld works just like in sci-fi movies; your avatar is instantaneously transported to the specified location with the accompaniment of a *beaming* sound effect! Teleportation has seriously warped the nature of geographical distance and accessibility as any location in the 429,000 km^2 expanse can be reached instantaneously from any other point in the world with no costs in terms of time or money. Consequently, every point in AlphaWorld is equally accessible. This is truly the death of distance (Couclelis 1996, Cairncross 1997). Teleportation is available to the user at any time as a menu in the browser; they just have to type in the co-ordinates of their desired destination and then they are whisked there in a second.

Distance may be dead in AlphaWorld, but the importance of location is alive and well. When people are choosing a location to visit or, more importantly, a place to build their homestead, they want a good location. A good location for most AlphaWorld inhabitants is determined by two factors, being as close as possible to Ground Zero, the center of the world, and having a location with memorable co-ordinates. Human nature, particularly when interacting with computers, means that people tend to select regular numbers for co-ordinate pairs, such as 256, 256, when teleporting. This has given rise to the star-shaped pattern of the urban growth, with radial spokes of development emanating from the city center along the principal compass axes apparent in Figure 11.6. The spokes are clearly evident in the December 1996 map, although in the second map, taken just over a year later, they have become less pronounced as fill-in development has taken place

The ability to teleport is a powerful feature, but interestingly it was not made available to users when AlphaWorld was first launched in the summer of 1995. It has only been progressively introduced for fear of its affects on the world. As the AlphaWorld newspaper, the *New World Times*, reported in November 1995:

> Teleportation! Yes Teleportation! The one most common request of AlphaWorld citizens has been teleportation... With teleportation more of AlphaWorld will become readily accessible. ... There is still some concern that teleportation will ruin the simulation of reality in AlphaWorld. In order to keep this simulation within bounds, teleportation will be implemented in a somewhat limited fashion. A 'Grand Central Telestation' located at or near Ground Zero will enable citizens to teleport to key locations, from which they can travel more easily to their destinations of choice. (*New World Times*, #4, p. 2, http://vrnews.synergycorp.com/nwt/).

AlphaWorld also warps conventional spatial norms and rules of physical movement at a local scale that have an impact on geographical accessibility. You are able to fly, unaided, above and even below the ground. To achieve this one simply presses the + and - keys to effortless float up and down. It is also possible to walk through any walls and structures by holding down the shift key, which has the effect of making all objects immaterial to your avatar. These two god-like *powers* have had a significant impact on the architectural design of buildings in AlphaWorld (Damer 1997). Visibility in AlphaWorld is also artificially constrained because it is only possible to see a maximum of 120 meters in any direction. This is due to the limits of graphics hardware and software on PCs to render a larger 3d landscape in real-time. However, the effect on the space is really quite unnerving, like walking around in an opaque bubble 120 meters across, where streets and buildings appear to end, with a sharp, artificial looking cut-off line. This impacts on local accessibility because it is hard to orientate and navigate with no fixed landmarks and distant vistas.

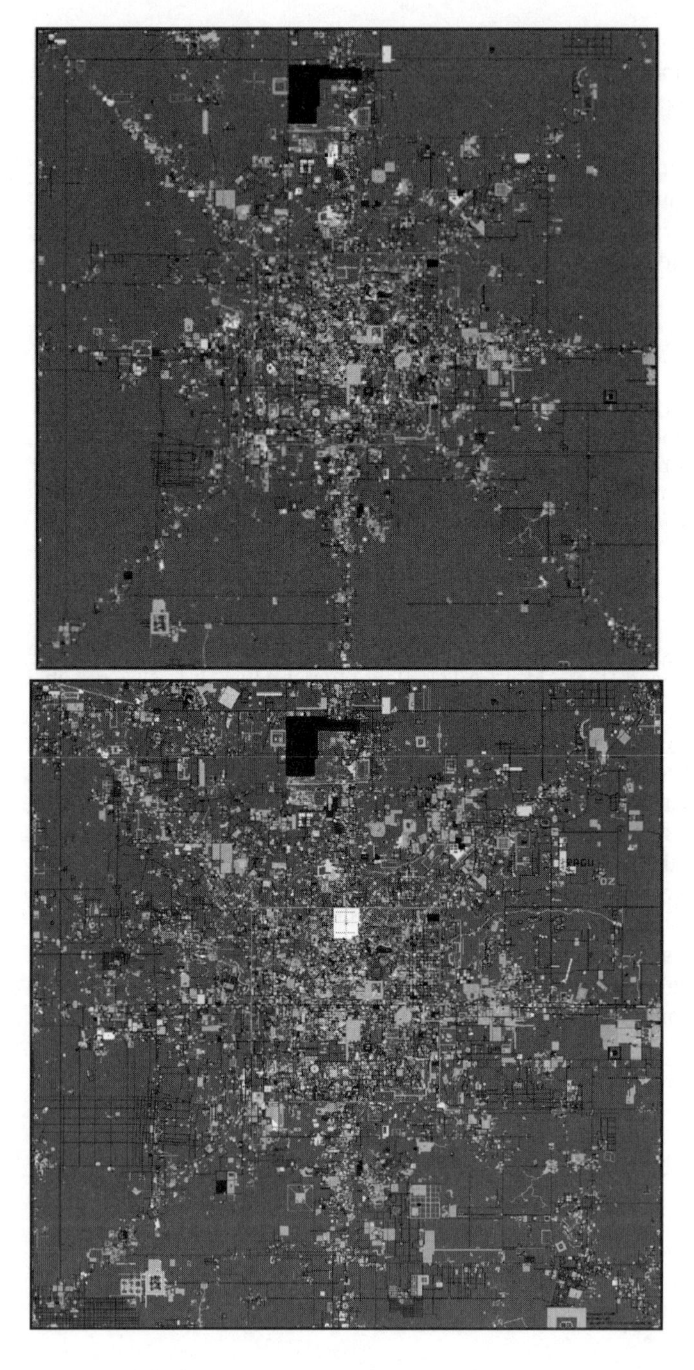

Figure 11.6. Maps of the city at the center of AlphaWorld in December 1996 (top) and February 1998 (bottom). Reproduced with permission from Vilett (1998).

11.6 Conclusions

In this paper I have been concerned with the idea of *information accessibility* within the virtual spaces of the global Internet. Two particular Internet-based information spaces, the Web and Virtual Worlds, have been examined as exemplar. One approach to measuring and mapping the relative accessibility of Web sites is to use the structure of hyperlinks between them to calculate measures of virtual distance. In contrast, virtual worlds provide a fascinating challenge to the conventions of geographic accessibility. Although the worlds have a tangible, simulated geographic environment that has many of the spatial characteristics of the real world, they can also warp the physical conventions of distance and travel. If, as some predict, these kinds of shared, 3d environments became prevalent as the next generation of information interface, then it will be vital to understand how accessibility effects the way people navigate and find places in the virtual worlds (Anders 1998).

In the Information Age the importance of access to information spaces, such as the Web or places in a virtual world, will increasingly take precedence over access to physical facilities in the real-world for certain important human activities. Being able to quantify and visualize accessibility to these virtual information spaces will be an important challenge in extending the notions of geographic accessibility to encompass Cyberspace.

References

Adams, P. 1998. Network topologies and virtual place. *Annals of the Association of American Geographers* 88:88-106.

Alderman, D.H. and Good, D.B. 1997. Exploring the virtual South: The idea of a distinctive region on 'The Web'. *Southeastern Geographer* 37:20-45.

Alderman, D.H. 1998. Finding the heart of Dixie in cyberspace: The Internet as a new laboratory for cultural geography, paper presented at *The Association of American Geographers 94th Annual Meeting*, March, Boston MA.

Anders, P. 1998. *Envisioning Cyberspace: Designing 3D Electronic Space.* New York:.McGraw Hill.

ASIS 1998. *American Society for Information Science (ASIS) '98 - Information Access in the Global Information Economy*, 24-29 October, Pittsburgh PA.

Batty, M. 1990. Invisible cities, *Environment and Planning B: Planning and Design* 17:127-30.

Batty, M. 1997. Virtual geography, *Futures* 29: 337-52.

Beckett D. 1997. 30% accessible ⇔ A survey of the UK Wide Web, *Proceedings of the Sixth International WWW Conference*, April, Stanford CA.

Berghel H. 1997. Cyberspace 2000: Dealing with information overload, *Communications of the ACM*, February 1997 40 2:19-24.

Bharat, K. and Broder, A. 1998. A Technique for measuring the relative size and overlap of Web search engines, *Proceedings of the Seventh International WWW Conference*, April, Brisbane Australia, 379-88.

Bray, T. 1996. Measuring the Web, *Computer Networks and ISDN Systems* 2:8 7-11, 993-1005.

Brunn, S.D. 1998. The Internet as 'the new world' of and for geography: speed, structures, volumes, humility and civility, *GeoJournal* 45: 1-2, 5-15.

Cairncross, F. 1997. *The Death of Distance: How the Communications Revolution Will Change Our Lives*. Cambridge MA: Harvard Business School Press.

Castells, M. 1996. *The Rise of the Network Society*, Vol. 1 of *The Information Age: Economy, Society and Culture*. Oxford: Blackwell Publishers.

Clemente, P. 1998. *The State of the Net: The New Frontier*. New York: McGraw-Hill.

Couclelis, H. 1996. The death of distance, *Environment and Planning B: Planning and Design* 23: 387-89.

Damer, B. 1998. *Avatar!: Exploring and Building Virtual Worlds on the Internet*. Berkeley CA: Peachpit Press.

December, J. 1995. A cybermap gazetteer. In Staple, G.C. (ed.) *TeleGeography 1995: Global Communications Traffic Statistics & Commentary*. Washington DC: TeleGeography Inc., 74-81.

Dodge, M. 1998. Journey to the centre of the Web. In Staple, G.C. (ed.) *TeleGeography 1998/99: Global Communications Traffic Statistics & Commentary*. Washington DC: TeleGeography, Inc., 150-54.

Dodge, M. 1999a. Explorations in AlphaWorld: The geography of 3d virtual worlds on the Internet, paper presented at *RGS-IBG Annual Conference - Geographies of the Future*, Leicester UK, 4-7 January.

Dodge, M. 1999b. *Atlas of Cyberspaces*. London: Centre for Advanced Spatial Analysis, University College London (http://www.cybergeography.org/atlas/).

Donath, J.S. 1997. *Inhabiting the virtual city: The Design of Social Environments for Electronic Communities*. PhD thesis, MIT (http://judith.www.media.edu/Thesis/).

Dyson, E. 1999. Search and searchability, *The Guardian*, Online section, 14 January, 11.

Fagrell, H. and Sørensen, C. 1997. It's life Jim, but not as we know it!, *Proceedings of the WebNet'97 Conference*, October, Toronto. (http://ifi.uio.no/iris20/proceedings/35.htm).

Gaines, B.R., Chen, L.L. and Shaw, L. 1997. Modeling the human factors of scholarly communities support through the Internet and World Wide Web, *Journal of the American Society for Information Science* 48:987-1003.

Henkel, T. 1998. An Internet artifact: place-concepts revealed in the virtual world. Paper presented at the *Telecommunications and the City Conference*, Athens GA, University of Georgia, March.

Hillis, K. 1996. A geography of the eye: the technologies of virtual reality. In Shield, R. (ed.) *Cultures of the Internet: Virtual Spaces, Real Histories, Living Bodies* London: Sage Publications, 70-98.

Holderness, M. 1998. Who are the world's information-poor? In Loader, B. (ed.) *Cyberspace Divide: Equality, Agency and Policy in the Information Society*. London: Routledge, 35-56.

Huberman, B.A., Pirolli, P.L.T., Pitkow, J.E. and Lukose, R.J. 1998. Strong regularities in World Wide Web surfing. *Science* 280(3 April):95-7.

Huberman, B.A. and Lukose, R.M. 1997. Social dilemmas and Internet congestion. *Science* 277 (25 July):535-7.

ITU 1997. *Challenges to the Network: Telecommunications and the Internet*. Geneva: International Telecommunication Union, September.

Jackson, M.H. and Purcell, D. 1997. Politics and media richness in the World Wide Web representation of the former Yugoslavia. *The Geographical Review* 87:219-39.

Jellinek, D. 1998. Door policy at Number 10, *The Guardian* Online section 7 May, 7.

Krol, E. and Hoffman, E. 1993. FYI on 'what is the Internet?' *Request for Comments: 1462* (May 1993), (ftp://ftp.ripe.net/rfc/rfc1462.txt).

Lawrence, S. and Giles, C.L. 1998. Searching the World Wide Web, *Science* 280(3 April):98-100.

Lesk, M. 1998. How much information is there in the world? Paper presented at the *Time & Bits: Managing Digital Continuity conference,* Getty Information Institute, February, Los Angeles (http://www.gii.getty.edu/timeandbits/ksg.html).

Murnion, S. and Healey, R.G. 1998. Modelling distance decay effects in Web server information flows, *Geographical Analysis* 30:285-303.

Network Wizards 1998. *Internet Domain Survey* (http://www.nw.com/zone/WWW/top.html).

Nielsen, J. 2000. *Designing Web Usability: The Practice of Simplicity*. Indianapolis: New Riders Publishing.

Norris, D.A. 1998. The developing world and the World-Wide Web: Search engine topical coverage, 1997. Paper presented at *The Association of American Geographers 94th Annual Meeting*, March, Boston MA.

NUA 1999. *How many online?* (http://www.nua.ie/surveys/how_many_online/index.html).

Paxson, V., End-to-end routing behavior in the Internet, *IEEE/ACM Transactions on Networking*, October 1997 5(5):601-15.

Pirolli, P., Pitkow, J. and Rao, R. 1996, Silk from a sow's ear: extracting usable structures from the Web, *Computer-Human Interaction (CHI'96) Conference,* April, Vancouver, Canada.

Pitkow, J.E. 1998. Summary of WWW characterizations. Paper presented at *The Seventh International World-Wide Web Conference*, April, Brisbane, Australia.

Quarterman, J.S., Smoot, C.M. and Gretchen, P. 1994. Internet interaction pinged and mapped, *Proceedings of the INET'94 Conference*, Internet Society, July, Prague, 522:1-4.

Quarterman, J.S. 1997. The Internet weather report. In Staple, G.C. (ed.) *TeleGeography 1997/98: Global Communications Traffic Statistics & Commentary.* Washington DC: TeleGeography Inc., 69-72.

Rafaeli, S., Sudweeks, F. and McLaughlin, M. 1998. *Network & Netplay: Virtual Groups on the Internet.* Cambridge MA: MIT Press.

Rickard, J. 1996. Mapping the Internet with Traceroute. *Boardwatch Magazine* 10 (December). (http://boardwatch.internet.com/mag/96/dec/bwm38.html).

Rickard, J. 1997. Backbone performance measurement, *Boardwatch Magazine Directory of Internet Service Providers* 2(Fall):22-31.

Rossney, R. 1996. Metaworlds. *Wired* 4.06:140ff.

Salus, P.H. 1995. Casting the Net : From ARPANET to Internet and Beyond... Addison-Wesley Publishing Company.

Schroeder, R. 1997. Networked worlds: social aspects of multi-user virtual reality, *Sociological Research Online* 2:4. (http://www.socresonline.org.uk/socresonline/2/4/5.html).

Shenk, D. 1997. *Data Smog : Surviving the Information Glut*. San Francisco: Harper.

Taylor, J. 1997. The emerging geographies of virtual worlds, *The Geographical Review* 87(2):172-92.

Vevo, 1999. *AlphaWorld Mapper* (http://awmap.vevo.com/).

Vilett, R. 1999. *AlphaWorld maps* (http://www.activeworlds.com/satellite.html).

Woodruff, A., Aoki, P.M., Brewer, E., Gauthier, P. and Rowe, L.A. 1996. An investigation documents from the World-Wide Web. Paper presented at the *Fifth International World Wide Web Conference*, May, Paris.

Wresch, W. 1996. *Disconnected: Haves and Have-Nots in the Information Age.* New Brunswick NJ: Rutgers University Press.

Wurman, R.S. 1997. *Information Architects.* New York: Graphis Press Corp.

12 Towards Spatial Interaction Models of Information Flows

Shane Murnion

Department of Geography, University of Portsmouth, Portsmouth, Hampshire P01 3HE, UK. Email: Murnions@geog.port.ac.uk

12.1 The Impact of E-commerce on Developed and Developing Economies

Internet related commerce (e-commerce) is going through a phase of rapid growth. The size of this business sector will undoubtedly have effects on the economies of many nations, particularly those, which rely heavily on 'invisible earnings' from the quaternary business sector. However, as yet, the scale of these effects cannot be predicted since there has been very little in the way of quantitative analysis of Internet information flows. It is not even possible to answer basic questions about the Internet, such as how many users there are or where they are located. In the absence of hard information and useful models, reliance is often placed on the predictions of Internet 'gurus' whose estimates often vary considerably.

The rapid growth of electronic commerce will undoubtedly raise concerns about its potential effects on world economies. At present, the vast majority of information services available on the Internet are free. However, the growth in e-commerce represents a gradual development of charged services, which will inevitably have an effect on national balance of payments. This will be of particular concern for countries that are net importers of information. This is certainly the case for the United Kingdom, as illustrated recently by JANET's (The UK academic Internet network) decision to start charging for access to US WWW information sources (JISC 1998). JANET will only be charging for Internet traffic flows from the United States to the United Kingdom since it is only the inward information flow that is congested. If an attempt is to be made to determine the financial effects of e-commerce on international trade, then it shall be necessary to develop models of information flows, to improve our understanding of how on-line information is accessed and used, and by whom. Furthermore the relationship between Internet-developed and Internet-developing nations is itself worthy of study since there is a strong relationship between Internet development and economic development, not only in the telecommunication sector, but in overall economic development as well. Figures 12.1 and 12.2 compare the telecommunication resources and GDP of various nations with the speed of the Internet connection between those nations and the United Kingdom. The strength of the relationship is both clear and striking.

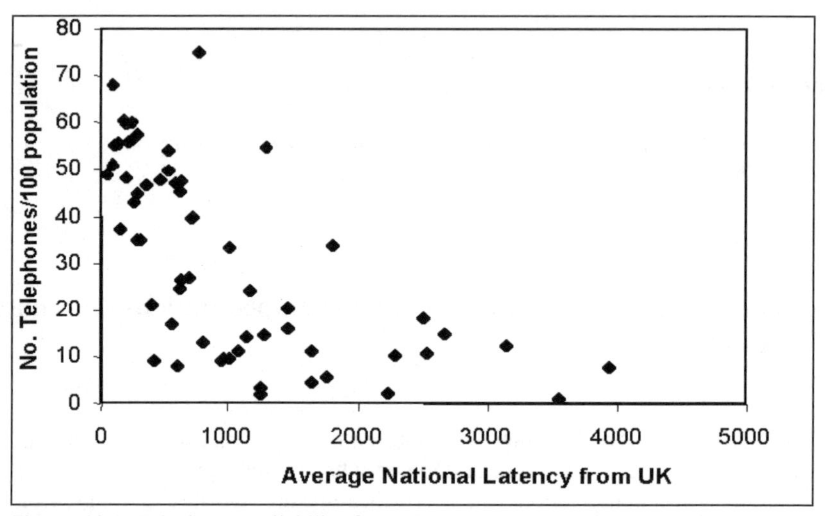

Figure 12.1. Telephone availability for various countries vs. the speed of the Internet connection (measured using latency) between those countries and the United Kingdom.

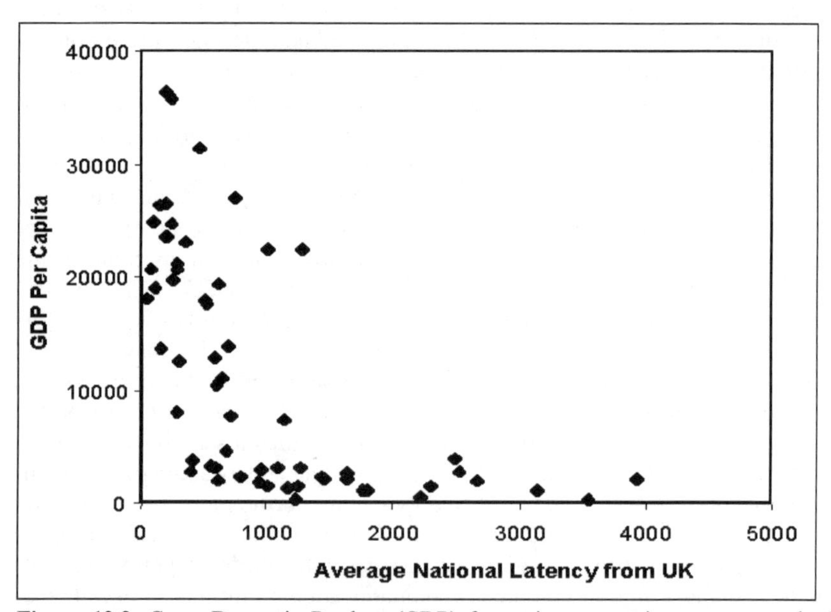

Figure 12.2. Gross Domestic Product (GDP) for various countries versus speed of the Internet connections (measured using latency) between those countries and the United Kingdom.

Thus there are two major areas where the growth of e-commerce will have a significant impact – on the developed economies that rely on the quaternary business sector and on developing nations. The effects on the first area are difficult to predict, however some insights into what may happen to developing nations may be drawn from previous studies in communication geography. Certainly the effects of the Internet on developing economies is starting to attract the attention of development consultants (Daly 1999). One study of telecommunication network growth from urban to rural regions in the United States and elsewhere showed that businesses based in the developed urban centers tend to dominate over rural-based business (Abler 1991). This pattern is likely to repeat itself in the relationship between businesses in the developed and developing regions. A further area of concern for developing regions is the asymmetry of latency space. Previous studies show that there is a connection between the length of time required to access Internet information and the amount of information accessed (Murnion and Healey 1998). Consider the hypothetical situation of a region *A* that has a poor Internet-connected network exhibiting high latency values. For customers within the region all Internet services delivered from sites on the local network will appear far away in latency space. If the network borders a highly developed network from a developed region *B* with very low latency values then services based in *B* will appear to be almost as close as their own internal services. Furthermore with the enhanced levels of Internet expertise and business experience available within developed regions, *B*'s services may well be superior in quality to *A*'s. As a result customers will tend to import services from *B* rather than use their own region's services. However, for customers within the *B* region any service originating from *A* will seem much farther away in latency space and of poorer quality than their own internal services and, as such, they are unlikely to use them. Thus it seems that the inequalities between developed and developing regions will be exacerbated by Internet commerce. To determine the effects of Internet services and the links between demand for these services and network quality as it varies regionally, we require some method of determining the quality of the network between two remote points. In this chapter we discuss previous attempts at determining the link between network quality and demand and examine how the methodologies used may be extended to facilitate the type of analysis required.

12.2 Previous Work

One method that obviously applies in the analysis of Internet information flows and their effects is that of spatial interaction modelling and indeed this method has been used in a closely related study of telecom information flows in Europe (Fischer and Gopal1998). However spatial interaction modelling can only be applied if space has some discernible effect on the flows studied. From an examination of possible metrics and measuring tools, Murnion (2000) suggested that latency, as measured using a utility called ping, might prove suitable in this type of analysis. Using this method Murnion and Healey (1998) undertook an analysis of information flowing from UK academic Web servers to the rest of the world. The aim of the study was to use a simple gravity model to determine whether or not information flows on the Internet were affected by space. The study showed that audiences accessing information from WWW servers are mostly located near to those Web servers in latency space. The main result from the analysis was that the catchment areas for WWW servers are regional in nature rather than global. Figure 12.3 illustrates the decay curve extracted from the analysis. The result is intuitive in that one might expect users would tend to choose a WWW server that can provide information rapidly over one that provides the same information more slowly. Although useful, the scope of the analysis was limited in some respects. In this work an attempt is made to expand upon this initial work.

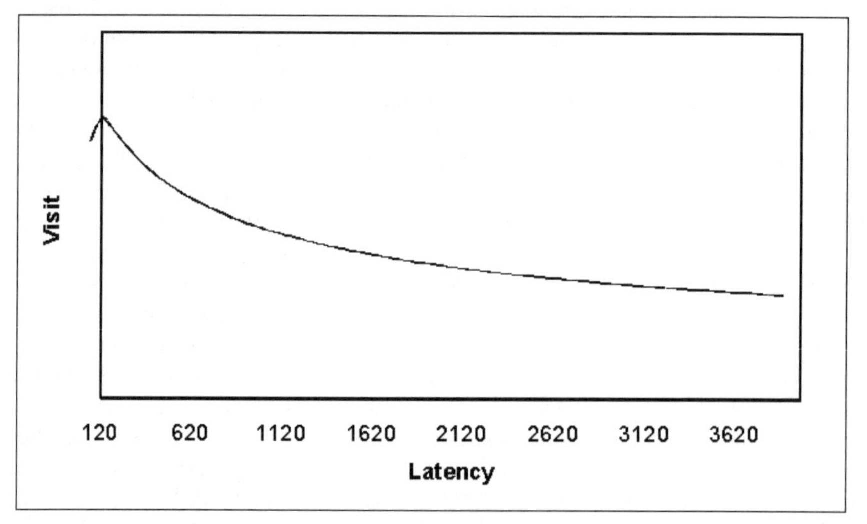

Figure 12.3. Latency distance decay curve derived in the Murnion and Healey (1998) study.

12.3 Extending the Information Flow Model to Include Many Sources and Destinations

The main limitation of the previous study was its scope. Only information flows from one sector, that of university academic information servers were examined. This type of service exists on a high-performance homogeneous network (JANET) and the types of services supplied were also highly homogeneous. One might expect that the wide variety of commercial services available might exhibit different modes of behavior. Furthermore the study only covered the reasonably simple one-to-many relationship of the information flows from the United Kingdom to other countries. Of much wider interest would be the flow of information from many countries to many countries. This type of study would allow the development of a full spatial interaction model rather than the simple gravity model used. By examination of the latency decay curves of different nations, it may be possible to determine the effectiveness of those nations' information services and obtain some measure of how well they may compete in the growing information economy. However, there are many major methodological obstacles to be overcome. The main barrier is the current inability to measure the latency between two remote computers. The current suite of measuring tools widely available for monitoring network congestion and latency are designed for measuring the latency between a local computer and a remote computer. This difficulty may explain why latency studies, which are increasingly common (University of Oregon 1998, MIDS 1998), tend to simply examine latency variation from a single source to many destinations. Some method is needed to allow remote measurement of latency. One possibility is the use of triangulation methods. Figure 12.4 illustrates how triangulation works. If distance measurements are taken by two range-finding stations $\boldsymbol{R}_1$ and $\boldsymbol{R}_2$ of a remote object $\boldsymbol{O}$, and the locations of $\boldsymbol{R}_1$ and $\boldsymbol{R}_2$ are known, then using simple geometry it is possible to calculate the position of $\boldsymbol{O}$. If the number of range-finding stations is increased it is possible to reduce the error in defining $\boldsymbol{O}$'s position.

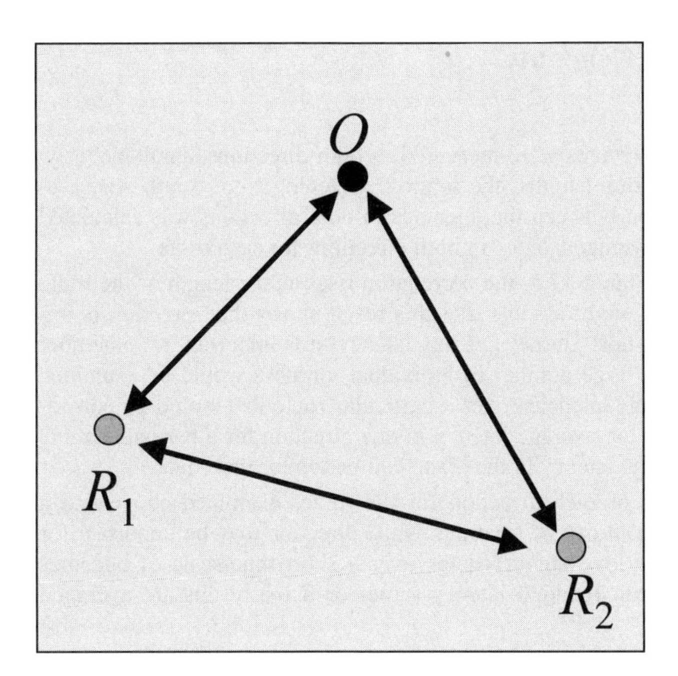

Figure 12.4. Locating a remote object *O* using triangulation.

If this method is transplanted to the Internet, it may be possible to use multiple computers $\boldsymbol{R}_1..\boldsymbol{R}_n$ to measure the latency between these range-finding computers and a remote computer $\boldsymbol{O}$. The measurements gathered could be used to calculate $\boldsymbol{O}$'s position in latency space. If two remote computers are located in this way, it may be possible to determine the distance between them in latency space. There is one flaw in this argument however, since the latency measurements taken do not relate to a straight-line distance, but instead relate to a distance along a network. This problem will certainly introduce errors in any attempt at positioning in latency space. However, since at this point the interest is in very aggregated flows of information between nations then the errors resulting may be acceptable. To test whether or not the method is feasible and to examine the magnitude of the errors involved, an attempt was made to estimate the latency between a computer in South Africa and one in Finland, using latency triangulation.

A further point of importance is whether or not the latency is affected by the direction of information flow. Intuitively one might expect that latency values would be commutative, i.e., that the latency between A and B would be the same as the latency between B and A. However this may not be the case since different routes may be traveled in each direction. A study was made of whether or not latency values commute.

12.4 Does Latency Commute?

Latency values along 10 routes were measured in both directions simultaneously. The routes were of various lengths, the longest stretching from South Africa to Australia. The correlation between the latencies in both directions was calculated and compared with the average latency in both directions for each route.

As can be seen from Figure 12.5, the correlation rises as the length of the route increases. Although one might assume that this result shows that direction is important, particularly for short journeys of low latency, it is important to remember that for most analyses a large number of individual journeys would be examined and, thus, it is the average latencies over a particular route that would be considered. Figure 12.6 shows the average latency in one direction for a particular route plotted against the average latency in the reverse direction for that route.

The average latencies in each direction for the routes examined above had a correlation of 0.99. Thus, it can be seen that while direction may be important for analyses of latencies for low latency routes over a short time-scale, it becomes increasingly less important for high latency routes or if the results are averaged over time.

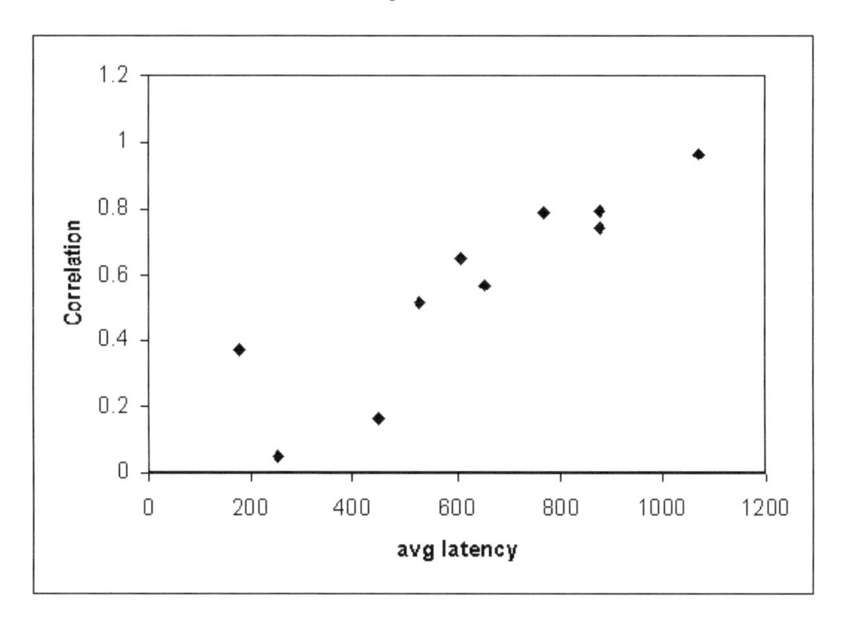

Figure 12.5. Correlation on ten routes between latency in each direction plotted against the average latency in both directions for that route.

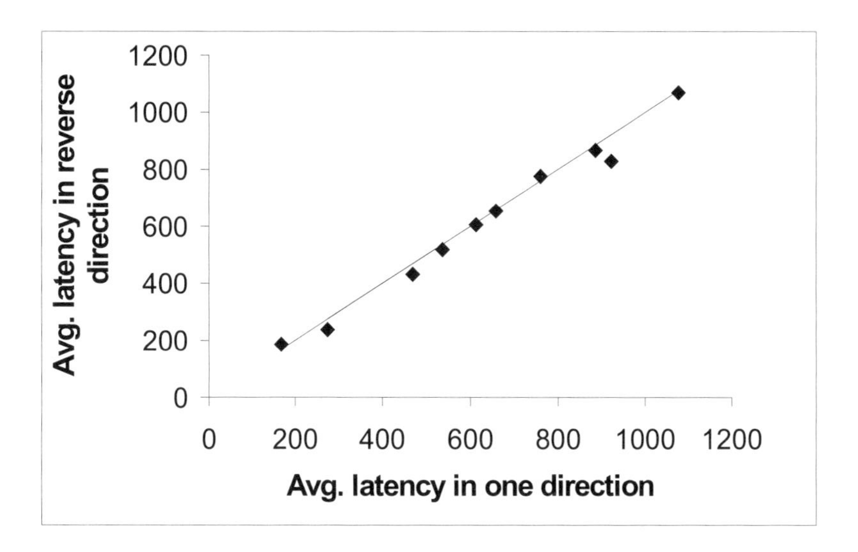

Figure 12.6. The average latency measured in one direction plotted against the average latency measured in the reverse direction for 10 separate information flow routes. The solid line represents a hypothetical perfect correlation.

12.5 Remote Latency Measurement Methodology

To attempt the latency triangulation, six computers were used, located in Santa Barbara, Portsmouth, Cape Town, Perth, Helsinki, and Tokyo respectively. The locations are shown in Figure 12.7.

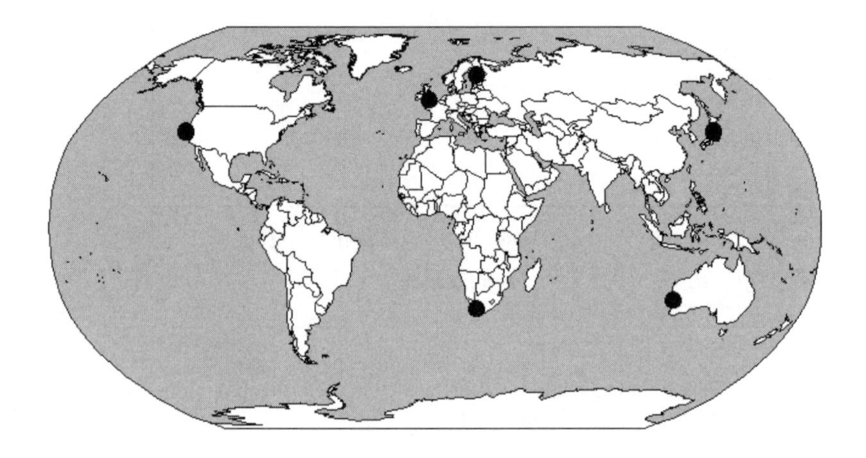

Figure 12.7. The locations of the computers used in the Internet triangulation exercise

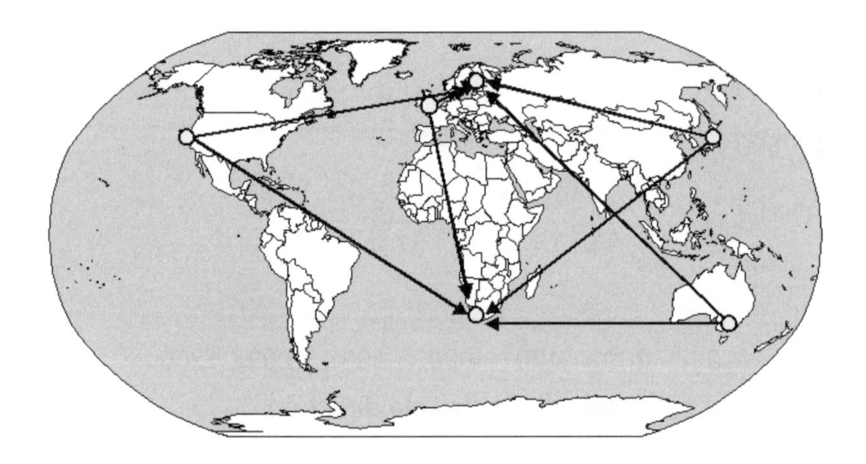

Figure 12.8. Simultaneous latency measurements taken for the attempted triangulation.

The computers at Santa Barbara, Portsmouth, Tokyo, and Perth were used as latency range-finding stations. Simultaneous latency measurements were taken by each of these stations to the computers in Cape Town and Helsinki as shown in Figure 12.8.

As each of these latency measurements was gathered, the true latency between the computer in Cape Town and the computer at Helsinki was also measured. The measurements were taken every 15 minutes over a period of three days. The complete data set was split into two parts. One third of the cases, randomly chosen, were used to train a neural network, which attempted to predict the latency between Cape Town and Helsinki from the measurements taken by the triangulation stations. The actual latency measurements for the training cases were used to determine the error in the neural network training. The details of the creation of the neural network are not included since the choice of modeling method is unlikely to critically affect the result. The trained neural network was then used in an attempt to predict the latency between Cape Town and Helsinki for the remaining, as yet unseen, triangulation measurements. The results of the exercise are given in the next section.

12.6 Triangulation Results

Figure 12.9 shows the measured latency against predicted latency for the unseen measurements.

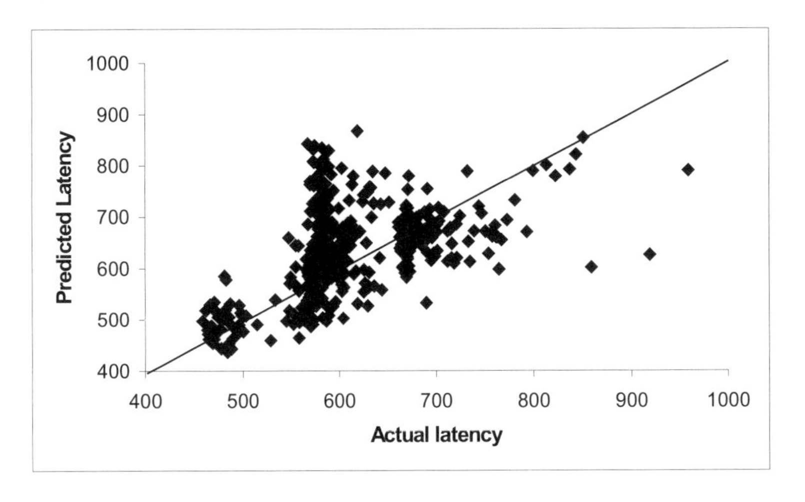

Figure 12.9. Latency predicted by the neural network against the actual measured latency. The solid line represents a perfect result.

Overall, the results are poor though statistically significant, giving an overall R^2 of 0.27 on 600 measurements. However it is noticeable that there is one large cluster of data points between 580ms and 610ms that contributes the majority of the error observed. If the data points between 580ms and 610ms are filtered out then the R^2 value rises to 0.56. Figure 12.10 shows the actual latency between South Africa and Finland measured over time.

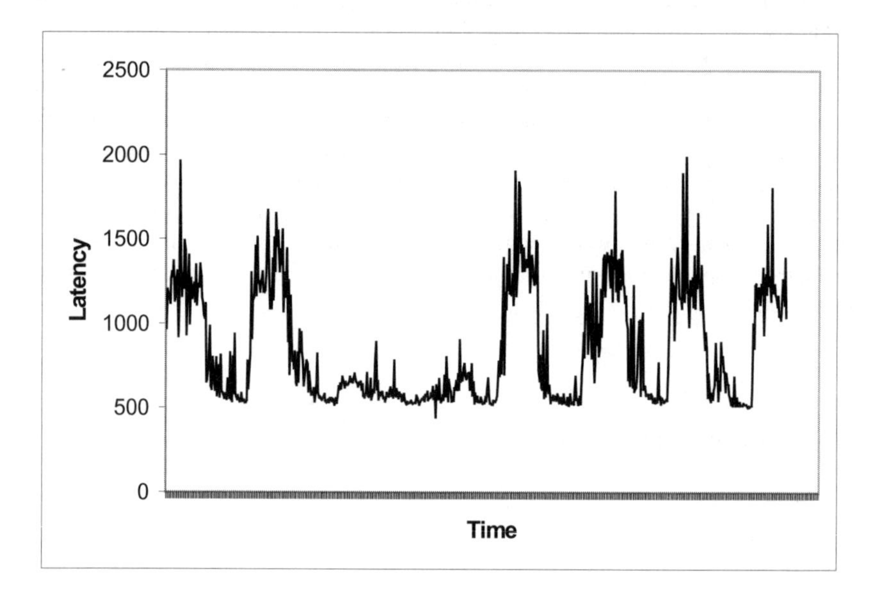

Figure 12.10. Latency between Helsinki and Cape Town (The time distance between each of the large peaks is one day).

Examination of Figure 12.10 shows the curious result that the latency values poorly predicted by the neural network represent approximately the baseline minimum latency values recorded. It may be that the triangulation method is better at detecting changes in network traffic then in absolute values.

12.7 Discussion

Quantitative analysis of Internet information flows is a very recent research area and, as shown here, there are many problems in conducting research in this area. These problems relate mainly to issues of scale and rapid temporal change. Since this is a new area, and our understanding of the behaviors seen are poor, it is critical that reliance is placed on the results of empirical testing rather than on intuitive assumptions as the commutation results show. The results of this study show that the triangulation method holds some promise as a potential method for measuring remote latencies. However the methodology needs further improvement to reduce the errors in the technique such that it could be used as a practical monitoring tool.

The importance of Internet triangulation goes beyond simply providing a method of measuring latency. For the method to work, it requires the location in cyberspace of each of the objects measured. If it is possible to obtain a dataset that contains the cyberspace location of a large number of objects for which geographical locations are known, then it may be possible to build a model that can map between latency space and geographical space. Using such a model, it would be possible to pinpoint in geographical space the location of a computer connected to the Internet, simply by triangulating on its IP or Internet address. If this is possible then it should be feasible to build dynamic maps of Internet usage and activity, locating users and domains in geographical space. The possibility exists then of building an Internet census that accurately reflects information use on a global scale.

References

Abler, R.F. 1991. Hardware, software and brainware: mapping and understanding communication technologies. In Brunn, S.D. and Leinbach, T.R. (eds.) *Collapsing Space and Time: Geographical Aspects of Communications and Information* London: HarperCollins Academic.

Daly, J.A. 1999. Measuring impacts of the Internet in the developing. World. *IMP Magazine*. http://www.cisp.org/imp/may 99/daly/05 99daly.htm..

Fischer, M., and Gopal, S. 1998. Artificial neural networks: A new approach to modelling interregional telecommunication flows. In Haynes, K.E. *et al.* (eds) *Regional Dynamics.* Cheltenham: Edward Elgar, 503-27.

JISC. 1998. Usage-related charges for the JANET network. *JISC Circular 3/98.* http://www.jisc.ac.uk/pub98/c3 98.htm.

MIDS. 1998. *The Internet Weather Report.* http://www.mids.org/weather/

Murnion, S. 2000.Cyber-spatial analysis: appropriate methods and metrics for a new geography. In Openshaw, S. and Abrahart R. (eds.) *GeoComputation.* The Netherlands: Balkema Publishers.

Murnion, S., and Healey, R.G. 1998. Modelling distance decay effects in Web server information flows. *Geographical Analysis* 30(4):285-303.

University of Oregon. 1998. *Internet Weather for News Hosts,* http://twin.uoregon.edu/iwr

13 Application of a CAD-based Accessibility Model

Paul C. Adams

Department of Geography, Texas A&M University, College Station TX 77843-3147, USA.
Email: adams@geog.tamu.edu

13.1 Introduction

Geographical understanding of accessibility usually proceeds macroscopically, from the vantage point of a remote and detached observer. Total minutes of telephone communication between a set of countries, presented as a network map, would be one form such knowledge might take. Frequency of flights between a set of cities would be another. While the macroscopic perspective provides a good sense of the overall degree of interaction between places and how such interaction varies spatially, it can obscure the way communication and transportation are incorporated in individual lives in real places, and hide much that is of interest from a cultural, social, or psychological viewpoint. This is perhaps even more true in the information age than in previous ages.

Time geography's seminal question 'where are the people in regional science' (Hägerstrand 1970), was a primary initiative in micro-scale geography. This interest in people is relevant today, as is time geography's ensuing focus on activity and authority patterns in time and space. This emphasis is best understood as a complement to macro-scale, aggregate approaches rather than as a substitute. We can ask: 'where are the people in accessibility studies' and, following Janelle (1973), use the term 'extensibility' to differentiate the focus of micro-scale interest from the macro-scale concerns that dominate studies of accessibility. The two are so thoroughly intertwined, however, being appearances of a single phenomenon from two perspectives, that insights into extensibility clarify accessibility and vice versa.

I have shown elsewhere (Adams 1995) that Hägerstrand and the other time-geographers oddly restricted their view of human activities to the movement of the body and the range of opportunities that people could encounter through physical proximity. The problem with this approach is that it reduces people to unbranching line-objects in time-space, whereas people are better understood as branching structures.

When we think of people not as bodies but as social agents and sensate beings, we immediately notice that agency and sensation are stretched out through space in various ways. One can both sense and act through the use of a telephone, which is obviously connection to a distant location. Less apparent, such a distant involvement depends on another technology or technique that is far older: language, a mechanism of spatial and temporal connection that goes beyond physical presence. If we ask how it is that the language spoken and understood by one person is the same language spoken and understood by a large number of people living throughout a region or regions, we arrive at the simple observation that language is always *shared.* Through sharing, something that is experienced in diverse places by multitudes of people (a philosophy, a car model, a movie star) is internalized as part of a particular individual's mental structure. Internalization of world as language, and externalization of that world as actions (including language), are the two manifestations of language as a space-transforming technique that foreshadows later space-transforming technologies. This ubiquitous technique is perhaps the archetype of all subsequent space-transforming technologies.

The implications for the individual are significant. To live a human life demands that we relate to others, and by doing so we must communicate, and by communicating we create linkages through time and space. To imagine an individual as capable of being lifted out of these connections is only natural – since that awareness is indeed part of the everyday sense of self in industrial and postindustrial society – but it is an illusion. Geographers are favored particularly in their ability to perceive people in their complex connectedness by grace of training in seeing spatial connections. The main adaptation that must be made is to see these connections as *part of* people rather than *between* people.

In the information age, it may be that only geographers can give an adequate account of the individual. With the development of transportation and communication technologies, as well as the increasingly dispersed production and investment networks, extensibility becomes dauntingly complex. The amoebic *pseudopods* through which people are present to others at a distance become increasingly important to self (Adams 1995). I may phone across town to ask my wife to put my umbrella in the car, or to remind her to put sunscreen on my daughter's ears; I may fax a request to a colleague to write me a letter of reference that will actually reach its destination via e-mail; I may direct a visiting friend to my house from the airport as he talks via mobile phone from the airplane or rental car. In all of these cases, my extended agency matters not only to me but to others, and hence to social organization in general. Although invisible, such extensions are constitutive of social structure no less than the interactions of co-present individuals. A fax, phone call, letter, memo, video image, or broadcast voice are all equally capable of affecting consumption, directing business activities, perpetuating a romantic involvement, or reinforcing political values. They can all provide a connection to the past or future, and ultimately help build communities of various types. A person cut off from a significant number of such connections would need to undertake a major redirection of his or her life. A person completely unable to communicate would be in danger of losing touch with reality, and hence with him or herself.

Information Technology (IT) – the technological vortex into which computing devices, telephones, computers, and a myriad of associated devices are whirling – appears at the macro-scale to be a global information system. This sense of a 'given' scale attached to the technology is derived from the rather obvious fact that the information technologies are capable of supporting international communications. This scale, once accepted as real, seems all the more ominous since the condition of *globality* is not uniformly accessible. Those most able to use and direct the system are the ones with the most money, education, and political power. Thus, we see a global system dominated by powerful groups, accessible to powerful groups, and potentially destructive of individual agency and freedom (e.g., Castells 1996; 1997; 1998; Graham and Marvin 1996; Curry 1995). Without fully analyzing our perspective, we have automatically applied a macro-scale lens to the phenomenon. Its implications, political and otherwise, are in fact scale dependent and look different at other scales.

To understand IT more clearly, we must include individual lives in our analysis, and particularly the routines people pursue in and through IT. Rather than simply asking what people *can* do, we must ask what they *do* do. A portrait of extensible individuals, though not self-sufficient, reveals meanings of IT that are missed in aggregate studies. In particular it reveals what I call the *spatial strategy* of each subject. This is a stance that is adopted relative to various geographical scales, a seeking out of certain scales of involvement through either sensation or agency, and an avoidance of other scales. Some people adopt spatial strategies that reach out beyond the locality, towards global economic, political, and cultural systems. Others seek to strengthen economic, political and cultural ties to locality. Within this general range of spatial strategies, from globalizing to localizing, there are a range of sub-strategies, aligned with various business or professional goals, personal interests, and life histories, attached to various scales of personal extensibility.

13.2 Social Organization and Scale

In the past decade, political geographers have begun to pay increasing attention to the social construction of scale. Scale is not, on this account, an inherent quality of interactions, but rather a social product growing out of the 'scale politics of spatiality' (Jonas 1994). A labor struggle, for example, is not local, regional, or national *by its nature* but, instead, takes on such scale characteristics through processes of social contestation. Parties to a conflict seek either to expand or reduce the scale of the conflict, depending on their social status. The expansive or contractive tactics of groups can be predicted on the basis of their situation in the conflict, such that movements resisting the dominant authorities will often try to broaden the scale of social involvement, while authorities will strive to constrain involvement to their territorial jurisdiction (Adams 1996).

Social conflicts make the constructedness of geographical scale abundantly clear, but more mundane and everyday social processes also manifest the construction of scale. Robert Sack explores this topic in both *Human Territoriality*

(1986) and *Homo Geographicus* (1997). In the more recent work, he argues that the familiar events of daily life, such as cooking a meal for guests or helping a child with school work, provide opportunities for exploration of meaning, nature, and social relations, and these in turn involve processes and phenomena at a wide range of scales. Likewise, Jurgen Habermas argues that communicative action constantly defines and redefines the 'horizon' of the lifeworld, the range of phenomena involved in a particular situation.

> Communicative action relies on a cooperative process of interpretation in which participants relate simultaneously to something in the objective, the social, and the subjective worlds, even when they thematically stress only one of the three components in their utterances (Habermas 1987, 120).

An argument, for example, may arise in which one participant justifies an action as a private choice (bounded) while the other participant criticizes it as a violation of universal moral principles (unbounded), or one conversant may justify an action as a response to international social conditions while the other justifies the action as a response to family problems. So speech constantly reconstructs and problematizes the issue of scale. It is clear that economic relations and political actions do this as well. In fact, all communication does so.

Communication constructs scale in several ways. The two I will explore here are distant sensation and indirect action. First, sensation is actively extended through various audio and visual media. I watch a television news feature on the war in Bosnia or current controversies in the nation's capitol, I read an English novel, or I listen to a friend in another state who tells me over the telephone about his new job. Second, action or agency is extended through institutional frameworks and media. I order a dozen widgets for my latest professional contract, call my wife to tell her to buy tomatoes for dinner, or sign a permission slip letting my child go on a field trip. In these cases, it is virtually assured that agency was guided by sensation, in the form of viewing a part description in a catalog, exchanging greetings with my wife, or reading the schedule of the field trip. Less obviously, sensation implies agency, since sensation is the basis of knowledge, and what we know (or think we know) affects our actions and the actions of others with whom we communicate.

This study applies two different micro-scale lenses to understand personal extensibility, and more generally, the social construction of scale. Both lenses are directed towards the lives of five people who live in the Albany, NY metropolitan area. First, a narrative lens provides a general *feel* of the subjects' different lifestyles, which range from a quiet retirement to a frenetic schedule juggling several high-level professional positions. Also included in this narrative are the intersections of these individuals' life-paths in the experience of one subject who integrates the other four to form a social network. This network is no more important or real than dozens of others that could be shown involving these persons with others. It is simply one particular *coming-together* of people's extended agency and sensation in an ordinary American city in the 1990s.

Second, *models* of the subjects' daily routines are shown. These models are in fact virtual objects constructed in the abstract representational space of a computer-aided design (CAD) program (see Appendix). Since they are stored as a set of objects in a database, the models can be rotated and examined from various angles and can be queried to reveal selected themes. For example, one thematic selection would show only two-way communications such as telephone, and disregard one-way communication links, such as radio and television; alternatively, one and two-way communications could be shown together but in a way that differentiates them by appearance. This extensibility diagram is like a GIS database in that it has no given appearance, but elements can be selected, oriented, and organized by the user. Furthermore, like GIS, the representation renders a large amount of data available for analysis rather than condensing the data into statistical measures, such as mean and standard deviation, which result in a loss of information.

The goal of the study is to show a way that emerging technologies can be linked to individual spatial strategies and interpersonal power relations. The sample is too small to provide more than a hint of what one might find with a more satisfactory sample. The methodology is entirely new, so it is a pilot study. Even so, it hints at how social power and extensibility are related. The relationship is not a simple matter of quantity, with more power implying more extensibility; although the most powerful subjects appear to maintain economic, political, or social power through extensibility, we cannot reduce that power to a simple metric, such as more time communicating with distant places or more frequent connections with distant places. Power is expressed in very different rhythms of extensibility. Furthermore, the less powerful subjects maintain a high level of involvement with scales of social integration beyond the locality – state, region, nation, and/or world. Their ability to act at a distance is limited and directed by others, but they use media to extend their ability to sense the world in ways that fit in with coherent overall spatial strategies.

The *powerless* are not simply receivers of distant information. They circulate perceptions in local social contexts that are drawn from distant origins on the Web, in newspapers, on the radio, and in books. Although minimally capable of affecting social processes at these scales, they are informed about non-local events and become primary sources of shared local information regarding non-local events. If this information is distorted by the news sources, as indeed it must be, people nonetheless are not passive dupes of the media.[1] They use media to establish a certain meaningful relationship to the world, and do so in active ways: combining sources, seeking out information related to interests, mixing news and entertainment, comparing news sources, and sometimes regarding the sources they depend on with a jaded and skeptical eye (Fiske 1987).

---

[1] I am assuming here that all *news* is a social construction and therefore perpetuates certain biases in the way it constructs world events.

13.3 Five Daily Time-space Routines

Five subjects were chosen for this study in a non-random way.[2] Although too small a sample to indicate general patterns, a range of intriguingly different time-space routines was indicated. This range sufficed to indicate some questions regarding the simple assumption of *information haves* and *information have-nots* that has driven other studies of accessibility and IT. In addition, practical aspects of the representation of extensibility in a CAD-based extensibility diagram could be explored. Such an approach cannot tell the whole story (even if expanded considerably), but it can help overcome the limitations of macro-scale studies.

Diann works in her home, primarily doing light assembly work for Thomas' company and caring for her 3-year-old daughter. In her spare time she designs and creates textile art for sale at a small gallery and invests in commodities. Mr. Worley, her neighbor, is a retired widower who spends much of his time at home. He spends his days reading the newspaper, exercising, taking long walks, and running errands in his 15-year-old station wagon. Martin Kroopnick is the general manager of a public radio station with branches throughout New England and upstate New York, who is also a college professor and the host of a weekly television show. Thomas owns and manages a small business that designs and produces promotional materials for police departments, primarily a child security kit that helps parents gather information that will assist police in identifying a child who is lost. Lisa is Thomas' secretary; she has a degree in religious studies and is looking for a job that better uses her skills and contributes more to furthering her social values. In her spare time she reads esoteric religious texts and does volunteer work for an environmental organization.

The five people are part of a social network with Diann at the hub. Thomas is Diann's employer; Lisa is Diann's most common work contact; Martin Kroopnick is the general manager of Diann's favorite radio station; Mr. Worley and Diann regularly greet each other when they pass on the sidewalk. The five have obviously different levels of social power. Martin Kroopnick and Thomas are both, in different ways, powerful social agents. They both direct the work activities of dozens of other people and identify their careers with the achievement of social power, influence, and success (though the three are mixed in different proportions). They both seem pleased with their career achievements, if also dissatisfied by some aspects of their lives. Diann, Mr. Worley, and Lisa are all near the bottom of conventional scales of social power. Mr. Worley is marginalized by his age and his previous employment status, as well as by his comparative lack of education. Diann and Lisa are professionally marginalized, serving the needs of a business

---

[2] The subjects were all known to the researcher prior to the study and were chosen on the basis of their willingness to participate and diversity of lifestyles. With a funded study a larger population could be studied and more reliable results could be obtained. All subject names have been changed, but are intended to indicate social status in a way that often occurs in society. Whether a subject is referred to by first name, last name, or both names mirrors the way the actual subjects were known to the *hub* subject, Diann.

process that they are not able to guide or direct. To the degree that their work entails communication, that communication is not so much a facet of their own extensibility as their bosses extensibility. In the terminology of Castells (1996, 244), these two are the *operated* rather than the operators, integrators, designers, researchers, or commanders of the information age. These are the *executants* who work for the executives. The three 'powerless' subjects differ in that while Diann and Mr. Worley do not use information technology at all in a professional capacity, Lisa's job is heavily dependent on IT. Here we have what Castells (1996, 244) calls a 'networked' worker, one who is 'on-line but without deciding when, how, why, or with whom.'

The five can be contrasted by several aspects of extensibility: (a) the frequency, duration and overall time devoted to travel, (b) the frequency, duration, and overall time devoted to incoming communication (e.g., reading), (c) the frequency, duration, and overall time devoted to outgoing communication (e.g., writing). These three general states combine to form a time-space rhythm that is unique to each individual. In addition, the geographical range of both incoming and outgoing communication varies, as does the social impact of the communications; working a crossword puzzle and editing a quarterly report are both ways of responding to incoming communications, but the former clearly has less social impact than the latter. While it is obvious that the powerful subjects enjoy their power, it is also evident that the less powerful subjects have qualitatively different spatial strategies, and this makes it hard to determine who is *better off* in an absolute sense.

Thursday

The following is a construction of a Thursday in Fall 1997. Although constructed from five different Thursdays, it could plausibly be a single day in the life of five persons. Martin Kroopnick wakes first, followed by Diann, Thomas, Mr. Worley, and Lisa, in that order. Waking is shown on Figures 13.1-13.7 by the jog in the *spine* of the time-space diagram where it expands from a narrow channel to the level labeled *proximate*. When they wake they are located throughout the Albany metropolitan area, except for Martin who lives in western Massachusetts. Three of them – Martin, Thomas, and Lisa – mainly use their early morning hours preparing to leave the house and going to work. Martin's routine also includes an exercise session, Thomas helps his wife with childcare, and Lisa writes in a dream journal. Diann, in contrast, begins working almost immediately after waking up, since she works in the home. Martin's extensibility also begins right away, but in his case the reason is that he obsessively listens to his own radio station. We will now consider each subject's day in turn.

Martin's day includes a diverse array of communication situations (Figure 13.1). He is the manager of a chain of public radio stations, who also has his own television program and provides a weekly commentary for a talk show on another radio station. He contributes a daily commentary to a network television affiliate, as well as editing a newsletter about the state legislature and teaching three college courses. Martin's extraordinary schedule suggests an exceptionally high level of extensibility. It is surprising, therefore, that much of his time is spent simply receiving one-way communications – reading.

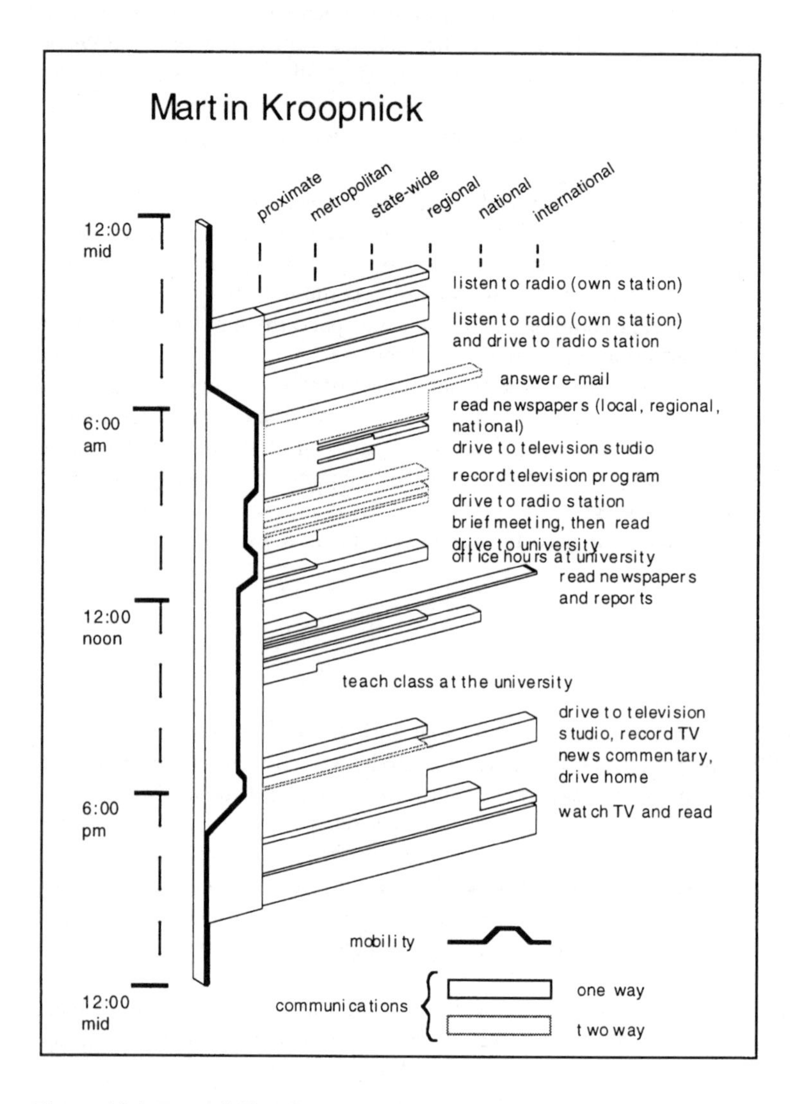

Figure 13.1. Extensibility diagram indicating the subject's mobility and communications for a typical Thursday in autumn 1997.

He wakes up to the sound of his radio station and continues to listen while eating breakfast and driving west across the New York state border to work. When he arrives at his office he reads the e-mail, which includes communications from radio station employees to him, and communications to station employees from persons outside the station. He forwards the in-coming e-mail and answers the internal e-mail. Next, he begins to scan a total of seven daily or weekly newspapers that have arrived at his desk that morning. A break occurs around 9:00 a.m. when he and his producer drive to a television studio in an adjacent municipality to record his Sunday television program. He returns an hour and a half later, handles details of the radio station administration, and continues skimming the papers. During the ride back to the station he listens to his own radio station, discussing a program and forming a link to a different space than the one his car moves through. The metropolitan space shifts to the background of his consciousness and he attends to the regional space covered by his radio stations.

The afternoon brings a shift of pace with office hours at the university and a lecture in his journalism class. At 5:30 p.m., when class is over, he drives to the television studio and prepares for his nightly political commentary. He waits during earlier parts of the 6:00 p.m. news broadcast, chatting with the photographer, the news anchor, and others on the set. He is on the air for two minutes, a span oddly out of proportion to the familiarity it brings him as a political commentator. He is in his car by 6:30 p.m., heading home to the small Massachusetts town where he sleeps.

Martin is in a small minority of the population whose outgoing communications shape public perception of issues, a role that cannot be measured strictly in quantitative terms, such as in minutes of communication. In essence, he serves a *gatekeeper* function (McQuail and Windahl 1981, 100-101; White 1950), influencing the kinds of issues that will be in the public eye. To be an *opinion leader* it is not necessary to hold public attention for long periods of time. What matters are the channels in which one communicates. Martin reflects this awareness in his obsession with his *audience* and potential ways to expand it.

Also noteworthy is his mobility. He drives for an hour to his office in Albany from his house in a small Massachusetts town, rides with his producer for 15 minutes to and from his appointment at the television studio where he records his weekly television program, drives for 10 minutes between his studio and the university campus, drives for 15 minutes to return to the television studio to record his daily news commentary; then drives for an hour back to his house in Massachusetts in the evening. It is not unusual for him to spend as much as three hours on the road during the course of a day.

With such a varied set of responsibilities and a home far from his work places, Martin's time away from home is elongated beyond the typical range. He leaves home around 5:30 a.m. and returns at 7:30 p.m. Martin explains that when he encounters members of his audience in the town where he lives they are surprised to find that he lives there. The time is short in which to enjoy the locational benefits that his income and power provides. In effect, his activity at the regional scale necessitates that he detach himself from local and metropolitan attachments.

Diann's working hours are even more unusual than Martin's (Figure 13.2). She begins work at 5:40 a.m. in her living room, assembling child security kits for Thomas' company while surfing the Web and listening to the radio. Oscillating between homework and childcare, her working hours extend clear to bedtime, but work is interspersed throughout the day with non-work activities, such as walks to and from her daughter's pre-school, errands, visits with her friend, and 'time out' to read a novel. The day of the study she takes a bath at 10:40 a.m. and attends a sing-along with her daughter at 4:00 p.m. More non-work time appears at a finer scale, in the numerous fluctuations between work and leisure that permeate her day on a minute-to-minute basis. Even during working hours, her attention constantly shifts between her work, her four-year-old daughter, and the Web or other media that she uses to occupy her eyes and mind while working on kits with her hands. The evening of the study day is a bit unusual in that her family runs several errands in the car: they attend a photography session so they can send personalized Christmas cards, they visit Thomas' office where Diann drops off the day's production, and they shop at the grocery store. Diann lets her husband drive because she hates driving in the city; she prefers to read a book and ignore the city entirely until they reach their destination.

Diann's involvement in paid work at the same time she is using telecommunication and broadcast media for personal pleasure and exploration indicates a rather unusual spatial strategy. She is not telecommuting, since her use of computer networks is for purposes other than work; but, like telecommuters, she is trading extensibility for mobility. She spends less time than any of the other study subjects communicating at a face-to-face range, and more time engaged in national and international scale communications. Her low mobility indicates a personal spatial strategy of localization, but her use of media such the World Wide Web and National Public Radio indicates a strategy of globalization. The home, traditionally private, domestic, and local, has become for her somewhat public, professional, and non-local.

Her social *influence* is primarily local, but this view of personal scale is problematized by her extensibility pattern, including the special role she serves as an information gatherer and community member. She compares the online news with stories on National Public Radio to get a better sense of non-local events. This active media use is important to her sense of autonomy and individuality, and adds validity to her information-gathering role in her immediate social network, which includes mainly family and friends. Even her boss, Thomas, has consulted her regarding the Y2K bug, the world economy, and other issues.

Her pleasure with this extensible lifestyle is indicated as she cites Web surfing as a reason she prefers to work at home. The choice to stay at home reduces the time she spends driving, but that does not reflect a complete withdrawal from her urban environment. In fact, an unusual proportion of her total travel time is spent walking, which allows her to attend to her surroundings more than someone who drives, particularly if they listen to the radio. In addition, she patronizes small shops in her neighborhood, helping maintain the nodes of the community's activity. A quantitative comparison of time spent physically moving through the city would obscure this qualitative difference. Summing up these characteristics we find a spatial strategy combining attachment to the local – home and neighborhood – and the global, with avoidance of the metropolitan and state scales of activity.

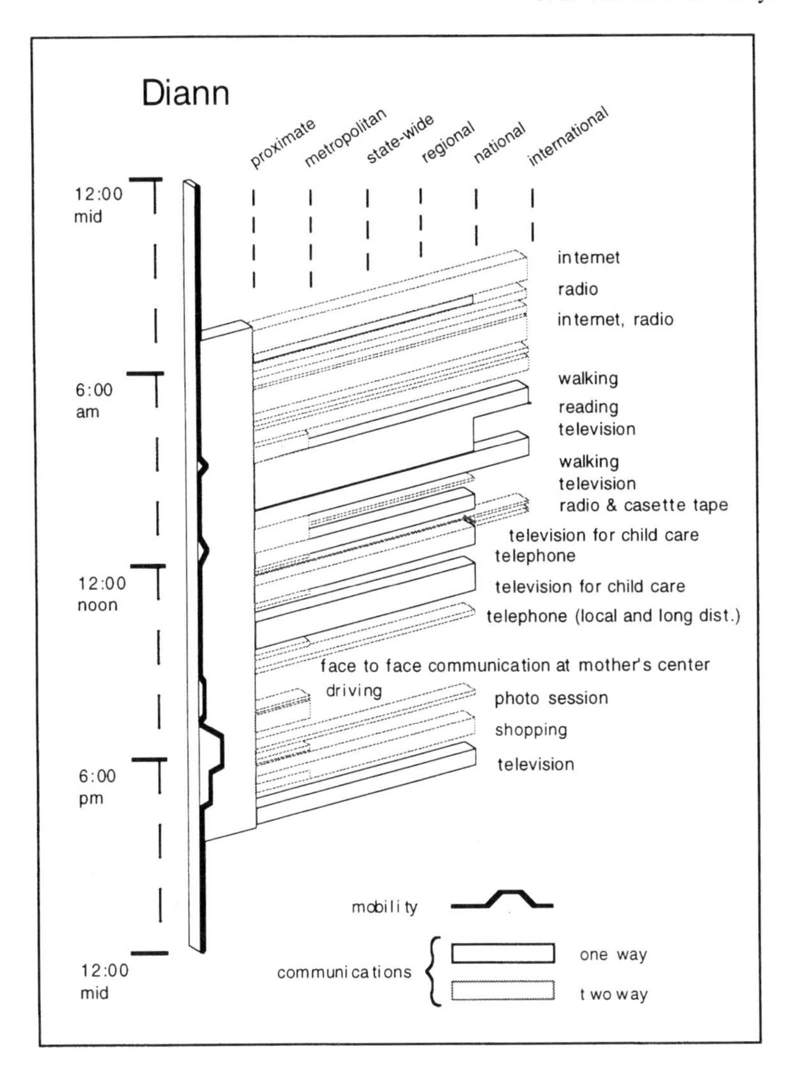

Figure 13.2. Extensibility diagram indicating the subject's mobility and communications for a typical Thursday in autumn 1997.

Thomas has a more 'normal' work life than Diann, with a clear phase of work-related extensibility starting shortly after 9:00 a.m. and ending around 5:00 p.m. (Figure 13.3). This 'window' of activity in time-space defines a great deal of his interactive (two-way) professional communication. The 9:00 a.m.-to-5:00 p.m. work period is evident in his routine in the form of frequent, short communications (mainly by phone) at the national scale, alternating with even shorter communications (by voice) at the proximate scale (in his office). He roams his office all day saying: 'Did you mail the order yet?' 'Is this your coffee?' or simply

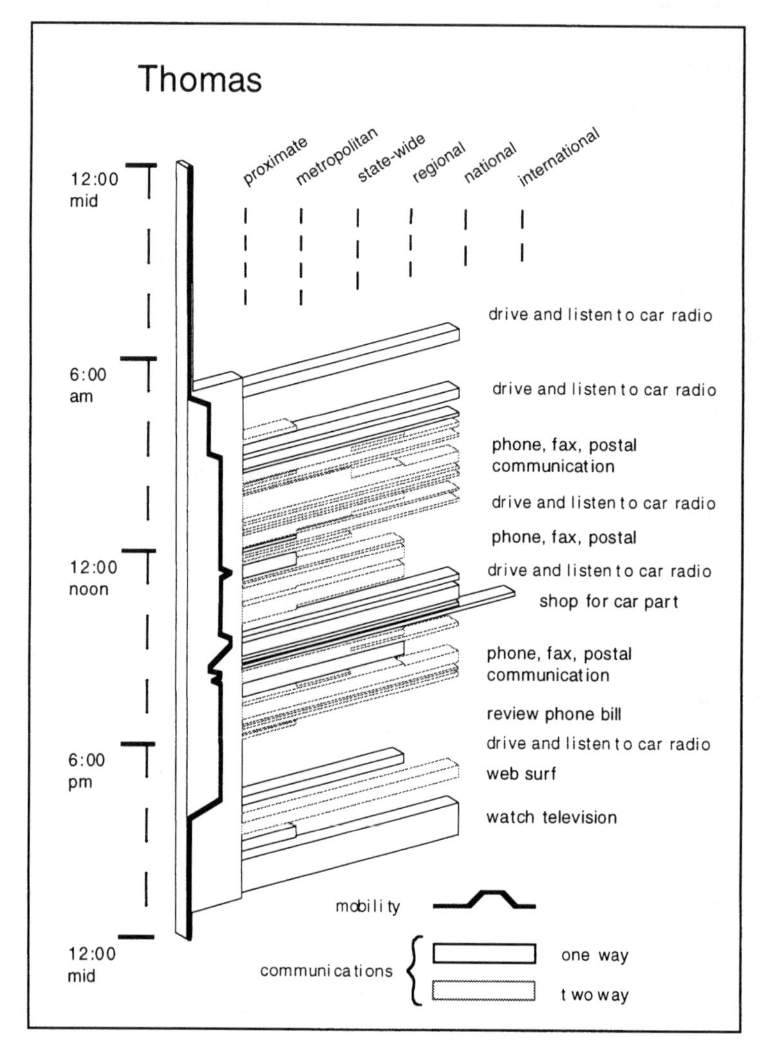

Figure 13.3. Extensibility diagram indicating the subject's mobility and communications for a typical Thursday in autumn 1997.

'What's up?' Work spills over into the rest of his day in the form of letters to read and accounts to check, either of which can keep him busy until 9:30 p.m.

Two-way communications at the national scale, phone conversations with clients, dominate his daily schedule. These are primarily routine negotiations with clients involving issues such as the layout of customized kit covers, production schedules and fees. While none of his communications are politically powerful in the way that Martin's are, their existence creates job opportunities for about a dozen employees and are therefore economically powerful. While Lisa and Thomas may spend equal amounts of time communicating at the national level, Thomas' communications are more closely tied to the exercise of personal power. Lisa's communications may be better represented as part of Thomas' extensibility.

Thomas is consumed by his work and knows almost nothing about political affairs, environmental problems, or the world economic situation. He knows Martin Kroopnick only by name. When he hears of a major world event he occasionally asks Diann or Lisa for their opinion, but otherwise does not search out information. Thus, he is active at a large geographical scale, but only in a very limited way, and with more agency than sensation.

His routine involves a high level of vehicular mobility. In the afternoon he drives his new Volvo proudly around town doing office errands. While he could send one of his employees, he values this chance to get away from the office and enjoys showing off his car and driving aggressively. His car radio is turned up high and tuned in to a local station with 'oldies' that provides a connection back in time to his youth. The station draws from a nationally shared list of 1970s hits thereby also connecting him outward in space to a national culture of 30-somethings. Although Thomas might be conscious of the city of Albany as he drives, his use of the radio limits the depth to which he is aware of his surroundings; he is as much *in* the music's ambiguous time-space as *in* the city (a string of unfortunate car accidents attests to that fact). The telephone at the national scale constitutes Thomas' professional persona while the car radio at an ambiguous spatio-temporal scale allows him to feel *at home*.

Thomas' secretary, Lisa, must adhere to a strict schedule, particularly in the mornings; she is usually the first person to arrive at the office and must be there to open the office at 9:30 a.m. (Figure 13.4). Still, she wakes up later than all but one study subject. Around 9:25 a.m. each morning she unlocks the door, turns on the light, and begins printing a computer report of the previous day's sales. While her duties are varied, the ringing of the phones is the most persistent claim on her attention, followed by the preparation of packages for UPS shipment. Strictly speaking, these activities are part of Lisa's extensibility; she speaks with people in hundreds of police offices around the country and sends them samples of child-security products and finished orders. But her actions are even more narrowly defined than are those of Thomas – aside from her personable manner on the phone, she has little control over what, when, or how she communicates. All of her actions are determined during work hours by office routines that have developed in an ad hoc way, or by the logic of trying to expedite orders, or (less often) by direct orders from Thomas. Her extensibility during work hours, therefore, is not entirely her own.

Nevertheless, unlike the automaton stereotype of the information age drone, or *protosurp* (Dear and Flusty 1998), Lisa appropriates this situation and retains her humanity in and through the technologies she must use for her job. She genuinely enjoys speaking on the telephone to the office's many clients, service providers, and employees. She brightens up their days with a bit of conversation, and remembers many people's names and personal trivia in this *telephone space*.

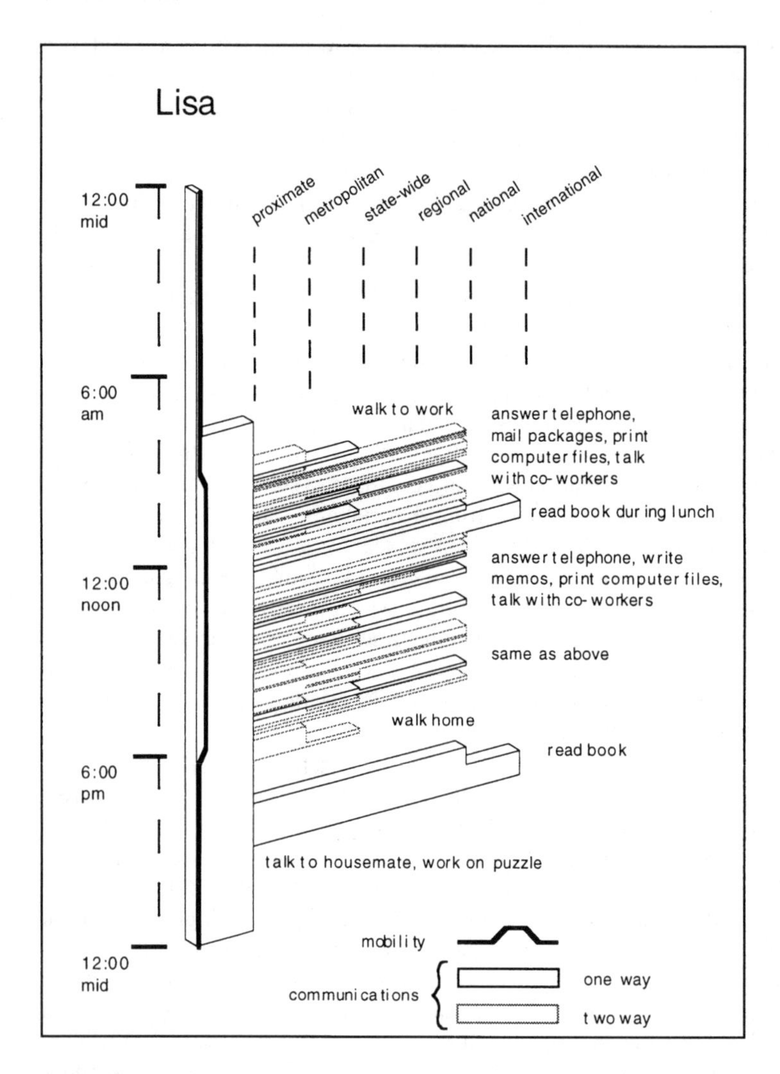

Figure 13.4. Extensibility diagram indicating the subject's mobility and communications for a typical Thursday in autumn 1997.

Outside of business hours and during her fugitive lunch periods (when she physically leaves the office and goes to the grocery store deli to avoid being interrupted), Lisa enjoys a very different kind of extensibility. She reads broadly from the texts of many different religions, expanding her knowledge in the subject in which she holds a Bachelor's degree. At present she is reading the *Hermetica* and a book of poetry about trees. She loves books, reading, and talking, and accordingly spends several hours a day engaged in conversation with her boyfriend about topics of interest from her reading. These emotional and communicational ties to realities outside the work routine could easily be overlooked as unimportant to the material fact of her status as secretary, but they provide intellectual sustenance to remain a friendly voice on the phone as she looks for a job that better uses her college degree.[3]

Her mobility is somewhat limited in speed and distance as she walks to and from work as well as to and from her lunchtime retreat at the grocery store deli. Qualitative issues are involved. She explains that the daily walks give her time to think about her poetry and her reading. She sometimes walks past her apartment or takes a longer route to extend this period of time. For her (like Thomas but in a slower framework) mobility is obviously more than simply a utilitarian concern and getting there is not the main objective. However, like Diann, her involvement with her surroundings seems to be higher that that of Thomas as she moves through town. Her spatial strategy is interesting.

She maintains strong ties to religious ideas originating in distant times and places, and to her home and the section of town where she walks. While her work-related extensibility is frequent and economically important, both to her and her business, it is questionable whether it should even be included in her time-space diagram or whether it perhaps should be shown in Thomas' diagram.

Mr. Worley has the most leisurely schedule (Figure 13.5). He rises any time between 6:00 a.m. and 7:30 a.m. most days, keeping a rather loose schedule. His days are filled with a small repertoire of activities: driving to the local *Friendly's* restaurant for breakfast, walking in the neighborhood, exercising on the mat in his TV room, reading the newspaper, watching TV alone or with his adult son, and buying groceries.

The pace is leisurely, as any one of these activities is drawn out to last at least half hour. The only activities of short duration are conversations with neighbors, such as Diann, whom he meets on the sidewalk. Even these can take half an hour if the other person is not in a hurry. His conversations serve a community-sustaining function, at least among the older residents of the neighborhood who have known him for years. In this sense, he manifests a localizing spatial strategy.

Nevertheless, Mr. Worley is not out of touch with the world. He spends several hours a day reading the newspaper and informs himself of key events in his local community and a smattering of national affairs that catch his attention. Most of his communication is one-way communication: reading the newspaper or watching television. His two-way communication opportunities are primarily in face-to-face situations, with neighbors, *Friendly's* employees, and his house cleaner. Unlike Diann, he does not act as an information source. Some of what he learns he keeps to himself. More often, he takes delight in bringing his discussions around to his cynical and ironic perception that the world is in a mess.

Less deliberately than Diann, he has traded mobility for extensibility. His housebound lifestyle increases his sensory involvement in affairs at the metropolitan, regional, and national scales. Attitudinally, however, he is not particularly

---

[3] By the time of this writing, she has found another job organizing the planting of 'champion' trees, which conforms more closely to her values and goals. Presumably her reading and writing were preludes to this career path.

receptive to such distant information, often summing up his view of the world with a few favorite expletives. This, too, is a spatial strategy. It is a brand of localism that draws parasitically on non-local media as a source of ironic critique of distant events.

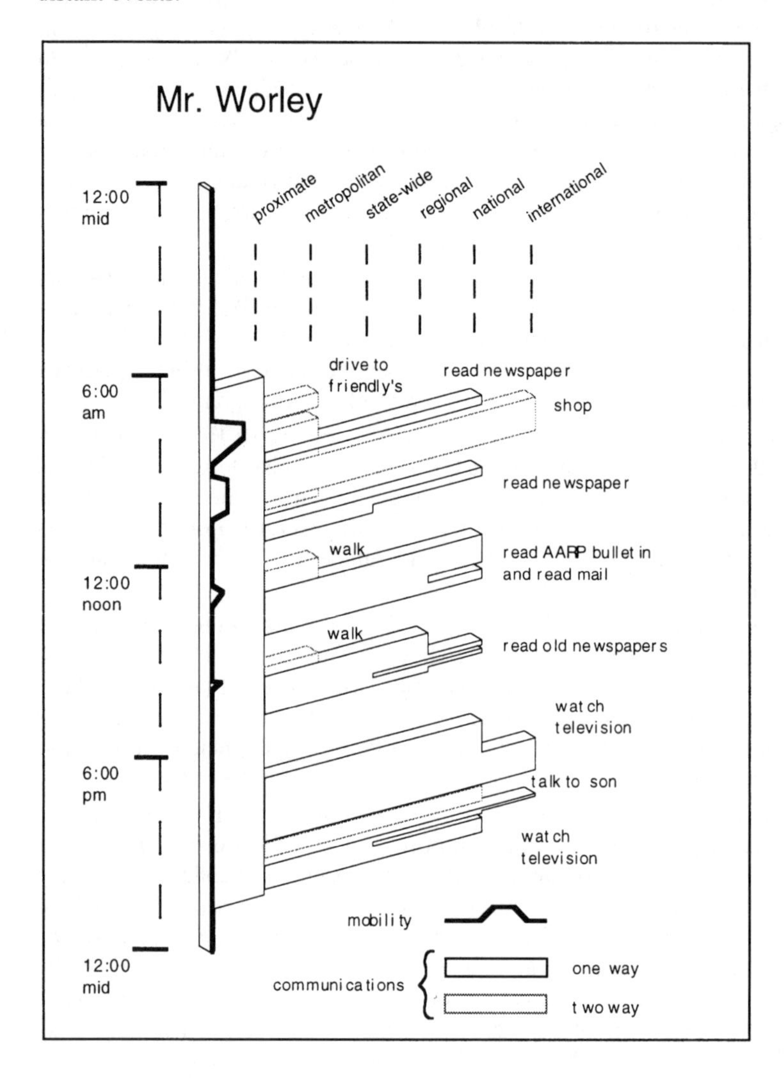

Figure 13.5. Extensibility diagram indicating the subject's mobility and communications for a typical Thursday in autumn 1997.

13.4 Discussion

These five people have greatly divergent patterns of extensibility. Their differences are only superficially tied to *opportunity*. Strictly speaking, Mr. Worley and Diann have similar opportunities to communicate but Diann spends much of her time listening to the radio and surfing the Web while Mr. Worley spends his time reading newspapers and chatting with neighbors. His lack of a computer is due not to a lack of funds or proper location, but rather a lack of interest. Nor can the difference be reduced to a difference in *cultural capital* without recognizing that Diann is not profiting from her use of the Internet but simply using it to pass the time and to situate herself in the world. Mr. Worley simply satisfies these objectives in a different way. To construe her greater degree of connectivity as an advantage is to misconstrue her purpose and rewards, and to impose an a priori judgment on these two forms of communication. In contrast, sometimes-similar levels of extensibility are not what they appear to be. Lisa and Thomas have much the same communication opportunities during work hours, but one determines what will be communicated while (with certain caveats) the other does not. Surely Thomas has an advantage in extensibility, although that advantage is difficult to specify.

What has been called accessibility is most often based on measures of opportunity derived from technological and economic patterns, but this is clearly an inadequate notion. Communications more closely reflect social power, but here again the connection is subtle. Most important are personal interests, goals, habits, and social connections, phase in the life cycle, gender, and other factors of the individual.

Communication is, in short, part of a time-space routine that is personal and difficult to generalize. The idea of 'opportunity' as typically construed in studies of accessibility must, therefore, be supplemented by actuality. Mr. Worley gets all the newspapers he needs, and what he misses on the Internet generally lies beyond his range of interests. Thomas' business depends on the telephone, but aside from purchasing several independent lines and a fax machine he has not yet felt a need to adopt more sophisticated telecommunications at work. He does have Internet access from his home, but uses it infrequently. Lisa's passion is communication, but not telecommunication; she prefers a good book that she can carry out of the office during her lunch hour, and discuss in the evening with her boyfriend as they sit in the kitchen or bedroom. Martin uses the Internet daily, but mainly for receiving e-mail; he keeps abreast of current events by reading national, regional, state, and local newspapers. Diann alone has built the Internet into her life as a serious passion, not because of her social status, but in part through the acceptance of a job with little social status.

Likewise, the most mobile subjects, Martin and Thomas, do not appear to be *in* the space they are so often driving through. They listen to the car radio, one to survey his own indirect labor, the other to bind past and present, dream and actuality. Physical mobility is, oddly, a constraint they both seek to overcome with extensibility. I would argue that what is overcome is the involvement with physical surroundings that is possible when one walks through an environment, like the apparently less mobile subjects. Again, qualitative issues complicate the picture: access to space is not merely measurable in terms of *distance*; one must also consider *depth.*

A map of communication opportunities, for example a map of Internet hosts and data transmission backbone, misses an essential point. Opportunities are determined not by an abstract calculus but by individually determined needs. Constraints are constraints only when they interfere with someone's self-defined needs. Only theoretical individuals can occupy the spaces of opportunity shown by traditional maps of accessibility. The five individuals in this study display five radically different lifestyles and five ranges of need, these correspond to five spatial strategies of real persons. Older media such as books and telephones easily satisfy some of their spatial strategies, while others require newer media, such as television and the Internet. The abstract metric of *accessibility* cannot capture the relativity of individual needs and goals.

It is notable that the connections between Diann and the other four subjects are brief. If one aggregates all such interactions, calls them interactions *in place*, and compares them with interactions at a distance, the two are roughly equal in time use. People who spend more of their time at home spend a higher proportion of total time involved in non-local communication situations. This points to a trend of localization within globalization, or perhaps globalization within localization. The opposition between global and local cannot describe the individuals in the study because they are folding the world into the space of the locality.

For some of the subjects, home has dwindled in functional importance because of a mobile lifestyle and heavy workload. In this category are Martin Kroopnick and Thomas. Perhaps, also, locality has dwindled for these people. While Thomas lives 15 minutes from his place of work, Martin lives an hour away. His lifestyle choice is not uncommon among the wealthy. The irony of this choice is that its costs, the personal loss of 16 hours per week and the public cost of automobile emissions, are *spent* on a choice that he does not appear to have much time or energy to enjoy. Thomas also suffers high costs in the form of the period of time he is away from home each day. Diann reveals an alternative strategy that does not provide the same benefits in social power, but satisfies goals relating to quality of life and the quality of others' (her child's) life. No doubt women are more inclined to emphasize these values. Even so, it is not necessary that they sacrifice career goals while adopting this spatial strategy (Helgeson 1998).

Surveying the different rotations of the extensibility diagram (Figures 13.6 and 13.7), we see a complex architecture of ties to the regional, national, and global scales. This architecture clings tenuously together at the local level, with momentary exchanges such as the 'hello' one says to a neighbor or the brief exchange of work-related information. The most intimate relationships are not shown, such as ties between husband and wife, best friends, and associates who work in the same office. Nevertheless the diagram suggests a truth about modern urban settings.

Almost any five citizens drawn randomly from the Albany metropolitan area will be held together tenuously if at all by communication links.

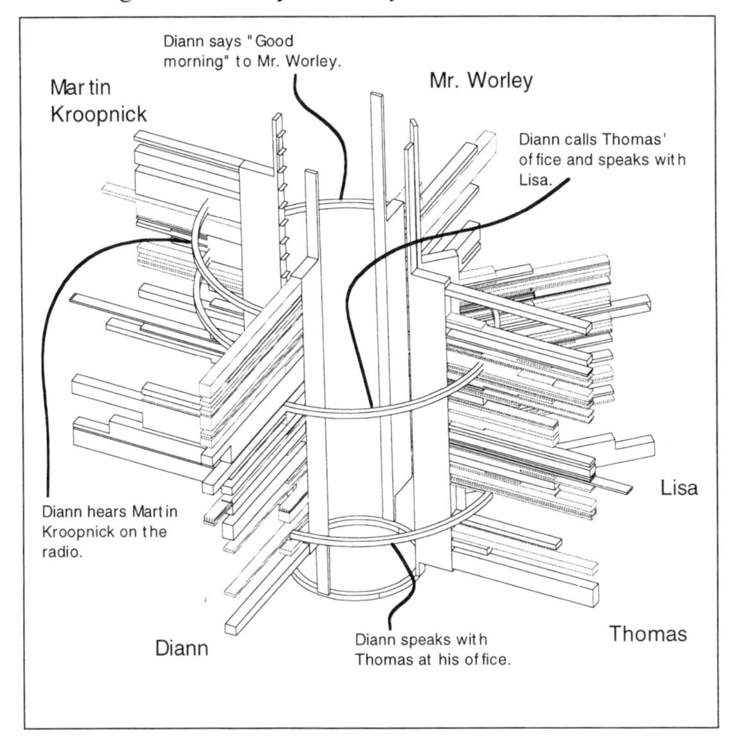

Figure 13.6. Extensibility diagram combining Figures 13.1 – 13.5, and also showing communications between Diann and other participants

In the information age, personal ties to non-local scales seriously rival ties to locality. What makes this situation perplexing is not its existence, because that could be attributed simply to the scale of the extant communication media. Rather, its mystery lies in the way people mix radically different spatial strategies in the pursuit of personal goals and seem to find mutually agreeable results in the process. There is a process of specialization that is entirely different than the specialization of skills employed in one's job: people also specialize in the spatial strategies they employ in the pursuit of personal and collective goals. In a way analogous to the merging of specialized skills to produce a diversified economy, we can observe a merging of spatial strategies to produce a diversified space-time fabric. It may be that the regional focus of one facilitates the local focus or global focus of another.

Ties at the macro-scale between accessibility and economic power must be acknowledged. The wealthy and powerful do have more communication tools at their disposal. However, we must not, as a consequence of this observation, overlook the free engagement and participation in the construction of communication flows by all strata of society.

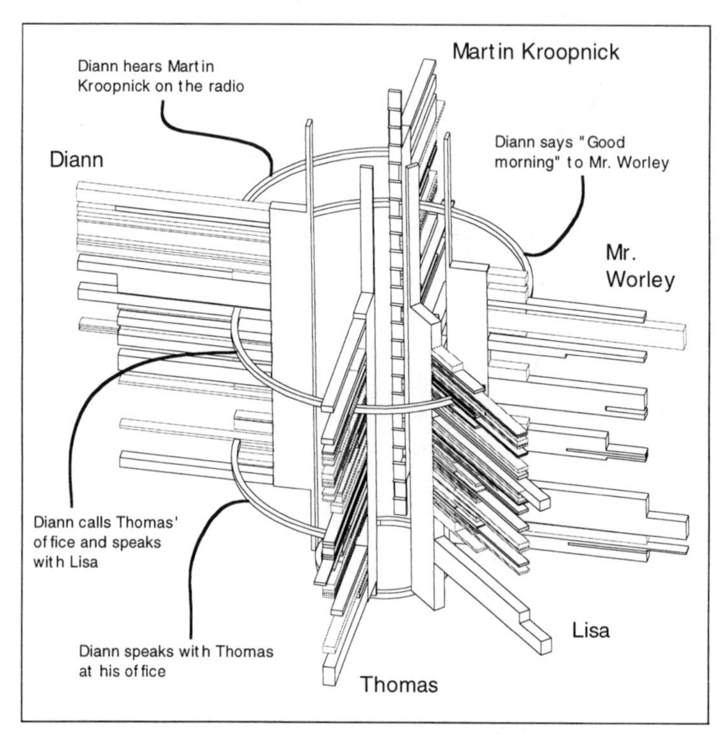

Figure 13.7. Extensibility diagram combining Figures 13.1–13.5, and also showing communications between Diann and other participants.

Appendix: A Note on Methodology

Information was gathered using a combination of interviews, time-diaries filled out by subjects, and direct observation. The mix of these techniques varied depending on the schedule and preferences of the subject. Significant attempts were made to approach the actual schedule of the subject, but the end result is intended as an illustration of a type of extensibility pattern rather than a precise replica of that subject's day.

To reveal the rhythms of daily communication, I eliminated the directional aspects of communication and retained the physical movements and the distance and duration of each communication. That distance was simplified to a more abstract concept of social scale by classification, as follows:

1 proximate (face to face, inside a building)
2 metropolitan (the Albany/Schenectady/Troy metropolitan area)
3 state (New York)

4 regional (New England and New York)
5 national (outside New England and New York but inside the United States)
6 international/global (outside the United States)

The communications of each subject over the course of a day are then arrayed within a unique plane in the virtual space. The rhythm (duration and frequency) and range (relativized distance) of communications are the primary attributes revealed. Standardizing distance measures helps reveal what is most socially relevant about distanciated (spatially extended) communications: their association with particular scales of social organization (see below).

The supporting software, Vellum®, is a CAD program by Ashlar, Inc. used most often for architecture and industrial design. Vellum converts two-dimensional shapes and lines to three dimensional 'wireframes' through 'extrude' and 'revolve' utilities, then transforms these into surfaces through the 'autosurface' utility.

Data was first entered as a spreadsheet chart (in MS-Excel), displayed in a chart, then exported as a 'picture' into Vellum where it was vectorized by hand. The classification of communication distances, and whether the communication is one-way or two-way, involved the following rule: if a person used more than one form of communication at a time, the code for the *farthest* appropriate range of communication was used. This was partly in the interest of revealing extensibility, and partly out of a conviction that a local or metropolitan sense of place can easily be overwhelmed by distanciated experiences.

Another decision was how to discern between one-way and two-way communication. The primary concern was to distinguish one-way communication for which the individual has no opportunity to reply (such as radio), from media that support discussion and comment (such as telephone and e-mail). For this reason, the *one-way* communications indicated are those in which the subject is a receiver rather than a sender. Any one-way communication in which the subject is a sender (e.g., writing a letter) is coded in the same category as spontaneously interactive communication (e.g., talking on the telephone), hence as two-way. The *out-going* communication is assumed to support interaction, and thus is part of a two-way exchange. Based on earlier research I have coded the use of computer networks as interactive. The typical use of Internet communication is active and exploratory, which results in a kind of dialogue.

Shopping is a particularly problematic type of activity to code. While it occurs in a bounded space, such as a supermarket or shopping mall, it is one of the most important ways people affect distant others. The *messages* they send are anonymous and abstract since they are transmitted by money, but they have real consequences. A farmer may switch from a subsistence crop to a market crop in response to perceived demand. His or her life has been affected by a message sent from distant markets in the abstract medium of money, so shopping is coded as communication, but in the chart with summations of interaction time (Figure 13.8) sums are shown with and without shopping included, A and B respectively.

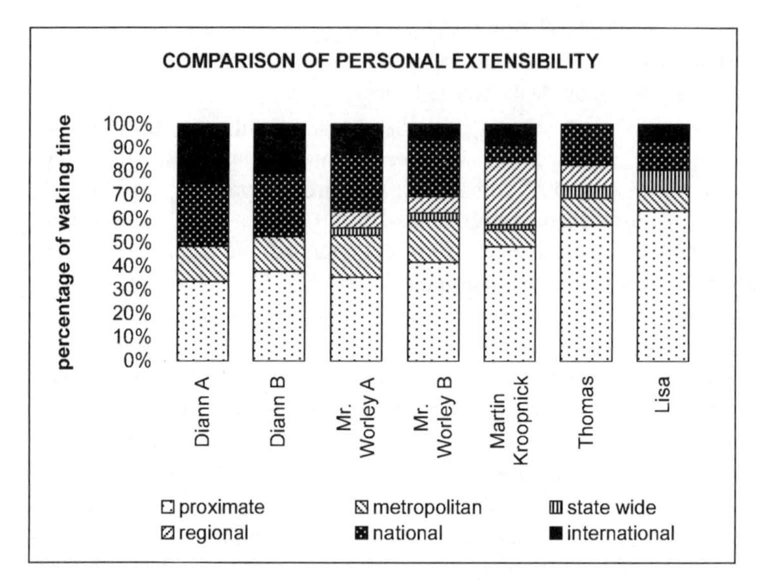

Figure 13.8. Approximate percentage of waking hours spent by each participant communicating at a given range, incoming and outgoing communications – combined. Reprinted with permission from *Urban Geography*, Vol. 20, No. 4, pp. 356-76. © V.H. Winston & Son, Inc., 360 South Ocean Boulevard, Palm Beach FL 33480. All rights reserved.

References

Adams, P. 1995. A reconsideration of personal boundaries in space-time. *Annals of the Association of American Geographers* 85(2):267-85.

Adams, P. 1996. Protest and the scale politics of telecommunications. *Political Geography* 15(5):419-41.

Adams, P 1999. Bringing globalization home: A homeworker in the information age. *Urban Geography* 20:356-76.

Castells, M. 1996. The Rise of the Network Society. The Information Age: Economy, Society and Culture, 1. Malden MA and Oxford: Blackwell.

Castells, M. 1997. The Power of Identity. The Information Age: Economy, Society and Culture, 2. Malden MA and Oxford: Blackwell.

Castells, M. 1998. End of Millenium. The Information Age: Economy, Society and Culture, 3. Malden MA and Oxford: Blackwell.

Curry, M. 1995. Geographic Information Systems and the inevitability of ethical inconsistency. In Pickles, J. (ed.) *Ground Truth.*, New York and London: The Guilford Press, 68-87.

Dear, M., and Flusty, S. 1998. Postmodern urbanism. *Annals of the Association of American Geographers* 88:50-72.

Fiske, J. 1987. *Television Culture*. London and New York: Methuen.

Graham, S. and Marvin, S. 1996. T*elecommunications and the City: Electronic Spaces, Urban Places*. London and New York: Routledge.

Habermas, J. 1987. The Theory of Communicative Action. Lifeworld and System: A Critique of Functionalist Reason, 2. Translated by Thomas McCarthy. Boston: Beacon Press.

Hägerstrand, T. 1970. What about people in regional science? *Papers of the Regional Science Association* 24:7-21.

Helgeson, S. 1998. Everyday Revolutionaries: Working Women and the Transformation of American life. New York: Doubleday.

Janelle, D. 1973. Measuring human extensibility in a shrinking world. *The Journal of Geography* 72(5):8-15.

Jonas, A.E.G. 1994. Editorial: The scale politics of spatiality. *Environment and Planning D: Society and Space*. 12(3):257-64.

McQuail, D., and Windahl, S. 1981. *Communication Models*. London and New York: Longman.

Sack, R. 1986. *Human Territoriality: Its Theory and History*. Cambridge: Cambridge University Press.

Sack, R. 1997. Homo Geographicus: A Framework for Action, Awareness, and Moral Concern. Baltimore and London: The John Hopkins University Press.

White, D.M. 1950. The 'Gatekeepers': A case study in the selection of news. Journalism Quarterly 27:383-90.

14 Human Extensibility and Individual Hybrid-accessibility in Space-time: A Multi-scale Representation Using GIS

Mei-Po Kwan

Department of Geography, The Ohio State University, Columbus OH 43210-1316, USA.
Email: Kwan.8@osu.edu

14.1 Introduction

With the increasing use of the Internet for getting information, transacting business and interacting with people, a wide range of activities in everyday life can now be undertaken in cyberspace. As traditional models of accessibility are based on physical notions of distance and proximity, they are inadequate for conceptualizing or analyzing individual accessibility in the physical world and cyberspace (hereafter referred to as *hybrid-accessibility*). To address the need for new models of space and time that enable us to represent individual accessibility in the information age, there are at least three major research areas: (a) the conceptual and/or behavioral foundation of individual accessibility; (b) appropriate methods for representing accessibility; and (c) feasible operational measures for evaluating individual accessibility. With the recent development and application of GIS methods in the study of accessibility in the physical world (e.g., Forer 1998, Hanson, Kominiak, and Carlin 1997, Huisman and Forer 1998, Kwan 1998, 1999a, 1999b, Miller 1991, 1999, Scott 1999, Talen 1997, Talen and Anselin 1998), it is apparent that GIS have considerable potential in each of these research areas. As shown in some of these studies, a focus on the individual enabled by GIS methods also reveals the spatial-temporal complexity in individual activity patterns and accessibility through 3D visualization or computational procedures.

Yet, even with the advent of 3D GIS tools, there are several difficulties when GIS methods are applied to represent or measure individual hybrid-accessibility. First, personal accessibility in the age of information involves multiple spatial and temporal scales (Hodge 1997), whereas current GIS are designed to handle only one geographical and/or temporal scale at a time. For instance, personal extensibility enabled by telecommunication technologies now allows an individual to access information resources at the global scale although the person's physical activities are still largely confined at the local scale. Further, the traditional temporal scale (hour/minute) is not adequate for studying cyber-transactions that may be accomplished within a few seconds. Second, GIS-based representational and computational methods, such as the space-time prism, are based on the sequential unfolding of a person's activities in the physical world. They are not developed to handle the simultaneity and temporal disjuncture that characterize many types of cyber-transactions. For example, a person may be talking over the phone and browsing a Web page at the same time. An email message sent out now may be read several hours later on the other side of the globe. These limitations of current GIS methods constitute a major challenge to any effort to represent and measure individual hybrid-accessibility in the information age.

As a preliminary attempt to address this methodological challenge, this paper explores how current GIS, given their limitations, can be deployed for the 3D interactive visualization of human extensibility in space-time. It develops and presents a method for the multi-scale, 3D representation of individual space-time paths based upon the concept of human extensibility (Janelle 1973, Adams 1995). Using geo-referenced activity diary data for an individual as an example and ArcView GIS software (© ESRI, Inc), the method is capable of revealing the spatial scope and temporal rhythms of a person's extensibility in cyberspace. It can also represent the complex interaction patterns among individuals in cyberspace using multiple and branching space-time paths within a GIS. Compared with the two-dimensional and/or cartographic representations in past studies, this method allows the researcher to interact, explore and manipulate the 3D scene (e.g., rotation, fly-through). This visualization environment not only greatly facilitates exploratory data analysis, but can also enhance our understanding of the patterns portrayed. It may provide the basis for formulating operational measures of individual hybrid-accessibility. In this paper, the nature of accessibility in the information age is first examined, and then alternative representational methods are discussed. Implementation of the GIS method using real activity diary data of an individual is described.

14.2 The Problem of Accessibility in the Information Age

In the physical world, the problem of accessibility is basically a problem of overcoming the impedance of physical separation between locations of demand and supply. Accessibility in physical space, therefore, depends largely on the spatial distribution of urban opportunities, available means of transport, and travel mobility (Burns 1979). Its foundation is the place- or location-boundedness of opportunities, where access to facilities and services is predicated on meeting the space-time co-location and co-presence requirements for spatial interaction (Giddens 1984). With distance between locations as the major impedance of movement, and with a given space-time distribution of opportunities in the urban environment, individual accessibility can be specified by these fundamental elements. In such an environment, the geometry or topology of the transportation network and the space-time constraints faced by individuals in their everyday lives are crucial for evaluating accessibility (Miller 1991).

In the virtual world enabled and created by telecommunication technologies and the Internet, however, the nature of access to opportunities or information resources is drastically different from that in the physical world. In cyberspace, access to resources and interactions between different individuals are mediated by communication technologies. It therefore depends more on the availability of these technologies to a particular person and the skills possessed than the time or cost necessary for overcoming physical separation. Except in the cases where the cost of access or interaction still depends on the physical separation of locations (e.g., long-distance phone charges), distance between locations or individuals would have little effect on individual accessibility and spatial interaction in cyberspace. When an electronic packet can travel around the globe within a second or so, time-space convergence is literally complete (Kwan 2000). Physical distance between the origin and destination of an electronic packet also seems to bear little relationship with the duration taken to traverse such distance (MCI 1998, MIDS 1998).

If physical separation between locations is playing a less important role in determining access and spatial interaction (as in cases where the Internet is used for a wide variety of activities), many fundamental determinants of accessibility in conventional models are no longer important. For instance, the *information superhighway* does not have many similarities with conventional transportation networks. Since locations are connected through worldwide computer networks that enable multiple access paths and that operate on optical fibers at light speed, the effects of physical separation between locations and of topologies of transportation networks are obliterated. Further, since telecommunication technologies and the Internet provide various means for moderating the space-time constraints of many activities, the space-time co-location and co-presence requirements for access or spatial interaction for many activities are also obliterated (e.g., voice mail, electronic bulletin boards, and the World Wide Web). It is possible for one individual to be at several (cyber) locations at the same time, and, thus, to violate the constraint that 'one individual cannot exist in two places at one time' (Carlstein 1982). The spatiotemporal configuration of resources or opportunities in cyberspace is, therefore, drastically different from what is available in the physical world.

With these complexities introduced by cyber-transactions, the problem of accessibility in the new hybrid physical-virtual world is far more complicated and difficult to deal with. This suggests that conventional models are inadequate for representing and measuring individual accessibility in such new hybrid spaces. Further, differences between accessibility in the physical and virtual worlds require new conceptual and analytical models since we are dealing now with two drastically different realms and their interface. To address the limitations of conventional methods for representing hybrid-accessibility, I examine the notion of the person as an extensible agent based on the work of Janelle (1973), Thrift (1985), and Adams (1995). Based on this concept of personal extensibility and Hägerstrand's (1970) time-geographic framework, I then describe a multi-scale three-dimensional representation of individual space-time paths using GIS.

14.3 Human Extensibility in Space-time

Janelle (1973) first formulated the concept of the individual as an extensible agent, where extensibility represents the ability of a person to overcome the friction of distance through space-adjusting technologies, such as transportation and communication. As the conceptual reciprocal of time-space convergence, which reflects the degree to which places are approaching one another in time-distance, human extensibility measures the increased opportunities for interaction among people and places (Janelle 1973). The development of communication and transportation technologies (or spatial technologies) and their associated institutions thus imply a *shrinking world* with expanding opportunities for extensibility (Adams 1995, Coucleclis 1994). Further, human extensibility not only expands a person's scope of sensory access and knowledge acquisition, it also enables a person to engage in distantiated social actions whose effect may extend across disparate geographical regions or historical episodes (Adams 1999, Thrift 1985).

Adams (1995) extended this notion of human extensibility through a new model of the person based on the structuration perspective (Giddens 1984), where the spatially contingent and socially embedded nature of human extensibility is emphasized. Inequality in human extensibility with respect to gender, race and other socially significant categories is understood in terms of the mutually constitutive relations between the individual experience of accessibility and macro-level societal processes. Adams (1995) captured the dynamic and fluid nature of personal boundaries through the notion of 'people as amoebas'. The body is reconceptualized as a dynamic entity, which combines

> a body rooted in a particular place at any given time, bounded in knowledge gathering by the range of unaided sensory perception, [and] . . . any number of fluctuating, dendritic, extensions which actively engage with social and natural phenomena, at varying distances (Adams 1995, 269).

This notion of human extensibility not only provides a useful point of departure for understanding individual accessibility in the information age. It also offers a theoretical foundation for overcoming many limitations in the traditional understanding of corporeality found in Hägerstrand's time-geographic framework. As Rose's (1993) critique suggests, depicting a person's trajectory in space-time as a linear and clear-cut path has many difficulties, especially when the framework is used to understand women's everyday lives. Further, since constructs of the time-geographic framework have been used to formulate accessibility measures in the past (e.g., Burns 1979, Lenntorp 1976, Villoria 1989), a representational device capable of handling this reconceptualized extensibility is an important first step in formulating operational measures of individual hybrid-accessibility.

For this purpose, Adams (1995) developed the extensibility diagram using the cartographic medium. The diagram, based on Hägerstrand's space-time aquarium, portrays a person's daily activities and interactions with others as multiple and branching space-time paths in three dimensions, where simultaneity and temporal disjuncture of different activities are revealed. Fuzzy zones surrounding the space-time paths represent the fluidity of personal boundaries. This method, as shown in Adams (1995) and expanded upon in Chapter 13 can be used to represent a diverse range of human activities in both the physical and virtual worlds, including telephoning, driving, e-mailing, reading, remembering, meeting face-to-face, and television viewing. Although the extensibility diagram is largely a cartographic device, most of its elements are amenable to GIS implementation. As a first step to improve the representation and measurement of individual hybrid-accessibility, the next two sections explore alternative GIS methods for implementing the extensibility diagram. The focus is on incorporating the multiple spatial scales and temporal complexities (e.g., simultaneity and disjuncture) involved in individual hybrid-accessibility.

14.4 Alternative Representational Methods

Early methods used to represent human extensibility as individual space-time paths are largely graphical devices (e.g., Burns 1979, Pred and Palm 1978). For the study of individual accessibility, these representations are useful for giving an idea about the size of or changes in the space-time prism resulting from particular activities. Due to the unavailability of the geoprocessing capabilities of GIS at that time, there were only a few attempts to implement these constructs in an operational sense. Those who have resorted to computational procedures for implementing time-geographic accessibility measures, however, encountered many difficulties. Their results may deviate from what might have been obtained if the original constructs were fully implemented (see discussion in Kwan and Hong 1998).

Recent application of GIS methods in representing and measuring individual accessibility in space-time has made significant progress. For instance, Forer (1998) and Huisman and Forer (1998) implemented space-time paths and prism on a three-dimensional raster data structure for visualization and computational purposes. Their method is especially useful for aggregating individuals with similar socioeconomic characteristics and for identifying behavioral patterns. On the other hand, Miller (1991, 1999), Kwan (1998, 1999a) and Kwan and Hong (1998) developed different network-based algorithms for computing individual accessibility using vector GIS procedures. Kwan (1999b) implemented 3D visualization of space-time paths and aquarium using vector-based GIS methods and activity-travel diary data. These studies demonstrated that GIS methods have considerable potential for advancing this research area. Further, implementation of 3D representations of human extensibility is a first step to the development of GIS-based computational procedures. The suitability of two possible methods in a 3D GIS environment is discussed below.

(a) Traditional Single-scale Representation. The simplest method for representing a person's space-time path is the space-time aquarium constructed first by Hägerstrand (1970). In a schematic representation of the aquarium, the vertical axis reflects the time of day and the boundary of the horizontal plane represents the spatial scope of the study area. Individual space-time paths can then be plotted as trajectories in this three-dimensional aquarium. Individual accessibility can be evaluated through deriving the space-time prism defined by fixed activities in a person's daily activity schedule. Recent examples of implementing this representation through 3D visualization using GIS are demonstrated in Chapter 5, in Huisman and Forer (1998), and in Kwan (1999b).

When transactions in cyberspace are included, visualizations using this method encounter one major difficulty: neither the spatial nor temporal scale of the traditional space-time aquarium is adequate for reflecting the full range of activities. For instance, the spatial scale at which local activities and travel can be visualized would render activities at the regional or global level out of range and invisible to the analyst. Further, the temporal intervals for recording activities in the physical world are too long for capturing the rhythm and pace of cyber-transactions, which have sub-second travel speeds (e.g., 0.4 second) and very short transaction durations when compared with physical activities. Using the zoom-in or zoom-out capability of a GIS also does not reduce this difficulty, as experimentation by the current author suggests. Other methods for representing multiple spatial and temporal scales are needed.

(b) Multi-scale Representation Using Linked Graphical Windows. One way to enable GIS-based 3D visualization of a person's space-time path at multiple scales is through the use of several dynamically-linked graphical windows. To implement this method, an individual's activities in the physical world and in cyberspace can be displayed using several graphical *windows*, each of which focuses on transactions at a specific spatial and/or temporal scale. For example, one window may show activities or travel at the local scale (e.g., the county of residence), while another may illustrate cyber-transactions involving Web sites or individuals located in other regions of the world. Each of these windows can be dynamically linked so that manipulations during the visualization process (e.g., rotation) in one of them will be automatically reflected in other linked windows. Further, the analyst can manipulate the graphical objects in any of these linked windows. A single-scale implementation of this method is found in the multi-panel 3D plot features in the S-Plus visualization environment of MathSoft, Inc. (1997).

This apparently attractive alternative, however, also has its difficulties. Since transactions involving different spatial or temporal scales are separately displayed, it is difficult for the analyst to identify the overall space-time patterns of and the interactions among the individuals involved. For instance, for an individual who has access to the Internet only at the workplace, Web transactions undertaken at home (and most likely in the evening) would be non-existent. In the case of an employee whose work phone number is not accessible to non-work purposes, distanciated communications with friends and relatives can only take place before or after work. These kinds of interactions between activities in the physical world and cyberspace would be difficult to identify when this method is used.

Because of the limitations of these two methods, another method for the multi-scale, three-dimensional representation of individual space-time paths in hybrid physical-virtual world is discussed in the next section. This method integrates transactions at different spatial scales in one graphical window, where the overall pattern or relationships among activities at different spatial scales can be easily identified. It can represent various types of transactions that take place at the same time (simultaneity). Further, the method allows differentiation of attributes for each transaction using graphical legends. For example, color codes can show transactions at different spatial scales (local, regional, global), with different temporal characteristics (synchronous, asynchronous), and undertaken through different communication modes (one-way incoming, one-way outgoing, two-way). Most of the data pre-processing was performed using ARCINFO while the visualization was implemented using ArcView 3D Analyst.

14.5 A Multi-scale Representation of Individual Space-time Paths: An Example

(a) Data. The activity data of an individual, Pui-Fun (a fictitious name), was collected and used to implement the GIS method. This person is a software engineer who works in a telecommunications company in Columbus, Ohio. Information about her activities in the physical world was collected in the form of an activity-travel diary. Data about her activities in cyberspace were compiled from the history file of her Web browser and email directory. As these data did not come with time stamps for computing the timing and duration of her cyber-transactions, the temporal information needed for constructing the space-time path was reconstructed through a personal interview, in which she also explained each of her activities recorded on the diary day (Table 14.1). This makes a GIS-based graphic-narrative of her activities on this day possible. Further, as several Web pages were browsed during each of her *visits* to the Web sites recorded, Web browsing activities are grouped into distinctive sessions identified by the site visited (instead of presenting details of each page browsed). Table 14.1 divides her activities in terms of the local, regional (15 northeastern states in the United States) and global scales according to the location of these transactions. The following account of Pui-Fun's cyber-transactions focuses mainly on the Internet since other forms of personal extensibility such as interactions via the telephone were not recorded.

Table 14.1. Activities undertaken by Pui-Fun on the diary day

Activity start time (hr/min)	Activity end time (hr/min)	Location involved: Local	Regional	Global
3:00	8:20	Home		
8:30		Work		
8:35	8:50		Chicago	
9:15	9:35		Chicago	
9:35	9:45		Charlotte	
9:45	10:00		Maywood	
14:00	14:20		Chicago	
16:15	16:25		Charlotte	
16:25	16:35		Chicago	
16:35	16:50		Maywood	
21:10	21:30		Chicago	
24:00	00:20 next day			Hong Kong
00:40 next day	3:00	Home		

Source: Activity diary and personal interview with subject, August 1998

Activity data provide information about Pui-Fun's activities in both the physical world and in cyberspace for a Saturday that she worked from 8:30 a.m. to 12 midnight (Table 14.1). This is an unusual schedule since she normally works only from 9 am to 5:30 p.m. on weekdays. On this day, Pui-Fun's husband dropped her off at work at about 8:30 a.m. As she was working with several co-workers in other branches of the company located in Chicago IL, Maywood NJ and Charlotte NC to meet the delivery deadline of a product, she regularly checked the company Web pages on which important news and updates about the project were posted. Further, she exchanged several email messages with these co-workers throughout the day since there might be last-minute debugging and testing tricks she needed to know for preparing the final shipment of the product to the client. On the diary day, Pui-Fun started her day at 8:35 a.m. with a brief session of Web browsing at the Chicago site. Then, shortly after, she browsed more extensively to make sure she did not miss anything important, covering the Chicago, Maywood, and Charlotte sites. After she got all the necessary information, she continued to work on the project. She only had a brief lunch break at her workplace at about 12:30 p.m.

Around 2:00 p.m., she came across a technical problem which required her to log onto the project's information site in Chicago again. In late afternoon, she conducted another round of routine browsing of the Charlotte, Chicago, and Maywood sites. Because time is so limited for meeting the project's deadline, Pui-Fun did not go out for dinner on this day. Instead, her husband brought her dinner from a fast-food chain at about 6:30 p.m. She stayed at her workplace for the whole day and was off at about 12:00 midnight. Because company policy restricts employees' private use of the Internet at work, she rarely reads personal email messages or browses her favorite Web pages during work hours. She usually does so before her formal work hours begin or in the twenty minutes or so after she is formally off from work while waiting to be picked up by her husband. On this day, she browsed some newspaper and magazine Web pages hosted in Hong Kong while her husband was on the way to pick her up.

(b) Procedures. Although Pui-Fun's diary day is an extreme example, which only involves a very limited range of activities, it can still be used to implement and illustrate procedures for constructing the multi-scale 3D extensibility diagram. As a preliminary attempt to represent activities in both the physical world and cyberspace, the 3D space-time paths constructed in this paper only include her Web activities. Email messages received or sent by her on the diary day are not represented in the diagram since the time at which these messages were read cannot be determined from the email directory.

The first step is to determine the most appropriate spatial scales and to extract the relevant base maps from various digital sources. For the case of Pui-Fun, base maps of three spatial scales were prepared. First, a map of Franklin County, Ohio, and a regional map of 15 U.S. states in the northeastern part of the country were extracted from Wessex's First Street geographic database. Franklin County is the home county of Pui-Fun, whereas the U.S. region extracted will be used to locate the three American cities involved in her cyber-transactions – Chicago, Maywood, and Charlotte. At the global scale, the world map layer was derived from the digital map data that came with ArcView GIS (many high latitude regions and islands in the world map layer were eliminated to improve visual clarity). As the best projection at a particular scale depends largely on the specific objectives at hand (e.g., minimum distortion in shapes or accurate distance between locations), coordinates in these three map layers are in unprojected decimal degrees.

As Pui-Fun's cyber-transactions were all undertaken at her workplace, these three map layers were registered to this location. The regional and world maps were then transformed into scales that allow them to be displayed in sizes commensurate with that of Franklin County. Results of these transformations in two dimensions are shown in Figure 14.1, where the location of Pui-Fun's workplace is identified by the epicenter symbol, which is the same point no matter which of the three map scales is used. At the local level, the end points of the line connecting Pui-Fun's home and workplace shows where she lives and works in Franklin County. At the regional level, her cyber-activities on the diary day involved Web sites located in Chicago, Maywood, and Charlotte (as indicated by the dashed lines). At the global level, she visited Web sites hosted in Hong Kong in the People's Republic of China. After the map-scale transformation, these three 2D map layers are converted to 3D shape files and added to an ArcView 3D Analyst scene as 3D themes. After preparing these map layers, 3D shape files for Pui-Fun's space-time paths were generated using Avenue scripts and added to the 3D scene. These procedures created the multi-scale extensibility diagram shown in Figure 14.2.

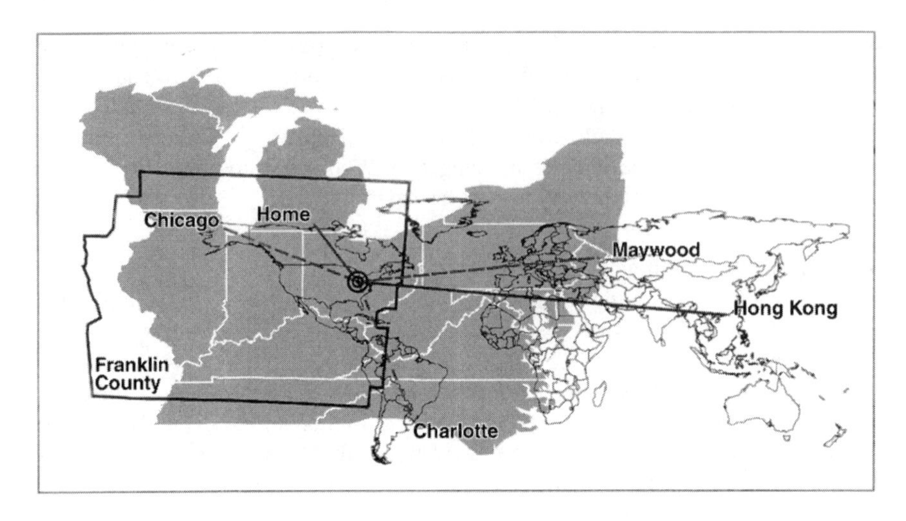

Figure 14.1. A two-dimensional representation of the three map layers after transformation.

(c) 3D Visualization. This multi-scale representation overcomes the limitations of the two methods discussed in the last section. Using this GIS-based extensibility diagram, the researcher can visualize all transactions at different spatial scales at the same time without need for multiple graphical windows (Figure 14.2). The visualization functions available in ArcView 3D Analyst also enable one to explore interactively with the 3D scene in a very flexible manner (e.g., the scene is visible in real-time while zooming in and out, or rotating). This allows for the selection of the best viewing angle and is a very helpful feature especially when visualizing very complex space-time paths. To focus only on one type of transaction or activity at a particular spatial scale, one can select the relevant themes for display while keeping the other themes turned off. Further, when the three sets of paths and base maps are displayed at the same time, they can be color coded to facilitate visualization. In the original color 3D scene, each segment of the space-time paths are represented using the same color as the relevant base map (e.g., blue for Franklin County and local activities), conveying a rather clear picture of the spatiality and temporal rhythm characterizing Pui-Fun's activities on the diary day. But, in the black-and-white version presented in Figure 14.2, spike lines are used to identify the location involved in each transaction.

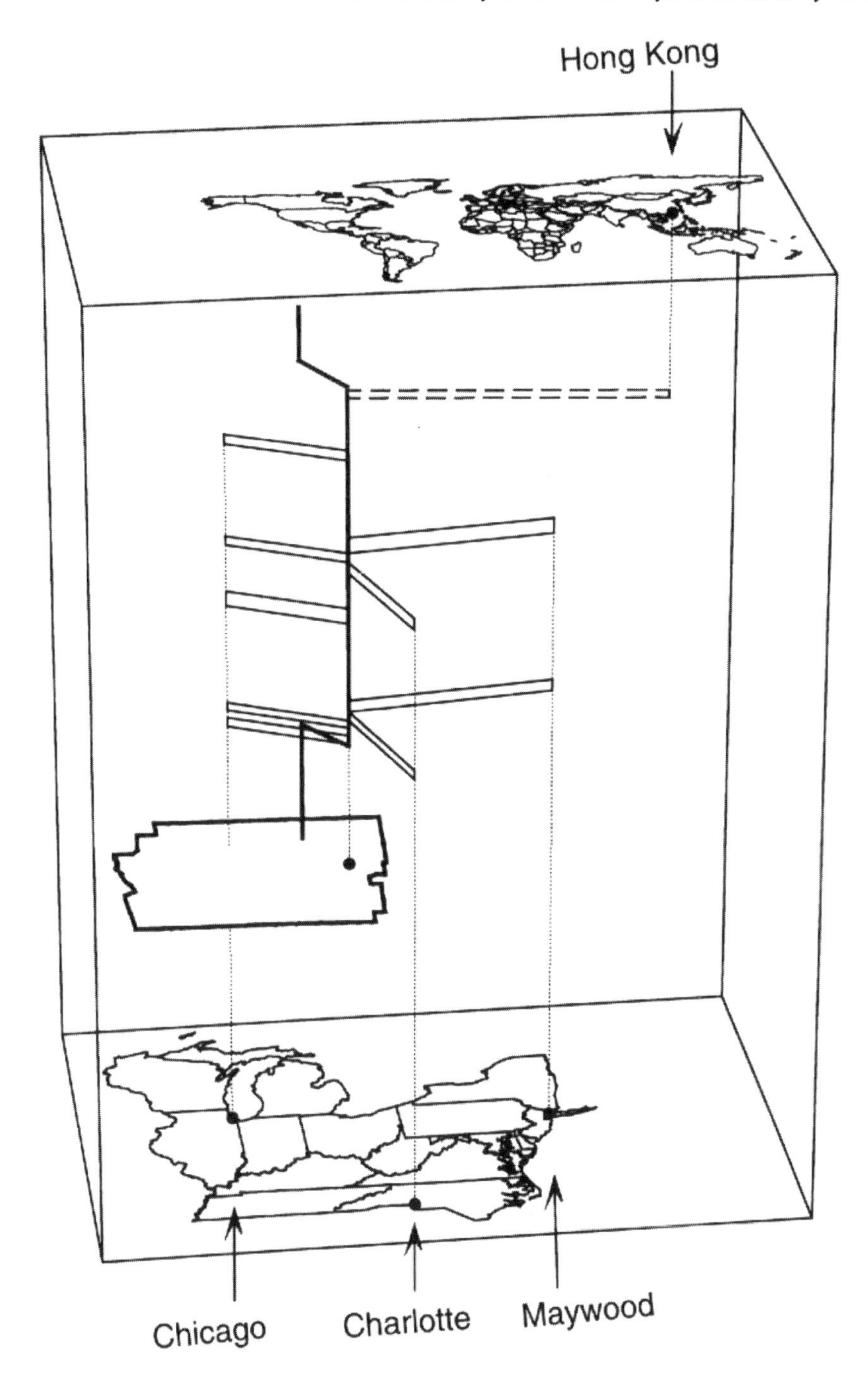

Figure 14.2. A multi-scale, 3D representation of the individual's space-time paths.

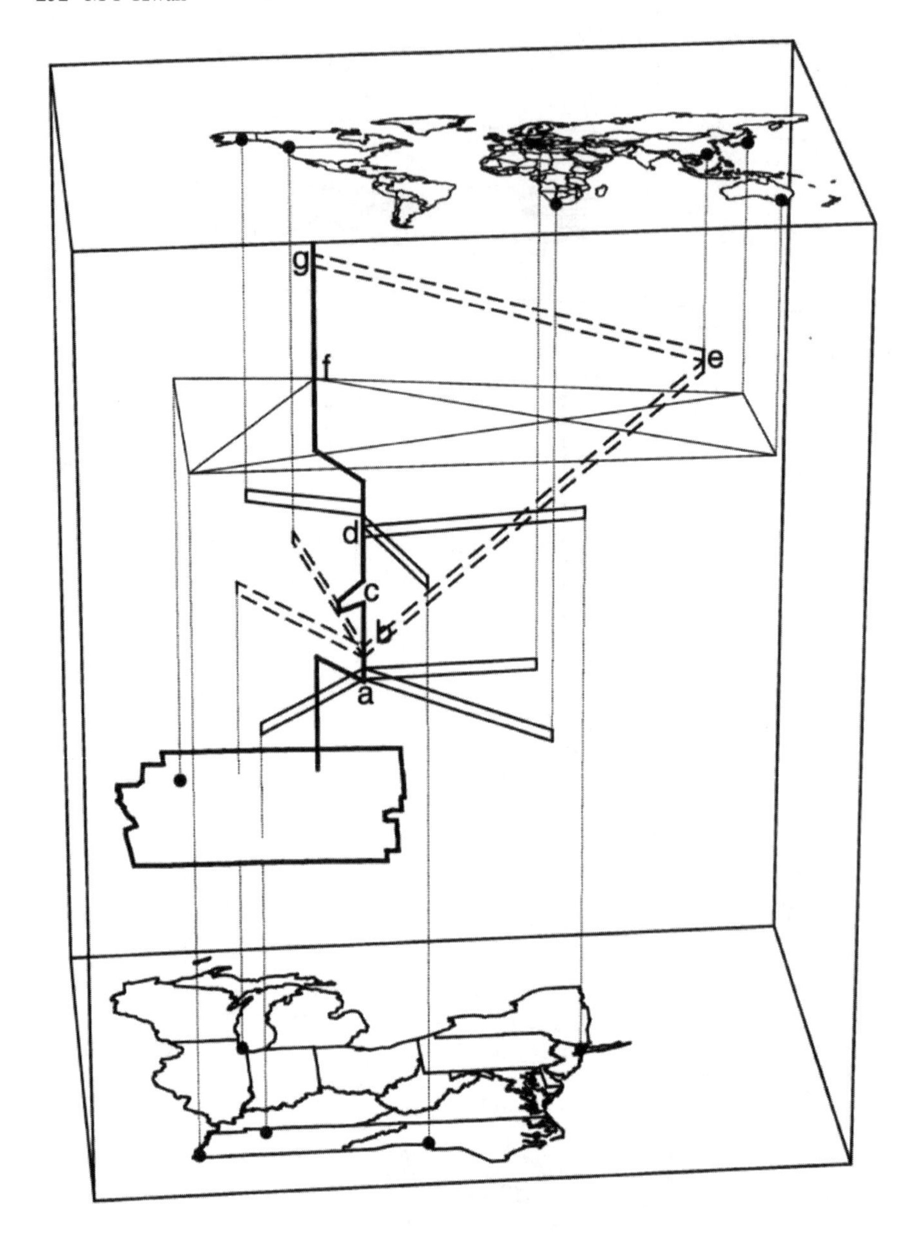

Figure 14.3. An extensibility diagram of a set of hypothetical activities.

Given the limited range of Pui-Fun's activities on the actual diary day, the potential of this GIS-based extensibility diagram is further explored using some hypothetical activities. The objective is to show how various types of transactions at different spatial scales can be represented. Using Pui-Fun as an example and partly following Adams' (1995) scheme, Figure 14.3 shows five types of activities undertaken on a particular day. On this day, Pui-Fun worked from 8:30 a.m. to 5:30 p.m., and had a one-hour lunch break at a nearby restaurant (c on the diagram). She subscribes to a Web-casting service where news items are continuously forwarded to her Web browser. On this day, she read some news about Yugoslavia, South Africa, and Nashville, TN (a on the diagram) before she started work. An hour later she sent an email message to three friends located in Hong Kong, Chicago, and Vancouver (b). The friend in Chicago read the email two hours later and the friend in Vancouver read the email five hours later. The friend in Hong Kong read the email 13 hours later and replied immediately (e). The reply message from this friend, however, was read at 2:00 a.m. at Pui-Fun's home (g). In the afternoon, Pui-Fun browsed Web pages hosted in New York, Charlotte, and Anchorage in Alaska (d). She was off from work at 5:30 p.m. and spent the evening at home. At 9:00 p.m. she started an ICQ (real-time chat) session with friends in Tokyo, Melbourne, Memphis TN, and Dublin OH (f on the diagram).

As shown in Figure 14.3, very complex interaction patterns in cyberspace can be represented using multiple and branching space-time paths. These include temporally coincidental (real-time chat) and temporally non-coincidental (e-mailing) interactions; one-way radial (Web browsing), two-way dyadic or radial (e-mailing), and multi-way (chat) interactions; in-coming (Web casting) and out-going (e-mailing) transactions (Adams 1998, Janelle 1995). The method is thus capable of capturing the spatial, temporal, and morphological complexities of a person's extensibility in cyberspace.

14.6 Conclusion and Discussion

Many characteristics of human extensibility can be represented using the multi-scale 3D GIS method presented in this paper. With appropriate time-space scaling, the extensibility diagram can reflect spatial relationships between different locations as a result of cyber-transactions. For example, the world can be as close as a few seconds away while it takes longer to reach one's next door neighbor. This kind of time-space inversion can be revealed and examined using this method. Further, although the focus of this paper is on the individual, the method itself can be used to represent interactions among many individuals. Thus, it allows for the study of social networks in space-time. The method, however, is limited by the capabilities of current GIS. For instance, current vector GIS can only represent objects as discrete entities with clear-cut boundaries, such as the straight-line representation of individual space-time paths. They cannot represent the fluidity of personal boundaries as fuzzy zones. It is also difficult to incorporate any qualitative information to account for the subjective experience of individuals in their everyday life. Future research on 3D GIS methods needs to explore how these limitations can be overcome.

There are other difficulties in implementing the method. First, since detailed data of an individual's activities in physical and cyberspace space are needed for constructing the 3D extensibility diagram, data availability will be a major issue. The problem is especially serious for transactions in cyberspace, as there are not only many different types of transactions to be recorded (e.g., e-mailing, Web browsing, Web casting, real-time chat, etc.), there is also no readily available means for recording these transactions. Data collected by commercially available server-side logging programs (used frequently by computer network administrators) are not adequate for this kind of study. Future research needs to investigate how to record these activities on the client side. This would involve a major difficulty regarding personal privacy: Will individuals be willing to disclose this kind of personal information in such detail?

Second, even when data about cyber-transactions are available, the location of a particular host on the Internet may be difficult to identify since IP addresses may not map onto geographical locations uniquely. Lastly, although individual space-time paths can be represented using this 3D GIS method, it renders the computation of space-time accessibility measures much more difficult. Given that cyber-transactions involve multiple spatial and temporal scales, and may include multiple and branching space-time paths, how can the space-time prism be identified? When fixed activities (such as work) may be ongoing with other flexible activities, which may involve far-away locations, how should space-time accessibility measures be computed? Each of these areas requires further research.

Acknowledgements

The support of an NCGIA Varenius seed grant for this research is gratefully acknowledged. I also thank the person who provided the activity data for this study.

References

Adams, P.C. 1995. A reconsideration of personal boundaries in space-time. *Annals of the Association of American Geographers* 85(2):267-85.

Adams, P.C. 1998. Network topologies and virtual place. *Annals of the Association of American Geographers*, 88(1):88-106.

Burns, L. 1979. Transportation, Temporal, and Spatial Components of Accessibility. Lexington, MA: Lexington Books.

Carlstein, T. 1982. Time Resources, Society and Ecology: On the Capacity for Human Interaction in Space and Time in Preindustrial Society. Lund Studies in Geography, Series B, Human Geography No.49. Lund, Sweden: CWK Gleerup.

Couclelis, H. 1994. Spatial technologies. *Environment and Planning B* 21:142-43.

Forer, P. 1998. Geometric approaches to the nexus of time, space, and microprocess: implementing a practical model for mundane socio-spatial systems. In Egenhofer, M.J. and Golledge, R.G. (eds.) *Spatial and Temporal Reasoning in Geographic Information Systems*. Oxford: Oxford University Press, 171-90.

Giddens, A. 1984. The Constitution of Socity: Outline of the Theory of Structuration. Berkeley, CA: University of California Press.

Hägerstrand, T. 1970. What about people in regional science? *Papers of the Regional Science Association*, 24:7-21.

Hanson, S., Kominiak, T. and Carlin, S. 1997. Assessing the impact of location on women's labor market outcomes: A methodological exploration. *Geographical Analysis* 29:281-97.

Hodge, D. 1997. Accessibility-related issues. *Journal of Transport Geography* 5(1):33-4.

Huisman, O. and Forer, P. 1998. Towards a geometric framework for modelling space-time opportunities and interaction potential. Paper presented at the International Geographical Union, Commission on Modelling Geographical Systems Meeting (IGU-CMGS), 28-29 August, Lisbon, Portugal.

Janelle, D. 1973. Measuring human extensibility in a shrinking world. *Journal of Geography* 72(5):8-15.

Janelle, D. 1995. Metropolitan expansion, telecommuting, and transportation. In Hanson, S. (ed.) *The Geography of Urban Transportation, 2nd edition*. New York: Guilford, 407-34.

Kwan, M-P. 1998. Space-time and integral measures of individual accessibility: A comparative analysis using a point-based framework. *Geographical Analysis*, 30(3):191-216.

Kwan, M-P. 1999a. Gender and individual access to urban opportunities: A study using space-time measures. *The Professional Geographer* 51(2):210-27.

Kwan, M-P. 1999b. Gender, the home-work link, and space-time patterns of non-employment activities. *Economic Geography* 75(4).

Kwan, M.P. 2000. Cyberspatial cognition and individual access to information: The behavioral foundation of cyber-geography. Environment and Planning B, in press.

Kwan, M-P. and Hong, X-D. 1998. Network-based constraints-oriented choice set formation using GIS. *Geographical Systems*, 5:139-62.

Lenntorp, B. 1976. *Paths in Time-Space Environments: A Time Geographic Study of Movement Possibilities of Individuals*. Lund Studies in Geography, Series B, Human Geography No.44. Lund, Sweden: CWK Gleerup.

MathSoft. 1997 *S-Plus User's Guide*. Seattle, WA: MathSoft, Inc.

MCI, 1998. Internet Traffic Report <http://traffic.mci.com/>

MIDS, 1998. The MIDS Internet Weather Report (IWR) <http://www.mids.org/weather/us/htmldir/>

Miller, H.J. 1991. Modelling accessibility using space-time prism concepts within geographic information systems. *International Journal of Geographical Information Systems* 5(3):287-301.

Miller H.J. 1999. Measuring space-time accessibility benefits within transportation networks: Basic theory and computational procedures. *Geographical Analysis* 31(2):187-212.

Pred, A. and Palm, R. 1978. The status of American women: A time-geographic view. In Lanegran, D.A. and Palm R. (eds.). *An Invitation to Geography*. New York: McGraw-Hill, 99-109.

Rose, G. 1993. *Feminism and Geography: The Limits of Geographical Knowledge*. Minneapolis, MN: University of Minnesota Press.

Scott, L.M. 1999. Abstract - Evaluating intra-metropolitan accessibility in the information age: Operational issues, objectives, and implementation. In Janelle, D.G. and Hodge, D.C. (eds.) *Measuring and Representing Accessibility in the Information Age, Research Conference Report*. Santa Barbara CA: NCGIA, 30-31.

Talen, E. 1997. The social equity of urban service distribution: An exploration of park access in Pueblo, Colorado, and Macon, Georgia. *Urban Geography*, 18(6):521-41.

Talen, E. and Anselin L. 1998. Assessing spatial equity: An evaluation of measures of accessibility to public playgrounds. *Environment and Planning A*, 30:595-613.

Thrift, N. 1985. Flies and germs: a geography of knowledge. In Gregory, D. and Urry, J. (eds) *Social Relations and Spatial Structures*. New York: St. Martin's Press, 366-403.

Villoria, O.G. 1989. An Operational Measure of Individual Accessibility for Use in the Study of Travel-Activity Patterns. Ph.D. dissertation, Graduate School of the Ohio State University.

Part III

Societal Issues

15 Accessibility and Societal Issues in the Information Age

Mark I. Wilson

Department of Geography/Urban and Regional Planning, and Institute for Public Policy and Social Research, Michigan State University, East Lansing MI 48824-1111, USA. Email: wilsonmm@pilot.msu.edu

15.1 Introduction

The preeminence of information as the foundation for the economies of most countries is often attributed to the technical possibilities available through computers and telecommunications. The information age is often presented as a product of the marriage of technologies and the triumph of advances in electronics and engineering. Increasingly apparent, however, is the need to incorporate social elements into our understanding of information technologies and the information age. By addressing the social context for these new technologies, discussion moves from the realm of what is technically feasible, to issues of access, equity, community, and identity. The theme of Part III is how to revisit the well developed theoretical foundation established for accessibility in transportation research, and to advance this foundation to understand the social impact of electronic media, such as computers, the Internet, and Geographic Information Systems.

The technical realm of the information age has tended to focus on electronic potential and possibilities. In many cases, the technical advance is accompanied by the assumption that technological constraints are the primary barriers to be confronted in the information age. This view is implicit in global maps that show Internet access by country (see Chap 7 by Sui), with only a few countries in 1999 not having Internet access (MIDS 1999). The advertising and rhetoric of the telecommunications industry offers a similar view, with low earth- orbit satellite systems allowing owners of Iridium telephones to call from anywhere on the planet. Global roaming options on mobile phones also *offer the world* in electronic convenience. In these cases, the technical possibilities are indeed impressive, and the ability to call home from almost anywhere represents a great advance on conditions a decade or two ago. The gulf remains, however, between the technically possible and the socially feasible.

Contributing to the scholarly challenge are the characteristics of the information technologies that we study. First, the pace of growth makes it difficult to identify the component elements and uses of technologies, such as the Internet, World Wide Web, and GIS; and, the information that is available tends to date quickly.

Second, information technologies and their services are not spatially confined, but are a global medium that appears to be aspatial by avoiding identity with a place or country. Third, as a new phenomenon, there is a lack of consensus about what to analyze and measure about the new technologies. Finally, much of the analysis and measurement of IT services is by individuals, firms, and agencies with a vested interest in the results, raising issues about standards and impartiality. Given these conditions, it is important for scholars to explore ways to understand, measure, and inform others about the nature of these new media.

In many ways the technical challenges are waning compared to the social challenges of access for the information society. World maps may show almost universal access, but there is a significant difference between access being possible and people being able to access the Internet if and when they wish. The non-technical challenges are many if access is considered integral to membership in the information society. Social factors that are important include the relevance of access to many whose daily life may not yet require access; and, education and experience in using information technologies. Economic factors, such as affordability; use in employment; and the growing importance of information-based economies shape access to information technologies. Political factors also loom large, with individuals having the freedom to access information; with regulatory environments supporting access and affordability of access; and government generating, managing, and using information effectively.

While not a unique element in this volume, Part III focuses specifically on the social context in which information technologies are used and developed. Implicit in this analysis is an awareness of the need to understand how society and technology meet and interact. In fact, the political, economic, and social dimensions of information technology use may well represent a far more complex and controversial arena for IT development than the technical barriers that scientists confront when advancing information technologies. As William Mitchell (1995, 5) notes in *City of Bits*,

> ...the most crucial task before us is not one of putting in place the digital plumbing of broadband communications links and associated electronic appliances (which we will certainly get anyway), nor even of producing electronically deliverable 'content,' but rather one of imagining and creating digitally mediated environments for the kinds of lives that we will want to lead and the sorts of communities that we will want to have.

Mitchell's comment is also a reminder that technology is not solely an external force, but one that is shaped by the actions of many in our society – individuals, corporations, and government -- and that technology is not necessarily inevitable, but a product reflecting many choices and decisions.

Part III introduces the major issues arising from the Asilomar meeting on *Measuring and Representing Accessibility in the Information Age*, with emphasis on the social dimensions that form the context for development and use of information technologies. Before identifying some of the specific social issues addressed, it is valuable to note some of the key questions that need to be considered when assessing the broader use of information technologies.

- How do information technologies change their host societies? The tendency to view technology as external to social forces misses many important issues, including the reverse relationship of how societies shape the technologies they adopt. This reverse relationship raises many normative questions and demands that we place more emphasis on understanding how people gather and acquire knowledge, and how they evaluate and act on that knowledge.
- What institutions shape the development of information technologies? The infrastructure of the information age stems from the actions of many actors with different and sometimes conflicting goals. The motivations of researchers, corporations (equipment manufacturers, software developers, service providers, etc.), government agencies, users, communities, and social movements can all vary both within and across these categories.
- What factors determine the use of information technologies? Use of IT can be necessary for work, or desired by individuals, for interaction (email, videoconferencing, and education) and information (employer intranets and the World Wide Web). Central to this factor is the ability of individuals to use information technologies, as well as a willingness to use IT. The rhetoric of technological determinism does not acknowledge the needs or wishes of those who do not like or want to use IT.

In the context of this volume, the core questions and issues revolve around access to information technologies. At the individual level, this type of access concerns (1) the availability of a nearby computer and the physical ability to use information technologies by knowing how to use a computer and to access the Internet; (2) having the resources to buy or rent a computer and to afford Internet access; and (3) having the freedom to interact with others electronically, or to view material of interest. As information technology evolves, some of these concerns will be accommodated. However, there will always be a need to focus on the broader implications of information technology to social change. The chapters in Part II of this book highlight five general constructs that weigh heavily in any consideration of the societal implications of IT – social context; equity; rights; time; and processes.

15.2 Constructs for Weighing the Social Implications of IT

Social Context

The power of information technologies needs to be addressed in social and historical context. In addition to the actors and processes shaping IT, already noted, there is also a set of issues associated with the value of access. What is the value and use of access, and, in particular, how useful is access to information technologies and for whom? We need to question the fundamental value of access to IT rather than moving directly to the assumption that it is or will be all encompassing. There are priorities and opportunity costs associated with improving access to information technologies. As part of that process, there is a need to identify and to understand the interests and voices that are heard about IT control, and those voices that are not heard. Also noteworthy is the role of information technologies in encouraging heterogeneity and diversity, associated with concern about the ability of IT and GIS to force homogenization and to undermine diversity. Do information technologies require a common language that produces a common mindset? And does it imply one frame of thought and one reality? In Chapter 19, Robert Mugerauer considers the possibility of alternative spatial configurations to capture qualitative elements of space. In particular, to harness the power of Geographic Information Systems to allow different groups and interests to define their own visions of space and location. He calls for empowerment though the ability to use GIS by each group or interest to define its own world and he notes that

> This does not mean that the conventional notions of accessibility no longer apply. It simply reflects the fact that accessibility is an intrinsically manifold notion, encompassing several definitions that can co-exist and not be reducible to each other.

This need for diverse interpretations of accessibility is also captured in Chapter 17, where Sylvie Occelli concentrates on the urban context for information technology and accessibility. She calls for an extension of the accessibility concept to include a temporal dimension, and for regarding access as a resource for urban populations. The commonly used concepts of accessibility are seen as being surpassed by increasingly complex urban societies that deserve more sophisticated conceptualizations of accessibility.

Equity

Equity considerations are increasingly important as the ability to gain access improves and the cost of access decreases. If access, control, and management of information are the foundations of economic growth and development, then the core equity considerations become: Who has access? And is access possible for those who desire access? While fundamentally an economic issue of affordability, the ability to access is also determined by government actions to shape the policies of service providers, or to make access possible through schools, libraries, and other public facilities. Emerging from equity concerns are research issues surrounding the relationship between access to information technologies, information, and economic and political power. Also relevant is an understanding of how inequalities change over time, and the ways in which IT affects the rate and direction of change in an information economy.

Beyond the affordability issues of access lie a number of geographic elements, as access has long been seen as a question of physical proximity. On one level, proximity to Internet service remains a crucial factor. Can I access the Internet where I live or work and at what cost? At a broader level, however, Internet access can reduce the physical barriers that have prevailed in the past. Information technologies may be able to remove the barriers that have defined peripheral locations to date. For example, Ireland's peripheral geography has been overcome in many ways through public and private investment in infrastructure, education, and training. The shift from peripheral geography to electronic centrality carries great benefits to the people and countries that are able to engineer relocation to electronic space.

Susan Hanson explores how spatial technologies affect equity in Chapter 16. In particular, she goes beyond issues of physical access to information technologies to address the social importance of information flows and their form. She calls for an understanding of how new technologies intersect with existing social relations in building and maintaining social equity, and she raises important questions about the social value of information technologies in comparison to face-to-face relationships and communication through existing community networks.

Rights

The discussion of rights recognizes a broader domain for the importance of freedom and the role of ethics in electronic interaction and access. Issues of right to access and use of technology and information are based in political and social context. The choices that societies make in relationship to utilitarian or communitarian systems raise several categories of core issues that warrant research attention.

First, the legal right to access information focuses on the existing constitutional and legal conditions in each country. The introduction of new technologies, however, raises again in many countries the internal political debate about what freedom is and what rights citizens should expect. Rights involve several elements: (1) right to speak; (2) should there also exist a right to be heard? (3) right to information access, and the shifting to private information from public information sources and control; (4) is there a right to receive a response from decision makers? and (5) what are the ethics of having decision-makers *lurk* but not respond to critical Web forums?

Second, the legal rights to access are not easily bounded, either legally or socially. Increasingly important is the boundary between public and private in an electronic world. Restriction of access to information may help or hurt people, presenting an obligation to balance needs for access and privacy, and to be aware of the information *haves* and *have nots*.

Third, rights of access and presentation of information also incorporate language/dialect and user community issues. The Internet offers the ability to express ideas in far more languages and dialects than possible using print or broadcast media. At the same time, the dominance of English and of a small group of online languages presents a reduction in choice for access to information.

Harlan Onsrud focuses on one of the important issues over rights in Chapter 18. He chronicles how the legal context for rights of access to information is changing in the United States. In particular, Onsrud is concerned about the erosion of access to public information sources for citizens. The diminishment of legal access comes from a series of legislative changes that are often buried deep within legislative bills. He observes (Chapter 18, 315) that the rights to access public information gained in the past are slowly being lost as '…publishers and government agencies use the threat of digital technology as an opportunity to limit the rights of citizens to access information'.

Time

The time dimension cuts across many of the chapters in this book. The ability of information technologies to reduce or end the friction of distance leaves time as one of the few remaining barriers to interaction globally. The importance of measuring and representing accessibility is to permit individuals and institutions to extract greater value from time, which can be expressed as the currency of the new economy, albeit a limited and finite resource. Andrew Harvey and Paul Macnab investigate an important aspect of this theme in Part II (Chapter 9), suggesting that time remains one of the key challenges to interaction now that distance can be overcome electronically. Using a case study of Canada and its six time zones, the constraints of interaction are defined clearly by temporal coincidence, and by the limited windows of real time communication possible at any one time. The societal increase in types of activities (both virtual and real) subjects individuals and institutions to allocation decisions.

Processes

The underlying theme of this research direction is to identify how social, economic, and political institutions in different places and times shape access to and use of information technologies. The importance of institutional players requires us to understand a range of factors. These include (1) how different types of institutions set agendas for information use and control; (2) how decision making by these institutions establishes the information infrastructure; and, (3) how relationships emerge between accessibility and social, economic, and political power. Social processes are also affected by information technologies, challenging scholars to explore the ways that people construct their social networks in an information age (a question addressed by Hanson in Chapter 16), and to investigate whether or not these new technologies require or generate new forms of social capital. Finally, it is important to seek understanding of how the spatial forces shaping the use of IT relate to and generate differences and similarities across places and spaces – a theme acknowledged throughout this book.

15.3 Conclusion

In many ways the technical challenges of information technologies are part of a larger challenge, with social factors of increasing importance. There may only be one way for physical properties to be seen or understood, but there are many differing interpretations of the way information technology affects our lives. One clear direction emerging from the chapters in Part III is the need to remind ourselves that social forces can and should shape the technologies; we must not accept a given system, network, or service without questioning its impact.

References

Matrix Information and Directory Services Inc. 1999. http://www.mids.org/

Mitchell, W. 1995. *City of Bits: Space, Place, and the Infobahn.* Cambridge MA: MIT Press.

16 Reconceptualizing Accessibility

Susan Hanson

School of Geography, Clark University, 950 Main Street, Worcester MA 01610, USA.
Email: shanson@clarku.edu

16.1 Introduction

One of my bad habits – one that I probably shouldn't confess to – is clipping newspaper articles and squirreling them away, solely on the basis of speed reading the headline and maybe a paragraph or two. If those few words suggest that something in the piece might possibly sometime be remotely related to anything I'm teaching or might ever teach, it gets filed away. One such piece, entitled 'Information Inequality' (1997, p. Ai4), appeared about a year ago on the editorial page of the *Boston Globe*. Thinking it dealt with unequal access to information technology (IT), I slipped it into my IT folder. When I finally read it carefully a short time ago, I was fascinated to find that the message touched hardly at all on the Internet. The editorial's touchstone was a statement that cultural critic Stanley Crouch had made that week on a local radio show: 'Talking about racial justice, or any justice, means talking about equal access to information.' (That was enough to activate the scissors.)

The editors go on to point out that the information people need access to relates primarily to people's immediate concerns like health ('what to do about a child's ear infection') and employment (what jobs are available and what training do they require). The editors' main point is that such information is exchanged most effectively through casual personal contact; *but* those contacts must cross the divide between information haves and information have nots if the information sharing is to 'dismantle information segregation.' For many reasons, including residential segregation, such divide-bridging personal contacts are unlikely in the course of everyday American life. After mentioning a few Boston-area initiatives aimed specifically at increasing the likelihood of such contacts, the editors conclude that these initiatives . . .

> need the resources to do more. Web sites can be packed with facts, but they haven't yet replaced the old-fashioned, labor-intensive – but more effective – approach of face-to-face talk. It's hard to champion a national conversation on ignorance. But talking about what we don't know is an inevitable first step toward breaking down nation-crippling information barriers.

I begin with this editorial because it highlights what I believe is the focus of this book – the role of information in access – and it raises (but doesn't really address) questions about the effectiveness of IT in reducing information inequality and in increasing access. In drawing attention to local initiatives for increasing personal interactions between information haves and have nots (all of these initiatives emphasize face-to-face contact) and in playing down the role of the Web, this editorial is a cautionary tale for technophiles. I want to use this cautionary tale as a starting point for rethinking models and measures of accessibility in an information age. My goals are

(1) briefly to review traditional (pre-virtual) conceptualizations and measures of accessibility, in order
(2) to point to key silences and omissions – especially the role of information – in these traditional approaches,
(3) to present the argument some scholars and policymakers advance that IT is the ideal way to fill these gaps,
(4) to examine how information relates to access and, in this context, how all information is not created equal, and
(5) to consider measures of accessibility that incorporate diverse, heterogeneous information sources.

I am particularly interested in people's access to paid employment, and most of the examples I'll use will have to do with access to jobs. I focus on employment because in contemporary North America, paid work is so central to issues of power, identity, and status – in short to economic, social, and political life. Moreover, employment relations are embedded in everyday life, including power dynamics within households, gender ideologies, cultural norms, and daily travel-activity patterns. Throughout this essay, my central concern is with equity and with how new spatial technologies might exacerbate socio-spatial inequities in access.

16.2 Pre-virtual Accessibility

The idea of accessibility has to do with reachability, obtainability, attainability. In the pre-virtual, non-virtual world, we geographers traditionally thought of access in terms of the ability of people to reach and use, or take advantage of, spatially dispersed activity sites providing jobs, goods, and services such as medical care, recreation, and entertainment. Access in the material world is closely linked up with mobility; as distances between activity sites in U.S. cities have grown, accessibility has come to require more (motorized) mobility. Accessibility has always been considered essential to living in a society where land uses are specialized and spatially separated. For this reason, accessibility has always been thought of as a good thing: the more accessibility an individual, group, or area has, the better. Society worries about those with limited access, and although some observers have begun to wonder if there might not be such a thing as too much mobility, I have not yet heard anyone voice the idea that some groups might suffer from too much accessibility.

Measures of accessibility typically involve counts of the number of opportunities (number of jobs, square feet of retail space) discounted by distance or some other measure of impedance. Such measures can be calculated for an origin zone, an individual occupying one or more trip origin points (home, workplace), or any number of people whose trip origins have been superimposed on a common point. In the pre-virtual sense of the word, then, accessibility refers to the theoretical ability of a person to reach and use dispersed points or zones on a plane (usually urban space); access measures capture the location of an individual vis a vis the location of a set of potential destinations. Such measures are really measures of interaction potential. All one needs to realize access, according to this view, is mobility.

16.3 Silences

Despite the utility of these measures in revealing access inequalities and inequities among groups and locations, the measures are limited by their inevitable silences – the dimensions of accessibility they overlook, neglect, and omit. Traditional accessibility measures provide rather narrow interpretations of core questions such as, Do you know what's available at potential destinations? Do you value what's there? Is it feasible for you to get to these locations and to participate in the activities there? Do you have social and cultural (not to mention geographic) connections to facilitate access to those sites?[1] In traditional accessibility measures, if a destination is located close by, it is considered accessible. As someone at the Varenius Conference noted, does someone who lives near a library but does not read or near an airport but does not fly have access? Traditional measures would answer yes.

Traditional measures of access neglect the fact that people are embedded in networks of social relations through which information is exchanged, networks that shape norms and values. That is, traditional approaches do not consider the informational, social, and cultural dimensions of accessibility. All of these affect a person's ability and willingness to leave an origin, ability and willingness to traverse distance, and ability to enter and participate at a destination. A job seeker needs information about current job openings, the locations of those jobs, and the

[1] Some of these and other silences are addressed in time geography. One example is the hours when stores and other activity sites are open.

qualifications (education, expertise, and experience) each employer is seeking; she might also want to know how women and visible minorities are treated in each workplace and how 'family friendly' each workplace is. For their part, employers might seriously consider only potential employees who possess certain social, geographic, and cultural attributes (e.g., some employers prefer female workers who are single, who do not live in certain stigmatized neighborhoods, and who do not have exceedingly long, multi-colored fingernails). Clearly, access requires more than proximity and/or mobility, and information is a key ingredient missing from traditional accessibility measures.

16.4 The IT-to-the-Rescue Argument

Recognizing the central role that information plays in access, some scholars and policymakers have heralded information technology (especially the Internet and the Web) as a, if not *the*, solution to the problem of poor accessibility. Harlan Cleveland (1985), for example, predicted that widespread availability of information would break down barriers of ignorance and secrecy, eroding hierarchies, increasing participation, and enlarging democracy. He saw the information revolution as undermining hierarchies of power based on control, hierarchies of influence based on secrecy, hierarchies of class based on ownership, of privilege based on early access to resources, and of politics based on geography. One of the main ways that IT acquires such power is by erasing the friction of distance, thereby providing access without mobility. IT is thus seen as the means to overcome the information segregation that derives in large part from residential segregation.

Stephen Graham (1998) and others (e.g., Hanson 1998) have debunked such technological determinism as unduly neglecting the lived realities of everyday life in a material world. In particular, utopian visions of equal access in cyberspace overlook the pervasiveness and power of the diverse place-based communities in which the majority of the globe's (and yes, even North America's) citizens live relatively grounded, even circumscribed, lives, with very real consequences for access to opportunities. Such utopian visions hint that IT will replace distance-based (and distance-biased) interactions, diluting their power to shape social life and ultimately rendering distance and geography obsolete. In my view, such visions distract attention from the real and difficult job of trying to understand how IT is complementing and interacting in complex and unforeseen ways with grounded social, economic, and political exchanges. Precisely because accessibility requires more than proximity and/or mobility, eliminating the friction of distance will not yield access.

16.5 Information and Access

Let us assume that everyone has access to the Web and therefore to almost limitless information. What does this mean for accessibility? Shannon and Weaver's classic information theory (1949) hinges on the idea that information reduces uncertainty. That is, an item of communication (a word, musical note, phrase, memo, medical test result) can be considered informative only when it reduces uncertainty. Shannon and Weaver show, moreover, that information's uncertainty-reducing role is context dependent: an additional musical note in a score reduces uncertainty only in the larger context of the particular musical passage.

This means that the same piece of information takes on different meanings in different contexts. Examples abound of words that mean entirely different things in different contexts. The Catalan word 'prou' means either 'yes, OK' or 'no, not OK' depending on context and intonation; the essence of Barbara Kingsolver's new book (1998) turns on a word in Kikongo, 'bangala,' which can mean 'precious dear' or 'poisonwood tree' depending also on context and intonation. It is obvious and widely recognized that virtual communication (VC) is not the same as face-to-face (F2F) communication precisely because it (VC) is decontextualized.[2] The medium of communication and the source of information affect the nature, quality, and reliability of what's communicated and therefore the degree to which it reduces uncertainty. Shannon and Weaver's insights about information raise questions about the information available on the Web. To what extent, and in what circumstances, is cyber information really informative in that it reduces uncertainty?

I believe that answering this question (and therefore understanding how information affects access) will require closely examining how IT intersects with other forms of communication in place-based communities. In the pre-virtual world, most people found jobs and most employers found workers through informal personal contact, not through formal information sources such as employment agencies or newspaper ads. And for good reasons: from the job seeker's perspective, key information about a potential work site (e.g., what it's really like to work there) simply does not appear – and never will appear – in formal job advertisements. From the employer's perspective, relying on word of mouth is cheaper than using formal advertising outlets and has a higher probability of resulting in the

---

[2] Sproull and Kiesler (1993) have studied how VC changes communication in work organizations and argue that it can lead to more egalitarian patterns of information sharing. But VC can also lead to more hierarchy and can create more barriers and balkanization (Jones 1995; Van Alstyne & Brynjolfsson 1996). These effects emerge not only because people's access to IT itself varies but also because VC allows us more choice regarding with whom we interact; it allows each of us to customize our social interactions based on shared interests, thereby narrowing the range of diversity with which we engage. As Van Alstyne and Brynjolfsson note, however, balkanization is not inevitable; we can 'use IT to select diverse contacts as easily as specialized contacts' (1996, 1480).

hire of a more productive worker. In each case, the screening function of personal networks (those of employers and existing employees) increases the likelihood that a particular employer-worker match is a good one. In short, personal contacts are effective at reducing uncertainty. Thus studies of labor market processes repeatedly find that the information that flows through social networks and everyday personal interactions plays a pivotal role in shaping people's access to jobs, affecting type of work (occupation, industry), location, and compensation (Granovetter 1974, Hanson and Pratt 1995).

The importance of these personal relations points to the need to recognize how social and cultural capital shape people's access to opportunities. Social capital 'encompasses benefits derived from relations of mutual trust and collaboration; it thus resides in the relations between members, not in the individuals who compose it' (Fernandez Kelly 1995, 216). Portes and Sensenbrenner (1993, 1323), who are interested in social capital primarily as it intersects with the labor market, define it as 'those expectations for action within a collectivity that affect the economic goals and goal-seeking behavior of its members, even if those expectations are not oriented toward the economic sphere.' They point to four sources of social capital: (1) value introjection (the socialization into consensually established beliefs), (2) reciprocity exchanges (the norm of reciprocity in face-to-face interaction), (3) bounded solidarity (common awareness), and (4) enforceable trust (rewards and sanctions linked to group membership) (Portes and Sensenbrenner 1993, 1323).[3]

Cultural capital consists of symbols and values that help people to make sense of their experiences; it is a byproduct of social capital because it is developed through the personal interactions that are the raw material of social capital (Fernandez Kelly 1995, 220). Social and cultural capital develop through social exchanges that until now have usually and necessarily been largely face to face. The social and cultural capital available to a person depend on the nature of these personal interactions, which cannot be ephemeral, fleeting, or singular; social capital requires repeated contacts and the expectation of on-going interaction.[4]

The expectation of repeated interactions points to the importance of geographic context in shaping social and cultural capital. The webs of social relations that sustain them develop and grow among the people living in a particular place and time. Social and cultural capital depend on the sustained contact that comes with residential rootedness (Hanson and Pratt 1995). Moreover, the geographic content of the information and knowledge exchanged through these social networks (e.g., concerning job opportunities or how to find a good physician) depends in large

[3] Fernandez Kelly (1995) and Porte and Sensenbrenner (1995) note that the concept of social capital has its origins in the classic sociological texts of Durkheim, Marx, Webber, and Simmel. Coleman (1988) was instrumental in reviving interest in social capital.

[4] A question that bears scrutiny is whether social capital can be built on the Internet. My suspicion is that the success of IT as a medium for building social capital is likely to depend on one's social position (e.g., gender and class) and the nature and amount of social capital one already has.

part on the geographic extent of the network members' experiences (Hanson and Pratt 1991, 1995).

This geographic dimension points to another crucial characteristic of social capital, namely the extent to which it embraces diversity. Do the social networks that yield social capital connect people across a range of interests and experiences and across lines of social cleavage (e.g., age, income, class, gender, race)? Putnam (1993) argues that the civic virtue of social capital lies precisely in its ability to do so. Putnam's point is that some interests that bring people in a community together (his examples include bowling in a league, singing in a chorus, or volunteering for the PTA) will necessarily cut across other lines of social cleavage (income, gender, religion, class, etc.). Fernandez Kelly's study of teen motherhood in a Baltimore ghetto suggests how seldom such cross-cutting ties are in fact part of the social capital of the poor.

Note the importance of embodiment in this argument: bowlers or singers may share a passion, but it is precisely their coming together in the flesh, as whole persons, that brings people in contact with difference and therefore bridges lines of social cleavage. Joining together with like-minded souls on the Internet to pursue a common interest does not necessarily have the same result. What is the probability that a person selected at random from one's social network will be the source of new or surprising information about employment opportunities? Are the people who supply one with information about jobs all working at the same level in the same kind of jobs? If so, any job information received from them is likely to be less useful for social mobility than if that information came from a network of diverse sources (Granovetter 1982, Hanson and Pratt 1991). Do networks extend beyond the immediate community, or are they socially and spatially confined? The amount of diversity that is built into the social capital of a place or person or group crucially affects accessibility. Because geographic mobility allows people to separate themselves spatially from those who are different, one serious source of unequal access in metro areas is the relative homogeneity – and impoverished homogeneity – in the social and cultural capital of the poor.

In sum, access to jobs requires more than having proximity/mobility and even more than possessing the needed human capital (education, skills, experience) for a particular type of employment. It requires information, but more importantly, it requires certain kinds of information – the kinds that inhere in social and cultural capital. Traditionally social and cultural capital have developed through networks of F2F contact, in which information has been contextualized and a basis for trust established. In thinking about accessibility in an information age, how might IT intersect with these often place-based and place-biased information networks? How might IT be used to intervene strategically to increase the access of those who currently lack it?

16.6 Accessibility = Grounded + Virtual

I have argued that traditional measures of accessibility are silent on the role of information and that any information-age concept or measure of accessibility must incorporate information, virtual and grounded, electronic and F2F. In this final section, I first sketch out how the pre-virtual information (exchanged F2F) embedded in social and cultural capital already does and might in the future interact synergistically with virtual or cyber information. Second, I consider desiderata in an information-age accessibility measure.

Some have seen IT as the perfect answer for those whose social capital lacks the diversity needed to connect them with 'good' opportunities, such as good jobs. Because most people find out about jobs through personal contacts, people with highly localized social networks are unlikely to hear about jobs that are located outside the immediate community.[5] Moreover, because previous co-workers are an important source of information about new jobs (more so for men than for women), having held few or no jobs in the past also constricts one's information about job opportunities. IT has been proposed as an ideal way to obviate these problems and supply people with the job information they need.

But this suggestion shows no appreciation for why social networks are so popular as conduits of job information in the first place and why digital job banks have not been very popular with either workers or employers. People value F2F information from known sources: the on-going nature of a social relation enables trust and sheds light on the veracity of the information; in this sense the source acts as a screen. The importance people accord the information they exchange face to face underlines the crucial role of context; information exchanged electronically means something different from information exchanged in person. Yet despite the power of F2F, the Web *does* have enormous potential for disseminating job and employment-related information and especially for bridging the divide between information haves and have nots. This suggests the possibility of a productive union between grounded and virtual information exchanges.

I've recently begun a study of entrepreneurship that focuses largely on how business start-ups and self-employment are related to and embedded in people's labor market experiences. A couple of anecdotes from entrepreneurs we've interviewed in Worcester illustrate how these people are combining IT with F2F to increase access. The point here is not so much that lots of small business owners are using the Internet; relatively few now do, although many more voice plans to

[5] Certainly not all social networks that connect people with jobs are localized around particular workplaces. Meir and Giloth (1985) found that word-of-mouth recruiting explained high unemployment and long commutes to low-wage jobs among Mexican Americans in a Chicago neighborhood that had good local job opportunities. The employers in this neighborhood used word-of-mouth recruiting, but their existing employees did not live locally and had no local networks. Neighborhood residents did not, therefore, have the needed network access to jobs in their own neighborhood.

use the Internet soon. What is striking is how use of the Web combines in interesting ways with personal contacts to serve the business owners' interests.

One woman, whose home-based business is marketing the products of high-tech firms to scientists and engineers, uses the Web extensively. A considerable share of her clients are in California and Texas. When we asked how she had penetrated those markets, she cited word of mouth and her firms' Website, two methods that she clearly sees as complementary. Having worked in the high-tech field for seven years before launching her own company, this woman had an extensive array of personal contacts among potential clients before start-up She now prepares a quarterly newsletter on marketing, which she faxes to all of her current clients (most of whom came to her via word of mouth) and to potential clients whom she's gleaned from the Web as well as from word of mouth with existing clients and others. The newsletter refers readers to her Website, which then prompts some personal contact and yields new clients.

A second example – of a woman who runs a flooring company – suggests how business owners use the Web strategically to broaden and diversify information sources well beyond what would be available through place-based personal contacts. When asked whom she relies on now for information and advice in running her business, she replied succinctly (and quite distinctly): 'My computer.' She uses the Internet extensively to learn what her competitors are doing as well as to identify potential clients. She has, in fact, been able to extend her market area by learning from the Web about potential jobs, *but* her floor installers have been willing to travel to these distant locations only because (and when) they have personal contacts (family, friends) there. These examples illustrate how small business owners weave together various forms of IT with F2F to promote their businesses.

Here are a few additional examples, with a more futuristic flavor. One involves setting up computer work-stations and Internet links in study rooms/labs for children in public housing projects or in neighborhood study centers in low-income areas. My sense is that most people rely upon their social networks in learning new IT and in negotiating the Internet; people often, for example, visit new web sites because they have heard about them in F2F exchanges. These facilities will need to be staffed with people who can help the children use the technology, a F2F/IT interface in itself, and offer them personal contacts that link them to the world beyond their immediate environments. Such adults might be college students, teachers, or retired persons whose experiences straddle the local community and other places and who can personally connect the children to opportunities outside their immediate experience. The synergistic nature of IT and F2F interaction will be evident, for example, in that staff in these facilities will often use the Web to identify opportunities previously unknown to themselves.

In the context of job search, it seems possible that intermediaries such as teachers, members of the clergy, volunteers, or members of non-profits could play an important bridging role. Perhaps both job seeker and intermediary would have access to job information on the Web and the intermediary could act as go-between with employers, helping to screen applicants. In this way IT might also help match teenagers with employers in internships and apprenticeships that would link them into work-based social networks. IT will be most powerful in expanding access if it enriches social capital by broadening and diversifying information exchanged face to face and by prompting the discussion of new questions.

How might information-age accessibility measures incorporate these informational dimensions of access; i.e., how might accessibility measures include aspects of access other than simple opportunities-discounted-by-distance? I suggest three components as desiderata for information-age measures of access. First, I think that we should not neglect the spatial arrangements captured in good old-fashioned spatial accessibility measures; these will remain important and should be retained in information-age accessibility measures. Second, as many other authors in this book have suggested, measuring access in the information age means including one or more measures of the access of people and places to IT itself – in the home, school, neighborhood, or workplace – and to locations in cyberspace. Third, access measures must incorporate measures of the social/cultural capital that individuals, groups, and areas have access to. This means devising measures that capture the collective information assets – and especially the diversity present in these group assets – that network members can tap into.

Operationalizing these concepts presents a challenge, but with GIS, devising such measures is not out of the question as long as the U.S. Census long form remains. For example, one could use the journey-to-work files derived from the long form to characterize for each person or household or area (1) the occupations and industries (measured by three-digit census codes) of employment and the wages earned per worker in that person's or household's place of residence and place of work, where places could be census tracts, block groups, or blocks and (2) the variance in occupations, industries, and wages by place of residence and place of work. The residence-focused measures could be weighted by length of residence. These are rough measures, to be sure, but the data are readily available and they do capture key aspects of social and cultural capital, i.e., the embeddedness of people in particular milieus. Another possibility that does not rely on the census long form but instead uses data from employers lies in examining the zip codes of residence of the current employees in a given work place (where place can be defined as an employer/establishment or as the collection of employers in an area). Mapping these would shed light on the spatial extent of information exchanges that people have access to at the workplace. I offer these as surrogate measures for the nature and diversity of information exchanged F2F through social networks.

16.7 Conclusion

I have argued that information-age concepts and measures of accessibility must go beyond incorporating the role of IT in providing access. Such concepts and measures must also acknowledge the powerful (though certainly not all-powerful) role of F2F personal interactions in shaping patterns of accessibility. IT *can* extend horizons, increasing access and helping to dismantle information segregation, but it will do this most effectively if embedded in webs of grounded, often place-based social relations that themselves bridge social and information cleavages.[6]

My main point is that IT and the Web are not simply technologies that are deployed in a vacuum or a materially neutral *cyberspace*; they invariably intersect with place-based processes, including patterns of communication, that strongly affect accessibility. So, even if everyone had equal access to the Internet and other spatial technologies, people's embeddedness in networks of social relations –and the likely continuing importance of this embeddedness – raises the two central research questions I've proposed: (1) How are people using IT (VC) and F2F to shape their access, either by increasing or constraining it? (2) How can information-age accessibility measures recognize people's embeddedness in the networks of social relations that play such an important role in determining access?

References

Cleveland, H. 1985. The twilight of hierarchy: Speculations on the global information society. *Public Administration Review* 45(1):185-95.

Coleman, J. 1988. Social capital in the creation of human capital *American Journal of Sociology* Supplement: S95-S120.

Fernandez Kelly, P. 1995. Social and cultural capital in the urban ghetto: Implications for the economic sociology of immigration. In Portes, A. (ed.) *The Economic Sociology of Immigration: Essays on Networks, Ethnicity, and Entrepreneurship*. New York: Russell Sage Foundation, 213-47.

Graham, S. 1998. The end of geography or the explosion of place? Conceptualizing space, place, and information technology. *Progress in Human Geography* 22(2):165-85.

Granovetter, M. 1974. *Getting a Job: A Study of Contacts and Careers*. Cambridge: Harvard University Press.

Granovetter, M. 1982. The strength of weak ties: A network theory revisited. In Marsden, P. and Lin, N. (eds.) *Social Structure and Network Analysis*. Beverly Hills: Sage. 105-30.

Hanson, S.. 1998. Off the road? Reflections on transportation geography in the infromation age. *Journal of Transport Geography* 6(4):241-9.

Hanson, S. and Pratt, G. 1991. Job search and the occupational segregation of women. *Annals of the Association of American Geographers* 81:229-53.

Hanson, S. and Pratt, G. 1995. *Gender, Work, and Space*. New York: Routledge.

---

[6] We might consider also how people use IT not only to increase their access to information and opportunities but also to control and limit the access of others to the self (e.g., via turning off the telephone, using caller ID and answering machines, checking e-mail sporadically).

Jones, S. 1995. Understanding community in the information age. In Jones, S.G. (ed) *Cybersociety: Computer-mediated Communication and Community*. Sage publications.

Kingsolver, B. 1998. *The Poisonwood Bible*. New York: Haprper Flamingo.

'Information Inequality,' *Boston Globe* August 23, 1997.

Meir, R. and Giloth, R. 1985. Hispanic employment opportunities: A case of internal labor markets and weak-tied social networks. *Social Science Quarterly* 66:296-307.

Portes, A. and Sensenbrenner, J. 1993. Embeddedness and immigration: Notes on the social determinants of economic action. *American Journal of Sociology* 98:1320-50.

Putnam, R. 1993. The prosperous community: Social capital and public life. *The American Prospect* No.13:35-42.

Shannon, G. and Weaver, W. 1949. *The Mathematical Theory of Communication.* Urbana, IL: University of Illinois Press.

Sproull, L. and Kiesler, S. 1993. Computers, networks, and work. In Harasim, L. (ed) *Global Networks: Computers and International Communication.* Cambridge, MA: MIT Press.

Van Alstyne, M. and Brynjolfsson, E. 1996. Could the Internet balkanize science? *Science* 274:1479-80.

17 Revisiting the Concept of Accessibility: Some Comments and Research Questions

Sylvie Occelli

IRES - Istituto di Ricerche Economico Sociali del Piemonte, Via Nizza 18,
10125 Turin, Italy. Email: occelli@ires.piemonte.it

17.1 Introduction

Almost all urban systems in developed countries are undergoing a number of institutional, socio-economic and cultural changes, pushing them towards a 'new' societal configuration that is generally taken to be more democratic, better educated, culture-based, and environmentally sensitive, i.e., the so-called *Post-Fordist society* (Amin1994). Space-adjusting technologies, and particularly the New Information Technologies (NIT), are playing a substantial role in this transition, since they affect both the range and time-related organization of activities offered in an urban setting, as well as the ways in which individuals participate in them (see Castells 1989, Graham and Marvin 1996).

Due to its intrinsic ability to provide the 'connections' between the pattern of activities and their interdependencies, accessibility is a very sensitive concept for analyzing these changes. In the current transition to a Post-Fordist society, however, the notions traditionally used reveal many drawbacks. A revision is needed to (a) improve understanding of the space-time changes taking place in modern cities (Couclelis 1996, Bertuglia and Occelli 1997) and (b) provide appropriate indications for coping with practical planning problems and, in particular, policies relating to accessibility.

While not exhaustive, this paper identifies a number of aspects likely to constitute the *accessibility question* in Post-Fordist urban development (see Rabino and Occelli 1997). The discussion builds upon recent work in which the need for this revision has already been advocated (Occelli 1998a). It was argued that, at least on substantive grounds, a major area for exploration is the impact of New Information Technologies (NIT) on the time dimension underlying the notion of accessibility. It was pointed out that the NIT gives accessibility a new role and new potential, making it not simply a time-space opportunity, but also a *resource*. This implies going beyond conventional definitions of accessibility and requires a broader perspective of analysis.

One implicit aim for this chapter is to show how *reasoning about accessibility* (i.e., developing an analytical framework, such as a modeling activity) can help both in disentangling accessibility problems and in defining better policy measures. The discussion is divided into four parts, which form the main building blocks of this reasoning.

The first part revisits the concept of accessibility, emphasizing how it relates to the main components of a spatial system, and the second part recalls classical definitions of accessibility proposed in the literature. It is argued that these reflect the features of the city that have emerged during the historical process of urbanization. A *meta-typology* of urban development is proposed as a way to improve understanding of accessibility in relation to the stages of urban evolution.

The third part shows that extensions in the analysis of accessibility do not result only from *phenomenological* issues associated with the transition to a Post-Fordist society. They are also related to a shift in analysis within the mainstream of quantitative geography. Implications of this new approach to the understanding of accessibility are illustrated.

The last part focuses on the formulation of some of the issues likely to be given priority in future research on relationships between NIT, urban evolution and accessibility. In this connection, a research project on accessibility in the Turin Metropolitan Area of Italy is described.

17.2 The Appealing Concept of Accessibility

Despite being an intuitive concept in everyday language as well as in geography, accessibility is difficult to translate into a single meaningful notion. One reason is related to the ontology of the concept, which is quite unique in geography. In fact, it does not refer to any clearly defined physical, social or economic entity, but to an hybrid *fuzzy* entity that shares features of the two fundamental components of any spatial system: the spatio-temporal pattern of activities and the spatio-functional pattern of interdependencies. In providing a bridge between these two components, the concept of accessibility acts as a *junction* between them[1].

Underlying any notion of accessibility is:

(1) An urban product--a bundle of activities, services or places--that is spatially distributed in a city or a region;
(2) A demand or need for the urban product. People and organizations are therefore motivated to gain access to a range of urban products, whose enjoyment is recognized as bringing some benefits;

---

[1] From a different point of view, the *junction* role of accessibility has already been recognized in *New Urban Economics*, where accessibility has been interpreted as an externality, thus sharing the properties of a *public good* (see Papageorgiou 1987).

(3) The effort necessary to reach an urban product to engage in an activity in a certain place at a given time. This effort may be seen as the monetary, temporal, travel, or psychological costs borne in order to have access to the product.
(4) A set of constraints, which depend on the behavior of the individual, usually associated with personal resources and household responsibilities, and on the opportunity pattern existing in a urban setting (i.e., the spatio-temporal ordering of urban activities, differentiated by type, location, characteristics, and form of organization).

Accessibility can thus be considered as the outcome and, at the same time, a component of the interaction process underlying the functioning of urban systems. It is intuitively evident, *ceteribus paris*, that the greater the importance given to an urban product and the lower the constraints on its enjoyment, the higher its accessibility. On the other hand, a high level of accessibility is likely to enhance the importance of an urban product and stimulate the demand, thus increasing its potential for interaction (i.e., the level of flows to and from the area where the product is located, and the number of contacts between people).

Empirically, the '*junction role* of accessibility has long been recognized in land-use and transportation analysis. Several measures of accessibility have been formulated and provide meaningful links within and between residence- and organization-based urban sectors. These relate the various activities available to individuals and organizations with the systems of transport and communications, which allows individuals to overcome the impeding distances and to participate in specific activities (see Hansen 1959, Wilson 1971, Morris Dumble and Wigan 1979, Leonardi 1979, Wachs and Koenig 1979, Koenig 1980, Hanson 1984, Hanson and Schwab 1987).

More recently, this intrinsic junction feature of the concept has been exploited, and measures of accessibility are explicitly included in the *performance indicators* developed for the analysis and evaluation of spatial structures (Clarke and Wilson 1987a, 1987b, Bertuglia, Clarke, and Wilson 1994). The junction role of the concept is, however, also responsible for certain elusiveness underlying the notion of accessibility. It explains the ambiguity, which often accompanies the common use of the term, associated primarily with travel-derived demand. Thus, accessibility is used to mean both the proximity and *ease* of interaction and the *possibility* of interaction.

As far as the former is concerned, its application in the field of transportation and land-use planning has generally involved an interpretation connected in some way with the spatial separation of human activities, usually expressed as a physical distance or a time. Time as *a locational and co-locational continuum* thus represents a basic dimension of accessibility (Carlstein, Parkes, and Thrift 1978).

In relation to the latter, the role of time associated with spatial separation makes it possible to relate the notion of accessibility to the *field of choices* of interaction available to an individual (Weibull 1980). The extension of this field depends on the individual's capacity and resources, and also on the patterns of opportunity he or she possesses in relation to the mix of activities, and their functional and spatial organization. Some typical factors characterizing the pattern of opportunities are the spacing of settlements, the range of urban activities offered, the modes of transport available, and the journey time.

An important consequence of this latter interpretation is that accessibility is not seen as an entity *per se*, but a property that an individual can take advantage of, given a certain pattern of opportunities. As this depends on the functional and spatial organization of human activities, the evolutionary path of an urban system conditions accessibility. Accessibility is, therefore, linked with the dynamics of cities.

The most direct impact of the new information technologies (NIT) is on the junction role of accessibility. This makes the concept of accessibility appealing for investigating the impacts of NIT adoption on city and spatial systems. This also explains why accessibility indicators are valuable tools for assessing the overall effects of NIT on both the spatio-temporal pattern of activities and the spatio-functional pattern of interdependencies. Not only has there recently been a new impetus to the application of accessibility notions in empirical analysis (see Helling, 1998, Bruinsma and Rietveld, 1998, Kwan, 1998), but the widespread diffusion of NIT is giving new scope for examining the junction role of accessibility.

Although they are difficult to identify the effects of the impact of NIT on accessibility are both substantive and methodological. On substantive grounds, one recently identified feature of NIT (and of any innovation; see Bertuglia and Occelli 1997) is related to its *hard* and a *soft* components, corresponding respectively to tangible and intangible parts. The former consists of the content of innovation itself (e.g., the network, information production, and communication); the latter is the *flux of change* in the interaction patterns – various forms of re-organization and the reshaping of activities at both individual and collective levels that result from its application.

Two features of NIT, which are particularly relevant for accessibility, are:

(1) The *time-space shrinking* potential: by substituting physical movement (travel) with virtual interaction (see point 3 above), NIT can affect the effort needed to reach an urban product; and
(2) The *enabling* potential: by allowing more flexibility in carrying out certain work and domestic responsibilities (or tasks), as well as in the provision of firm and household services, NIT can help relax the *set of constraints* an individual faces in everyday life (see point 4 above). In addition, NIT can also modify the *urban products* available in an urban setting by improving its relative competitiveness (see point 1 above).

While making the junction role of accessibility more complex, this endows it with new potentialities. Accessibility is no longer just a *property* to take advantage of, but a *resource* that a city as, a collective social entity, makes available[2]. It is the increased potential in its junction role that makes accessibility so relevant in the new information society. Although not yet fully recognized, this is foreshadowed in a number of recent studies where accessibility is referred to as a meaningful analytical notion for investigating the impact of innovation on the functional and spatial organization of urban activities. For example, Kobayashi, Sunao and Yoshikawa (1993), use an accessibility representation to describe the interdependencies created by knowledge exchanges. Martellato (1993) considers transport and telematic accessibility as distinct inputs to the production function of firms. He analyzes how changes in the price of such services impact on the locations of firms. Bertuglia, Lombardo and Occelli (1995) develop a model for simulating urban scenarios in which the location choices of firms depend on both transport and telematic accessibility.

On methodological grounds, acknowledgement of the hard and soft components of NIT changes our way of conceiving *accessibility* and hence its role in relation to Post-Fordist urban development. As discussed later in this chapter, there can be no single notion of accessibility. There will be many different definitions and measurements of accessibility (see IRES 1995, Kwan 1998 and Helling 1998). Depending on how it is represented, the differing knowledge of accessibility held by different individuals mirrors the complexity of the concept[3].

17.3 Accessibility and the City

The implications associated with the junction role of accessibility have never been fully explored in the literature. Nonetheless, they have inspired a number of definitions, which have shed valuable light on many urban phenomena and spatial

---

[2] It has been posited (Bertuglia and Occelli 1997) that this soft component is by far the most important aspect of present-day innovation. It can be related to economically based notions of *knowledge, experience* and *learning*, or to the circuits of communicative interaction that form the social fabric; it can even be associated with the so-called *software* network (including education, the arts, and science) that embellish human infrastructure (Andersson et al. 1993). More importantly, however, by changing interaction patterns, it can affect a whole range of behavioral (i.e., goal seeking, explorative, imitative, hyper-selective) processes. These can act at both the individual and collective level and affect spatial systems in a number of different ways to feed its evolution. One relevant implication is that everyone can be an agent of innovation. By pointing at the ordinary agent - whether an individual, household, organization, or collective society – as repository of innovation, we have a less élitist view.

[3] Acknowledging the individual ordinary agent (person, household or organization) as the main repository of innovation, implies that his/her notions of accessibility are increasingly important - how they are derived from individual agents' representations of their daily activity patterns and how these representations are continuously adjusted, modified, and re-created.

changes. Most of these definitions can be considered *by products* of the theoretical and methodological approaches to urban interactions developed in urban and regional studies over the last thirty years.

Table 17.1. Some definitions of accessibility

Definitions	Approaches
As the potential of interaction opportunities (Hansen 1959)	
Intrinsic characteristic (or advantage) of a place relative to the spatial friction encountered in gaining access (Ingram 1971, Dalvi and Martin 1976)	Physical-deterministic (gravity type approach)
Appreciation of the quality of transport conditions and availability of supply in a place, relative to a certain need (Vickerman 1974)	Economic-functionalist (entropy approach)
The outcome of choice between a set of alternatives (Ben-Akiva and Lerman 1979)	Economic-behavioral (micro-economic approach, random utility theory)
An aspect of the freedom of action of individuals, depending on physical and time constraints (Hägerstrand 1975, Burns 1979)	
An availability created in daily activity, measurable by the effort necessary to generate or maintain it (Pirie 1979)	Spatio-temporal (behavioral approach)
A resource associated with the many webs of interaction established by people and organizations in urban systems	Informational

Table 17.1 lists a few of these definitions and, although not exhaustive, gives an idea of the evolution of the accessibility concept. Also mentioned are the main approaches to urban interactions from which the definitions are derived or to which they can be related.

Occelli (1998a) observes that these definitions have progressed:

(1) from a concept of accessibility in physical-deterministic terms, according to which interaction factors are represented by the mass of the localized activities and by the impedance, which is a function of the physical distance between them;
(2) towards an economic conceptualization, where factors of interaction are the economic opportunities existing in zones and the impedance is a cost, expressible in monetary terms to cover the movement necessary to reach the various zones;
(3) involving a probabilistic-behavioral notion, where accessibility is the outcome of interacting individual behaviors resulting from a process of choice among a set of different alternatives; and
(4) arriving at a conceptualization in which the fundamental essence of accessibility is linked with the type, intensity, and forms of interaction established, or likely to be established, by individuals and organizations

An interesting exercise would be to put the definitions and approaches shown in Table 17.1 in a historical perspective and to link them with the features that mark the urbanization process. For reasons of space, it is not possible to undertake this here, but Table 17.2 gives suggestions for a kind of meta-typology to serve as a reference for future research into the changes that have occurred in urban systems over time.

The labels used (Pre-industrial city, Fordist city, Post-Fordist city) and the three stages of urban evolution described are not new. They are drawn from the existing literature, which has used them extensively to deal with issues of socio-economic growth, and to describe the consequences of growth in terms of spatial patterns, technological innovation, wealth, sustainability, and quality of life (Brotchie 1986, Freeman and Perez 1988, Batten 1995, Bertuglia and Occelli 1995, Brotchie, Anderson, and McNamara 1995).

What is more unusual is that the meta-typology matches a set of commonly recognized socio-economic and institutional features with a set of spatial features. This association relies on the assumption that spatial patterns cannot be fully understood without taking into account the underlying socio-economic and interaction structures. The spatial features relate to the spatio-temporal patterns of activities and to the spatio-functional pattern of interdependencies. They concern (a) the spatial distribution of activities (i.e., the spacing of human settlements and population density) that supports the socio-economic growth of the system, and (b) the type of the interaction (i.e., the number and average length of trips between demand and supply centers) between residence-based and organization-based activities. The bottom part of Table 17.2 indicates those correlates of accessibility that seem most relevant for the various stages of urban evolution. Also mentioned are the major urban issues for the different urban types.

The proposed typology is, of course, highly simplistic and ignores other determinants of urban change or, more importantly, continuities over time. It is however detailed enough to show that, in the evolution of urban systems, an increasing complexification of cities has taken place. This is particularly evident when we contrast the descriptions of Fordist and Post-Fordist types of urban development.

The Fordist City is generally seen as a place of industrial production, whose evolution is driven by the growth of (export oriented) industrial sectors, which in turn trigger the growth of the resident population and services. These services include all urban activities that are oriented to the local market (i.e., the resident population of the city). The spatial interactions consist mainly of flows of employees who travel between the many places of residence to the few places of work. The structure of the city is intrinsically stable and its path of growth is continuous and unvarying. The most relevant determinants of accessibility are associated with the costs of commuter movements and the opportunities within zones (i.e., the range of urban activities and vacant land), and with transportation (e.g., mass transit, road capacity, and speed). By contrast, the Post-Fordist city is seen as an information-based spatial system in which the kind and characteristics of relationships become increasingly important (Rabino and Occelli 1997).

Not only are urban relationships more numerous and varied (by type, time intensity, frequency, and number of actors involved), but they are also evolutive and self-organizing, based on different spatial and temporal scales (the world-wide and local) and interacting with the relationships of other sub-systems (e.g., the environmental and socio-cultural sub-systems, etc.). The most salient feature of the Post-Fordist City is, therefore, the pattern of networking relationships. Not unexpectedly, accessibility has a multi-dimensional nature related to the interaction opportunity (which can be social, functional, physical, and virtual) of individuals and organizations, resulting from their capacity to enter various fields of urban interaction.

From the definitions of accessibility in Table 17.1, none of them are associated with the pre-industrial city. While the definitions rooted in the physical deterministic and economic-functionalist approaches can be related to characteristics of a Fordist city, those derived from the economic-behavioral, spatio-temporal, and informational approaches clearly allude to a Post-Fordist type of urban development. The so-called informational approach, in particular, accommodates the role of the New Information Technologies in our understanding of the present-future continuum and of planning for the future (Miles and Robins 1992).

Tables 17.1 and 17.2 recognize the increasing complexification of cities. That accessibility, too, is becoming more complex is also acknowledged. As noted, the time-space shrinking possibilities and enabling potentials associated with NIT, are major determinants of this complexification.

Table 17.2. A meta-typology of the urban system in an evolutionary perspective

FEATURES	***Pre-industrial city (merchant and agricultural based society)***	***Fordist city***	***Post-Fordist city***
Socio-economic and institutional aspects			
Production sectors	Agriculture, electrical machinery, steel ships	Cars, armaments, consumer durables, petro-chemicals. Mass production	Computers, capital goods, optical fibers telecommunications,
Tertiary sectors	Domestic services, state and local bureaucracies, growth of transportation and distribution	Growth of social and financial services. Decline of domestic services	Expansion of information services. New forms of craft production
Infrastructures	Canals, railways, roads	Electrical cables, highways, airlines, airports	Digital communications network, satellites
Social organization and population trends	Rigid class divisions. Urbanization, high population turnover	Unified class formation and parties. Concentration of population in urban areas	Pluralistic class formations, multi-party system, regional diversification. Counter urbanization and aging of population
Aspects of regimes of regulation	Craft unions and early social legislation	Welfare state and its crises	New-style participatory decentralized welfare state
Spatial aspects			
Settlement pattern and urbanization processes	Isolated, small settlements with stable population	Formation of polarized, high-density agglomerations. Marginalization of peripheral areas	Metropolitanization, edge cities, dispersed, polycentric settlements of various size, network of cities
Type of interaction	One-to-one. Open non connected network	One-to-many. Radial network	Many-to-many. Interconnected network
Determinants of accessibility	Physical distance and transport. Place-based determinant	Cost of movement, centrality, transport means. Person-based determinants	Interaction opportunities, physical vs. virtual interaction. Field-based determinants
Major urban issues	Housing and health conditions	Employment, cost of opportunities, resource allocation , urban growth	Environmental sustainability, quality of life, urban performances, city competition

17.4 *A Million or So* Notions of Accessibility: Towards Socially Agreed Definitions

The discussion so far has provided arguments on the need for revisiting notions of accessibility. However, this need does not result solely from phenomenological issues that mirror the transition to a Post-Fordist society. Increased complexity in the notion of accessibility also depends on a broader shift in the approach to analysis. This is related to several changes in the modeling field (and more generally in quantitative geography) over the past twenty years. Some of these changes, which are also major topics of enquiry in the Varenius Project, have been addressed more extensively elsewhere (see Rabino and Occelli 1997) and are summarized below.

(1) A first source of change stems from the epistemological background. The acknowledgement of limits to rationality and the need to develop a new philosophy for social action has fostered an interest in the cognitive interpretation of modeling. In this approach, the model is a means for hypothesis testing, targeting ill-defined problems and yielding alternative visions of likely futures. This is distinct from the structuralist interpretation of models that seek a more rigorous understanding of the workings of the system. Whereas differences between structuralist and cognitivist interpretations are becoming more noticeable, the complementarily of their roles in dealing with urban phenomena is also more evident. If, for complex systems, such as cities, 'the multiplicity of disciplinary viewpoints and paradigms is the norm' (Batty and Xie 1996, 202), then one major challenge for urban modeling and geographical analysis is how to reconcile the connections between the two interpretations. In this direction, some possibilities lie in:

- the 'identification and judicious description of relatively invariant factors that can be expected to constrain the observable system states' (Couclelis 1984, 322), i.e., a system's historical and structural prior information;
- the possibility, useful from an operational point of view, of defining a number of windows of observation through which certain system features and properties can be described (see Rabino 1996); and
- reference to the concept of virtual system organization, which, in a self-organizing system, can enhance our prescriptive capacity for addressing urban problems (Turoff 1997).

(2) A second source of change is the new technological backcloth resulting from the introduction of NIT and, in particular, from the increasing power of desktop computing. Three related consequences are:

- a shift in the role of computers. Whereas, previously, a computer was simply a computing tool and the level of technological sophistication set the benchmark for the application potential, the computer is now an integral part of spatial analysis – GIS applications, simulation experiments, interactive visualizations, and virtual environments. Most current empirical studies could not even be conceived of without a computing device. Computer environments serve as templates for describing the real world, and for prescribing self-organizing environments;
- the increasing efficiency of operational procedures. Thanks to the advance in information technologies, it is now possible to deal with problems that were impossible or prohibitively costly to tackle earlier. By enabling alternative procedures of implementation through parallel computing and computer vision, an updating of existing approaches is also possible – through re-engineering; and
- enhancement of communication possibilities. This results from the improvement of both user and method interfaces, as well as from the diffusion of computing potentialities and communication devices among a wider and more diversified public. As far as modeling is concerned, the possibility of carrying out experiments locally broadens the scope of spatial analysis. In this respect, model applications providing the means for sharing experience gained elsewhere offer essential support in learning how to learn about urban processes.

(3) A final source of change concerns the socio-cultural context. As the cultural and information levels of society as a whole are rising, the socio-cultural context is becoming more demanding and selective in the kind of knowledge expected (Knight 1995). First, a new awareness is emerging about the multiplicity of processes that combine to produce overall changes in urban systems. The conceptual unity of the urban system is no longer considered an axiomatic entity. Rather, it results from a continuous re-definition that gives emphasis to the interplay of knowledge-driven actions (behavior) of a variety of actors. Planning questions also need to be put in a new perspective. For instance, besides the need to disentangle the key questions to be answered by policies, new needs are emerging for devising pro-active policies to anticipate problems[4]

---

[4] A different way to conceive relationships between the observer (the analyst) and reality (the modeled system) is advocated. The urban modeler is part of the observed reality and, as such, he or she is an agent of urban change like any other agent. Furthermore, his role as a *maven* no longer holds. On the one hand, as a *problem-solver* his role does not differ from that of any other practitioner. On the other, as a *problem-definer,* the role is likely to be enhanced. Modeling activity (in both the cognitivist and structuralist domains) makes it possible to set up *intelligent interfaces* based on NIT that favor the communication of the various *system descriptions* and to arrive at a collective shared description. This idea follows arguments in policy analysis and management science that point to the need to link

To give an idea of the likely consequences of these changes in the analysis of accessibility, Figure 17.1 shows how two different approaches to understanding accessibility compare. The diagrams are inspired by arguments discussed in Turoff 1997. The first, called the conventional approach, illustrates the traditional view of accessibility, with its roots in the positivistic assumptions held in mainstream social sciences. The second, referred to cautiously as beyond the conventional approach, reflects the conceptual shift alluded to above, and gives primary emphasis to the role of modeling. In both approaches the four following elements are emphasized, which correspond to the main aspects of abstraction the modeling process:

- observable components or determinants of accessibility;
- theories or metaphors – the abstractions that steer our understanding of models of accessibility;
- mental models of accessibility – an individual's representations and tacit knowledge of accessibility, which ultimately steer his/her day-to-day interaction patterns; and
- system models – the shared descriptions or explicit knowledge of accessibility that is understood by communicating individuals.

The circular links between these elements (indicated with Arabic numbers in Figure 17.1) relate to the process of abstraction underlying any approach to accessibility. The cross-links, labeled 'validation and experimentation' refer explicitly to specific features of the modeling activity. Although the methodological underpinnings of the contents of Figure 17.1 require more detailed discussion, this presentation addresses only the main differences between the two approaches and their likely implications in extending the accessibility concept.

- A first set of differences concerns the role of theories. In the conventional approach theories are the depository of the abstractions of reality –scientific truths that guide the formation of individuals' mental models. In contrast, in the unconventional approach theories have a more pragmatic role, being a means of validating models. Second, the role of 'system models' is modified in the unconventional approach to account for the 'cognitivist orientation' in the modeling process. A major difference in this respect can be seen in the role of simulation, as implied in the *What if?* enquiry. In the conventional approach the what was the main focus of attention. However, in the unconventional approach the if becomes the major focus of interest. In this approach, the possibility of exploring alternative courses of action, adjusting them, and assessing their viability is crucial for achieving shared descriptions of acces-

---

action and organizations more effectively to enhance the capacity of local planning institutions (see Bennett and McCoshan 1993, Bryson and Crosby 1998).

sibility models. Third, the 'observable' is also different in the two approaches. In the conventional approach, accessibility is assumed to be physically determined by the spatial and functional properties of the system. However, in the unconventional approach, accessibility is the outcome of a multiplicity of changing individual perceptions – collectively shared and evolving entities that may appear as policy issues. Finally, as far as the mental models are concerned, 'awareness and consciousness' are features in the unconventional approach, which seem unaccounted for in the conventional approach.

- A second set of differences lies in the links. It is assumed that encoding of the understanding process begins with an 'observation activity'. As shown in Figure 17.1, the analytical path moves clockwise in the conventional approach (observables → theories → mental models → system models) and anti-clockwise in the unconventional approach (observables → system models → mental models → theories). In then unconventional approach greater emphasis is given to the shared descriptions of models (the so-called negotiated accessibility) that result from an analytical activity (experimentation) to connect the representations that underlie 'negotiated accessibility' and individual mental models.
- A last point relates to transformation in the activities of experimentation and validation in the unconventional approach. Experimentation becomes an activity aimed primarily at seeking agreement between the representations of accessibility held by individuals in a changing environment and the 'notions of accessibility' that have been agreed upon. Validation does not necessarily need 'physical accessibility' to be carried out. It relies on a kind of pseudo-scientific process based on the relationships established between 'negotiated accessibility', theories and system models, in which societal issues (e.g., equity, sustainability, and quality of life) set the fundamental yardstick.

Figure 17.1 posits that, even from a strictly methodological point of view, accessibility is a manifold concept. Three distinct analytical dimensions for representing the notion of accessibility are suggested:

- *representations of individuals' and organizations' interaction patterns*, given system conditions that include economic, social, spatial, cultural and institutional context as well as personal and organizational constraints);
- *systemic* representations of accessibility for the city system as a geographical, social and collective entity are the kinds of representations associated usually in a standard GIS approach); and
- *policy* representations of accessibility from perspectives of goals, norms, and policy measures for 'prescribed courses of action' related to societal issues.

A. Conventional approach to the analysis of accessibility

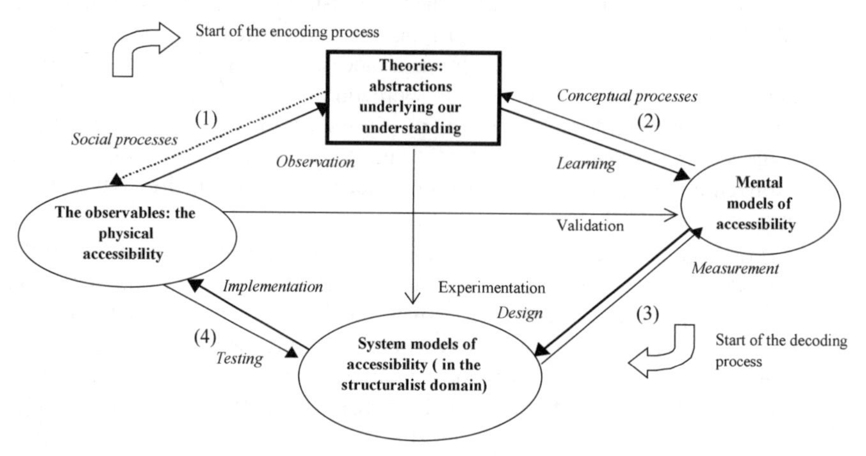

B. Beyond the conventional approach to the analysis of accessibility

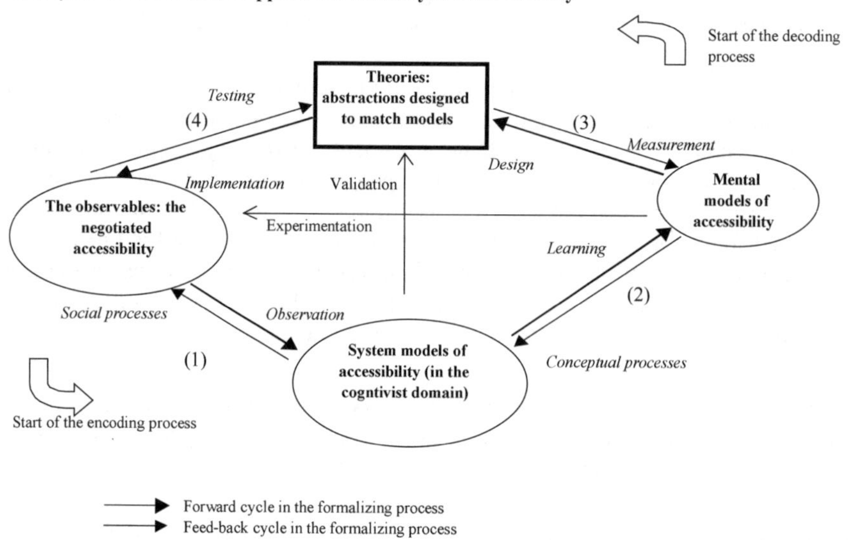

Figure 17.1. Approaches to the analysis of accessibility (adapted from Turoff 1997)

While these three representations of accessibility do not coincide, yet none of them can exist without the others. Furthermore, because of the intrinsic 'complexity' of the accessibility concept (see Casti 1984), they cannot be derived from each other in a simple way. Conventional notions of accessibility fail to recognize this distinction. Furthermore, this drawback introduces serious biases in the formulation of policy measures for accessibility, with the risk of negating from the outset the expected benefits.

From a methodological standpoint, there is a role for NIT (particularly for its 'soft' component) in creating, updating, and innovating these threefold representations of accessibility. As suggested elsewhere (see Bertuglia and Occelli 1997), NIT's effectiveness is strictly associated with the awareness (perception, information, and knowledge) of innovation potential (i.e., the technological self-referentiality mentioned by Sui (Chapter 7). This means that the expected impacts of NIT on these representations are likely to have no lesser role than the appearance of NIT itself (i.e., the hard component of NIT).

A major question concerns the potential of NIT in making it possible, and easier, to link the representations with different analytical dimensions. This extricates us from a discussion of complexity criteria (i.e., by improving our evaluation of the unpredictability and discontinuity of urban changes) and moves our attention to aspects likely to have many practical implications, such as those regarding the 'reducibility question'. To deal with this question, we make two assumptions:

- Knowledge about accessibility can be (socially) increased or improved by moving between representations obtained for various analytical dimensions.
- The possibility of 'informational convergence' of the representations cannot be excluded. As argued, this implies re-establishing the connections between the 'structuralist' and 'cognitivist' approaches to geographical analysis. This acknowledges the existence of 'historical and structural prior information' and 'windows of observation' of certain system features, and recognizes the possibility of setting up a virtual system to enhance our prescriptive capacities for the formation of urban futures.

In the creation of various representations, NIT can improve knowledge of accessibility through use of analytical tools, such as indicators, models, and visual representations (see Occelli 1998b). Because NIT contains hard and soft components, both the rationalizing and creative components underlying any of these tools will be deeply affected. This entails taking into account two fundamental and interrelated perspectives, which in the recent debate about accessibility have been overlooked. These are:

- **an evaluative perspective**. This is necessary to make it possible to carry out the 'informational convergence' of the representations of accessibility. We should be able to: (a) compare our representations of accessibility according to both diagnostic and socially relevant aims and (b) assess the representations in relation to given societal targets that are agreed upon or prescribed (see Toulemonde 1995);
- **a planning perspective**. As accessibility is a resource to be exploited, maintained, and regenerated, we need to recognize it as a distinct policy issue that does not depend exclusively on transportation. The intrinsic multiplicity of accessibility concepts therefore needs to be matched by a corresponding 'set' of policy measures. This recalls what is posited in Ashby's principle of 'req-

uisite variety', often mentioned in the literature (see Friend and Jessop 1969, Casti 1986, Batty 1995).

17.5 Concluding Remarks

In this paper the importance of the 'junction role' of accessibility has been emphasized, bridging the spatio-temporal and spatio-functional component of spatial systems. To cope with the space-time changes occurring in the transition to a Post-Fordist type of development, it is suggested that this junction role should be revisited and our notion of accessibility extended. An effort has been made to justify this extension on conceptual, phenomenological, and methodological grounds. The role of NIT has also been addressed. From all these points of view, the increasing complexity of the concept of accessibility has emerged as a main justification for its extension.

This does not mean that the conventional notions of accessibility no longer apply. It simply reflects the fact that accessibility is an intrinsically manifold notion, encompassing definitions that co-exist, but are not reducible to each other. In particular, three distinct analytical levels were mentioned as having relevance for representing accessibility – the individual, systemic, and policy levels. Although this may appear a trivial result on speculative grounds, it is certainly not simple from a policy point of view. Most definitions of accessibility currently used do not consider this distinction and usually assume that the same indicator of accessibility can be applied and have the same meaning in very different planning contexts.

In dealing with methodological aspects, it was emphasized that the many different notions of accessibility also depend on the kind of representation we have. It was argued that NIT could have a substantial role in the formation and updating of representations and it was suggested that NIT, in particular its soft component, might be helpful in allowing 'informational convergence' between the different representations of accessibility, thus providing a bridge between the various analytical levels.

Recent research suggests a number of questions and issues that need consideration (see Couclelis 1996, Handy and Niemeier 1997, Helling 1997). Building upon suggestions in Occelli (1998a), the following issues and questions are identified for future research:

(1) The first relates to the difficulty in identifying the appropriate level of definition for a given representation of accessibility. The problems are both conceptual and empirical. Conceptually, the need to formulate an appropriate time-space frame of reference has been pointed out (see Holly 1978). Also, the kind of description of the 'effort necessary to get at an urban product' at the individual and systemic levels can be quite different as a consequence of the ways distance and travel impedance are incorporated conceptually and em-

pirically (see Couclelis 1996, Handy and Niemeier 1997). Morris, Dumble and Wigan (1979) identified empirical problems in constructing of indicators whenever the two temporal dimensions of accessibility (the locational and co-locational continuum and the 'ladder' of time are taken into account. Do we wish to examine the opportunities available (through process indicators) of what is potentially offered for different types of individuals, independently of their observed behavior? Or, do we wish to focus on the properties of outcome indicators, such as observed journeys to activities, travel times for journeys-to-work or for leisure, and so forth? What should be investigated by means of process indicators at one spatio-temporal scale, however, would require outcome indicators at another. Different spatio-temporal scales reveal different issues, which can have contrasting implications for policy. A crucial point in this respect is the management or control of the so-called 'border effects' – variations in accessibility between different time-space levels. For example, improving regional accessibility in an international context will not necessarily have positive effects on all local areas. Conversely, higher accessibility levels in a local area do not necessarily guarantee the improvement of its connections with regional or international markets.

(2) A second set of questions is raised by two major features of 'Post-Fordist' type urban development, namely the growing importance of interactions (the demand for mobility and communication) and the effects that such interactions may have on sustainability and the quality of life. An important implication is that any definition of accessibility should be accompanied by an evaluation of the associated benefits accruing to the individual, an organization, or the city as a whole. In particular, considering accessibility as a resource implies that attention should be paid to:

- the type and quality of urban products, such as the various kinds of activity relative to their temporal organization and location. This means that 'what' is to be accessed and also 'how' are relevant in determining the benefits. This in turn raises questions relating to scarcity and efficiency in the provision of a range of urban services, as well as the co-ordination of the different activities (e.g., the opening times of services and transportation availability);
- the ways of overcoming spatial separation, such as the kinds of communication links involved in various human interactions. In this connection, attention needs to be given to the 'value' of time associated with the communication links, integrating the analysis of both cognitive and practical aspects of valuations;
- the kind of trade-offs likely to be involved: the positive effects at the individual level and any negative externalities at more aggregate levels (e.g., increased traffic congestion in some areas of the city). One further implication, particularly relevant in planning, concerns the relationships between accessibility, mobility, urban form (the spatial distribution of activities and patterns of land use), and the environment. A less myopic

view is required, and an assessment should be made of accessibility changes that result from growth in the Fordist-type city. This could lead to a revision of the mechanisms with which accessibility is provided in relation to the spatial expansion of settlements and the daily engagement in urban activities (e.g., the possibility offered by the new communications technologies for substituting certain trips by other forms of interaction).

(3) A further set of issues concern the representation of accessibility and the recognition that the kind of 'knowledge' held by individuals and decision-makers about accessibility is a fundamental determinant in the use of the accessibility resource, as well as in its preservation and regeneration. In the current transition to a Post-Fordist urban development, there is a risk that gaps in the different representations of accessibility (particularly between those of the general public and decision-makers) might exacerbate accessibility needs, raising problems of equity and social justice. Improving information about accessibility, therefore, should be an essential component of any policy strategy, since it can improve not only accessibility, but also social equity.

This last concern is a major focus of the IRES survey. The purpose of the project is to answer two main questions which, due to the provisions set out in a national law on local government enacted in 1992, are also relevant from a policy point of view: (1) how is accessibility perceived by the residents in metropolitan area? (For instance, what knowledge do people have of their accessibility and how do they value it?); and (2) what is likely to be the social acceptability of alternative bundles of accessibility measures? The underlying thesis of the project is that knowledge about accessibility, and the representations that individuals and decision-makers have of accessibility, is likely to be no less significant than the accessibility policy measures themselves.

Building upon observations made previously, Figure 17.2 shows a conceptual framework for defining the questionnaire[5]. Two main levels of definition of accessibility are considered: (1) a systemic level, and (2) an individual level. Accessibility at the systemic level reflects the functional and spatial organization of the activities provided by various transport agencies and service authorities (the mix of opportunities, travel times, transport services, opening

[5] The IRES survey is a pilot survey for a larger research project on accessibility in the Turin Metropolitan Area. The survey was conducted in November and December 1998 by means of home interviews. The individual is used as the unit of analysis. About 400 persons were interviewed. Due to resource constraints, the study area was limited to the western sector of the MA. The questionnaire was divided into three sections. Section 1 covers basic information about the individual and his/her family. Section 2 covers information about the action space of the individual in his/her urban environment. The respondent was also asked for an assessment of this action space. Section 3 requested an evaluation of a set of alternative measures that could be introduced to improve accessibility.

times of services, etc.) observed at an aggregate level. Accessibility at the individual level considers the action-space within which an individual (a person belonging to a household or a collective actor belonging to an organization) currently operates. Besides recognizing the existence of a multiplicity of representations of accessibility, co-existing at both individual and aggregate levels, the diagram emphasizes their changing nature, resulting from a process of learning. Emphasizing the notion of accessibility as a resource means that the information made available not only modifies the individual's action-space, but can also make it possible to achieve some social improvements in accessibility.

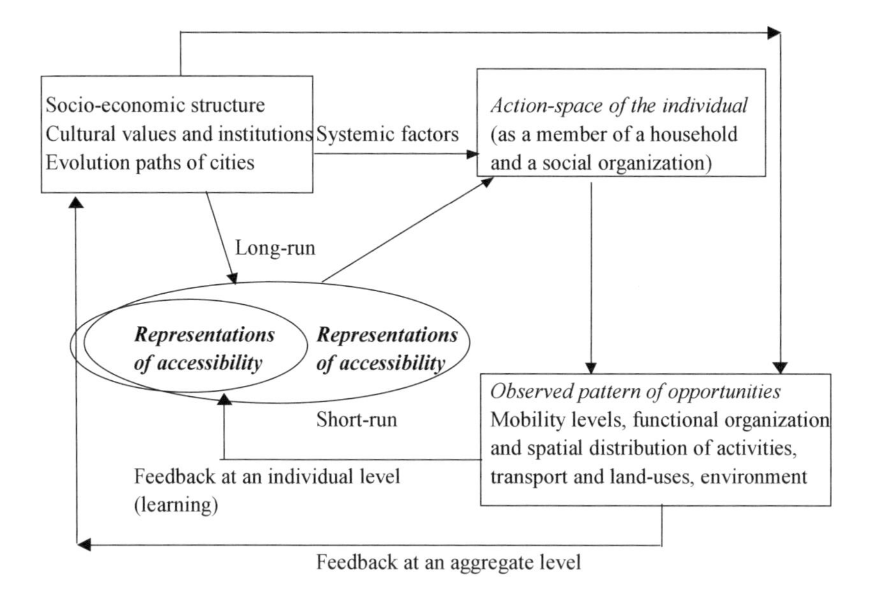

Figure 17.2. Conceptual framework for the IRES survey on accessibility

(4) The last question relates to the representation of accessibility and its measurement. Experience so far in the definition of accessibility measures suggests that an approach to their formulation should (see IRES 1995, Handy and Niemeier 1997):

- incorporate descriptors that reflect changes in the performance of the spatio-temporal and spatio-functional component of spatial systems (e.g., the possibility of substituting physical with telematic interactions);
- have a sound basis in theories of behavior of individuals and organizations;

- be technically feasible and operatively simple; and
- be easy to interpret, especially for decision-makers and the general public.

Experience indicates, however, that these requirements may be in conflict. The need, for example, to define measurements that are theoretically acceptable can be made impracticable by the lack of adequate information or perhaps cause greater difficulty of interpretation. Furthermore, the kind of accessibility extensions called for in this paper (and implied in the 'unconventional approach' described in Figure 17.1) contrast the desire for a more comprehensive holistic approach and the desire for a more focused one. While this makes the possibility of developing 'one single behaviorally-based, policy-relevant and socially-agreed measure' seem less feasible, it emphasizes the need for a frame of reference within which a variety of accessibility measures can be developed and compared. In this connection, it is reasonable to ask whether NIT will provide the appropriate 'environment'. So far, great promise and enthusiasm has been raised by the 'hard' component of NIT. The soft' component, however, is still largely unexplored and poses an important challenge for the analysis of accessibility.

References

Amin, A. (ed.) 1994. *Post-Fordism.* Oxford: Blackwell.

Andersson, A.E., Batten. D.F., Kobayashi. K. and Yoshikawa, K. (eds.) 1993. *The Cosmo-Creative Society,* Berlin: Springer.

Batten, D.F. 1995. Network cities: Creative urban agglomeration for the '1st Century. *Urban Studies* 32 (2) 313-27.

Batty, M. 1995. Cities and complexity. Implications for modeling sustainability. In Brotchie, J. Batty, M., Blakely, E., Hall, P. and Newton, P. (eds.) *Cities in Competition. Productive and Sustainable Cities for the 21st Century.* Melbourne: Longman Australia, 469-86.

Batty, M. and Xie, Y. 1996. Possible urban automata. In Besussi, E. and Cecchini, A. (eds.) *Artificial Words and Urban Studies*, Venezia: Daest, Convegni, no.1, 176-205.

Ben-Akiva, M. and Lerman, R. 1979. Disaggregate travel and mobility: Choice models and easures of accessibility. In Hensher, D.A. and Stopher, P.R. (eds.) *Behavioural Travel Modeling*, London: Croom Helm, 654-79.

Bennett, R.J. and McCoshan, A. 1993. *Enterprise and Human Resource Development.* Liverpool: Chapman.

Bertuglia, C.S., Lombardo, S. and Occelli, S. 1995. The interacting choice processes of innovation, location and mobility: A compartmental approach. In Bertuglia, C.S., Fischer, M.M. and Preto, G. (eds.) *Technological Change, Economic Development and Space*, Berlin: Springer, 118-44.

Bertuglia, C.S., and Occelli, S. 1995. Transportation, communications and patterns of location. In Bertuglia, C.S., Fischer, M.M. and Preto, G. (eds.) *Technological Change, Economic Development and Space,* Berlin: Springer, 92-117.

Bertuglia, C.S., and Occelli, S. 1997. The impact of the new communications technologies on economic-spatial systems. An agenda for future research, Paper presented at the International Seminar *The Impact of the New Communications Technologies on Economic-Spatial Systems*. Pisa: September, 12-13.

Bertuglia, C.S., Clarke, G.P., Wilson, A.G. (eds.) 1994. *Modeling the City: Performance, Policy and Planning*, London: Routledge.

Brotchie, J.F. 1986. Technological change and urban form, *Environment and Planning A* 16:583-96.

Brotchie, J.F., Anderson, M. and McNamara, C. 1995. Changing metropolitan commuting pattern. In Brotchie, J., Batty, M., Blakely, E., Hall P. and Newton, P. (eds.) *Cities in Competition. Productive and Sustainable Cities for the 21st Century*. Melbourne: Longman Australia, 382-401.

Bruinsma, K. and Rietveld, P. 1998. The accessibility of European cities: Theoretical framework and comparison of approaches, *Environment and Planning A* 30:499-521.

Bryson, M.J. and Crosby, B.C. 1998. Policy planning and the design and use of forums, arenas and courts, *Environment and Planning B* 20:175-94.

Burns, L.D. 1979. *Transportation, Temporal, and Spatial Components of Accessibility*. Lexington MA: Lexington Books.

Carlstein, T., Parkes, D. and Thrift, N. 1978. Introduction, in T. Carlstein. In Parkes, D. and Thrift, N. (eds.) *Time and Regional Dynamics*. London: Edward Arnold, 1-4.

Clarke, G.P.and Wilson, A.G. 1987a. Performance indicators and model-based planning: 1. The indicator movement and the possibilities for urban planning, *Sistemi Urbani* 9:79-127.

Clarke, G.P. and Wilson, A.G. 1987b. Performance indicators and model-based planning: 2. Model-based approaches, *Sistemi Urbani* 9:137-69.

Castells, M. 1989. *The Informational City*. Cambridge: Blackwell.

Casti, J.L. 1984. On the theory of models and the modeling of natural phenomena. In Bahrenberg, G., Fischer, M.M. and Nijkamp, P. (eds.), *Recent Developments in Spatial Data Analysis*, Gower, Aldershot, 73-92.

Casti, J.L. 1986. On system complexity: identification, measurement, and management. In Casti, J.L., and Karlqvist, A. (eds.) *Complexity, Language and Life: Mathematical Approaches*, Berlin: Springer, 146-73.

Couclelis, H. 1984. The notion of prior structure in urban modeling, *Environment and Planning A* 16:319-38.

Couclelis, H. 1996. From cellular automata to urban models: new principles for model development and implementation. In Besussi, E. and Cecchini, A. (eds.) *Artificial Worlds and Urban Studies*. Venezia: Daest, Convegni n.1, 165-75.

Couclelis, H. (ed.) 1996. *Spatial Technologies, Geographic Information, and the City*, Technical Report 96-10. Santa Barbara, CA: NCGIA,:

Dalvi, M.Q. and Martin, K.M. 1976. The measurement of accessibility: some preliminary results, *Transportation* 5:17-42.

Freeman, C. and Perez, C. 1988. Structural crises of adjustment: business cycles and investment behaviour. In Dosi, G., Freeman, C., Nelson, R., Silverberg, G. and Soete, L. (eds.) *Technical Change and Economic Theory*. London: Pinter, 38-66.

Friend, J.K, and Jessop, W.N. 1969. *Local Government and* Strategic *Choice*. London: Tavistock.

Graham, S. and Marvin, S. 1996. *Telecommunications and the City*. London: Routledge.

Hägerstrand, T. 1975. Space time and human condition. In Karlkvist, A., Lundqvist, L and Snickars, F. (eds.) *Dynamic Allocation of Urban Space*. Farnborough Saxon House, 3-12.

Handy, S.L. and Niemeier, D.A. 1997. Measuring accessibility: an exploration of issues and alternatives. *Environment and Planning A* 29:1175-94.

Hansen, W.G. 1959 How accessibility shapes land use. *Journal of the American Institute of Planners* 25:73-6.

Hanson, S. 1984. Environmental cognition and travel behaviour. In Herbert, D.T. and Jonhston, R.J. (eds.) *Geography and the Urban Environment*, New York: Wiley, vol. VI, 95-126.

Hanson, S. and Schwab, M. 1987. Accessibility and intraurban travel, *Environment and Planning A* 19:735-48.

Helling, A. 1998. Changing intra-metropolitan accessibility in the U.S.: Evidence from Atlanta, Diamond, D. and Massam, B.H. (eds.) *Progress in Planning*, London: Pergamon, 55-107.

Holly, B.P. 1978. The problem of scale in time-space research. In Carlstein, T., Parkes, D. and Thrift, N. (eds.) *Time and Regional Dynamics.* London: Edward Arnold, 5-18.

Ingram, R.D. 1971. The concept of accessibility, *Regional Studies* 5:101-07.

IRES 1995. Un'analisi dell'accessibilità in Piemonte. Studio di supporto alla valutazione delle politiche del piano regionale dei trasporti. Occelli, S.and Gallino, T. (eds.) Turin: Quaderni di Ricerca Ires, 74.

Knight, R.V. 1995. Knowledge-based development: policy and planning implications for cities, *Urban Studies* 32:225-60.

Kobayashi, K., Sunao, S. and.Yoshikawa, K. 1993. Spatial equilibria of knowledge production with 'meeting-facilities.' In.Andersson, A.E., Batten, D.F., Kobayashi K. and Yoshikawa, K. (eds.) *The Cosmo-Creative Society,* Berlin: Springer, 219-44.

Koenig, J.G. 1980. Indicators of urban accessibility: Theory and application. *Transportation* 9:145-72.

Kwan, M-P. 1998. Space-time and integral measures of individual accessibility: a comparative analysis using a a point-based framework. *Geographical Analysis* 30(3):191-216.

Leonardi, G. 1979. Introduzione alla teoria dell'accessibilità, *Sistemi Urbani* 1:65-88.

Martellato, D. 1993. Reti di interazione e tecnologia della informazione. Un'analisi dei loro effetti sulla localizzazione. In Lombardo, S. and Preto, G. (eds.) *Innovazione e trasformazioni della città. Teorie, metodi e programmi per il mutamento*, Milano: Collana AiS-Rre, Angeli, 130-47.

Miles, I., and Robins, K. 1992. Making sense of information, in Robins, K. (ed.) *Understanding Information, Business, Technology and Geography*, London: Belhaven, 1-26.

Morris, J.M., Dumble, P.L. and Wigan, A.R. 1979. Accessibility indicators for transport planning. *Transportation Research* 13A:91-109.

Papageorgiou, G. 1987. Spatial public goods, 1: Theory, *Environment and Planning A* 19:471-92.

Occelli, S. 1998a. Accessibility and time use in a post-Fordist urban system. Some notes for a research agenda, Paper presented at The International Conference on Time Use, Luneberg, 22-25April.

Occelli, S. 1998b. A New Perspective for methodologies in spatial and urban analysis. In Bertuglia, C.S., Bianchi, G. and Mela, A. (eds.) *The City and its Sciences*, Berlin: Springer 851-72.

Pirie, G.H. 1979. Measuring accessibility: A review and Proposal, *Environment and Planning A* 11:299-312.

Rabino, G.A. 1996. Complessità, scienza della complessità e modello dei trasporti, *Le Strade.*

Rabino, G.A. and Occelli, S. 1997. Re-thinking urban system modeling: New features and a proposal, Paper presented at the ICCS Conference, Cortona, 22-25 September.

Toulemonde, J. 1995. Should evaluation be freed from its causal links, *Evaluation in Program Planning* 1(1):179-90.

Turoff, M. (1997) Virtuality, *Communications 40(9):*38-43.

Vickerman, R.W. 1974. Accessibility attraction and potential: A review of some concepts and their use in determining mobility, *Environment and Planning A* 6:675-91.

Wachs, M.and Koenig, J.G. 1979. Behavioral modeling, accessibility, mobility. In Hensher, D.A. and Stopher, P.R. (eds.) *Behavioural Travel Modeling.* London: Croom Helm, 698-710.

Weibull, J.W. 1980. On the numerical measurement of accessibility. *Environment and Planning A* 12:53-67.

Wilson, A.G. 1971. A family of spatial interaction models and associated developments. *Environment and Planning A* 3:1-32.

18 Legal Access to Geographic Information: Measuring Losses or Developing Responses?

Harlan J. Onsrud

Department of Spatial Information Science and Engineering, University of Maine, Orono ME 04469-5711, USA. Email: onsrud@spatial.maine.edu

18.1 Introduction

A major proposition prevalent in the geographic information research community is that better models, tools and techniques are needed to measure and represent changes in access as more and more of the interactions and transactions of our daily lives occur electronically. Measures or representations from these tools and techniques will purportedly help us identify the winners and losers in society as we move to electronic social interaction environments. However, the losers quite often are obvious and a focus on measuring losses in such situations seems misplaced when energies might be better spent on lessening or reversing such losses. Further, measurement tools are often used by those in power positions in attempts to refute that losses are actually occurring. This is because many of the benefits of access and costs of lack of access (such as missed opportunities) are very difficult to measure or otherwise quantify in a convincing manner. In addition, by focusing on the scientific reliability of tools and measurements, those in power positions often are able to divert attention and energy away from the goals of opponents that otherwise would undermine their control over access.

In many instances, it is far more important for those advocating increased access or more equitable access to identify the processes by which losses in access are occurring, publicize that the losses are occurring, explore alternatives for halting or reversing the losses, and seek solutions for expanding access or providing more equitable access. Measuring, mapping, and modeling the nature and extent of changes in access are important but not if such efforts divert attention from arriving at solutions for enhancing access.

In this chapter the argument is made that the foundations of legal rights of citizens to access information are being undermined as we move into networked digital data environments. As a result, widespread loss of access to information and works of knowledge in U.S. society is occurring. The flood of data provided by emergent technologies is being channeled rapidly through legal mechanisms to provide wealth and power to very limited sectors of society. While citizens may be looking forward to more meaningful dialogue among each other and with government in future electronic environments, they should also be aware that past gains made in the ability to access and build upon the works of others and gains made towards increasing the transparency of government operations are being eroded. Measuring and modeling the who, what, when, and where of such losses isn't as important as understanding the manner in which such losses are occurring and exploring alternatives by which such losses might be avoided.

18.2 Background

The United States is unique in the world in the broad access to information that its laws support. Areas of the law influencing access to information, geographic or otherwise, include intellectual property law, freedom of information law, privacy law, electronic contracting law, and antitrust law, as well as several other areas of the law. Two generalizations about the interoperation of these laws appear germane to the topic of this book.

The First Generalization. *The forms that these laws take in the United States allow greater access to government information at the local, state, and national government levels and use of that information than is generally allowed in other nations.* For instance, few nations have national freedom of information laws that allow citizens broad general access to the public records of government.[1] Even in those nations that do allow such access, citizens are not allowed typically to add value to such information and resell it without the permission of government as they are allowed to do in the United States. In addition, the United States goes much further than other nations since it actually imposes affirmative obligations on federal agencies to actively disseminate their information as defined by the provisions of OMB Circular A-130 (June 1993). Agencies are particularly encouraged to disseminate raw content upon which value-added products may be built by the private sector and to do so at the cost of dissemination, with no imposition of restrictions on the use of the data and through a diversity of channels. The core provisions of OMB Circular A-130 were incorporated into the *Paperwork Reduction Act* of 1995 (PRA) and that act additionally encourages the use of information technologies by agencies for providing public access, rather than relying on cumbersome *Freedom of Information Act* (FOIA) processes. With the expanded use of World Wide Web servers by federal agencies, the cost of dissemination for many

[1] Fifteen nations that have general open government records laws, several of which are recent, are listed at http://www.cfoi.org.uk/foioverseas.html.

federal government data sets has become negligible and, thus, these data sets are now freely available to anyone with the ability to access them over the Internet.[2]

Actions have also been taken at the federal level specifically related to spatial information and agency contributions to building the National Spatial Data Infrastructure (NSDI). The Office of Management and Budget (OMB) established the Federal Geographic Data Committee (FGDC) in its 1990 revision of Circular A-16, *Coordination of Surveying, Mapping, and Related Spatial Data Activities.* FGDC is now composed of representatives from 17 Cabinet level and independent Federal agencies. In April 1994, President Clinton signed Executive Order 12906 that called for the establishment of a coordinated National Spatial Data Infrastructure (NSDI) as part of the evolving National Information Infrastructure (NII), and FGDC was charged with coordinating the federal government's development of the NSDI. In this executive order, FGDC was given a mandate to involve state, local and tribal governments, academia and the private sector in coordinating the development of the NSDI. The roles of various parties and their relationships in moving towards a common NSDI vision are being developed over time.

Similar to the federal situation, open access laws exist in most of our states that impose similar broad principles of access by citizens to the records of state and local government agencies. Because of this atmosphere of openness, many local municipal, county, and state governments have voluntarily been making geographic data sets available on the web for general use by for-profit businesses, not-for-profit organizations, and citizens generally.[3]

Rights of access to government information and the atmosphere of openness that these rights engender are currently very significant. We should not allow these rights to be chipped away at through growing numbers of legislative exceptions for geographic information databases and other government databases.

The Second Generalization. *U.S. law grants individuals greater leeway to use the work products of others without permission than is typically granted by the laws of other nations.* The law of copyright in the United States grants fewer ownership interests in intellectual works and greater access to the work products of others than in perhaps any other industrialized nation. Again, we should not aban-

---

[2] Data sets available from U.S. federal agencies may be traced typically through their official web sites indexed at http://lcweb.loc.gov/global/executive/fed.html. Examples of spatial data sets available from federal agencies include those found through http://www.usgs.gov/themes/info.html, http://www.census.gov/ftp/pub/www/tiger/, http://www.epa.gov/epahome/data, and http://gcmd.gsfc.nasa.gov/.

[3] See http://recorder.maricopa.gov/recorder/imaging/ for an example of open web access to deeds and plats; see http:// www.ci.ontario.ca.us/ for an example of access to community geographic information; or see http://fgdcclearhs.er.usgs.gov/ for state and community clearinghouse nodes.

don this high level of access to the information sources and products produced by others without substantial social benefit reasons for doing so.

The United States has productive scientific and commercial database communities that are the envy of the world. The vitality of these two communities go hand-in-hand. The government, commercial, and not-for-profit sectors have all benefited by the past balance of legal policies that has minimized the need to pay for access and use of data drawn from other government, commercial and not-for-profit sector sources. For-profit private sector creators of derivative databases have gained as much or more from this policy balance than any other societal sector. Although both new innovations and past investments are protected under U.S. law, the current complex balance in the law provides a tension in favor of new innovations over past investments. This keeps competition and the need to innovate high. The level of activity and growth of the database industry in the United States in comparison to other nations is evidence that a flexible ability to draw from and build upon the data sets collected by others is highly desirable for economic development. The strong health and high activity of the U.S. database industry also suggests that there is no crisis at hand that would warrant granting the private commercial sector immediate greater control over the data it collects.

Geographic data, like many other forms of scientific and technical data, possess the classic characteristics of public goods. That is, geographic data are typically *nonrival* and *nonexcludable*. In short, a *nonrival* good is one that may be consumed without detracting in the least from consumption of the same good by others. By example, the use of digital data for finding one's way does not make the same data any less useful to others for finding their way. A *nonexcludable* good is one whose benefits are available to all once the good is provided. Once nonexcludable benefits are provided they may be very difficult or perhaps impossible to exclude from others even though others may not have helped pay for the good. Due to the ability to copy data electronically and transfer it over networks at negligible cost, data made available to small numbers of persons are often transferred to much wider audiences regardless of the contractual provisions stipulated and the technical protection techniques used, such as electronic watermarking or encryption. Geographic data, as well as many other forms of factual data, are additionally nonexcludable in that others may recollect the same or substitutable data. It is because of these public goods characteristics that much geographic data in the United States has been collected in the past by government and access provided to all as a general public benefit.

The commercial sector has little incentive to collect data possessing the characteristics of a public good unless some form of subsidy is provided. Copyright law, by example, is a subsidy established by law that provides an incentive for creators to select and arrange data to make it more useful for society. Any consumer of the original and creative selection or arrangement is required to subsidize the creator (or owner) by paying a higher price than would otherwise be required. Because education and research also have socially desirable outcomes, those copying an authored data set for education and research activities are not required to pay the subsidy under certain circumstances (i.e., see the *fair use* provisions of the *U.S. Copyright Act*). Similarly, any other person or concern may extract factual information from a data set without paying a subsidy or asking for permission as long as that person doesn't copy creative original aspects of the work and as long as that person is not breaching a contract with the owner. Further, the *first sale* doctrine of copyright law (17 U.S.C., section 109) specifically authorizes the owner of a legally acquired copy to *sell or otherwise dispose of* the copy and thus subsequent sales of a copy do not require further subsidy payments to the creator (e.g., resale of a typically purchased book or CD). These examples illustrate that copyright owners have never had the ability to extract the full market value from their works. Protection for copyright owners has been established by the law over time at a level that provides strong incentives to produce works of authorship but not at a level so high that author rights significantly impinge upon society's interest in allowing the public to make reasonable use of authored works. Society benefits more in terms of advancement of science, the arts, and the overall economy if reasonable leeway is provided to allow each of us to draw from and build upon the works of others.

It is noteworthy that the United States gained significant economic strength and dominance in information technologies, as well as in research and technology generally, at a time when its information laws were very different from those of other nations. The role of U.S. laws and policies in supporting an open environment of access to scientific data for the commercial and science sectors and the role of U.S. laws in ensuring access to government data sets should not be overlooked when exploring the competitive success of U.S. businesses and scientists.

18.3 The Diminishment of Legal Access

Observation of recent actions in Congress and legislative actions at the state and local government levels suggest that the nation is back tracking on its openness principles rather than extending them. Some observations of recent lawmaking in action include the following.

(1) Cost Recovery Legislation. Restricting access to public records is contrary to the plain letter language of most state open records laws in the United States and, therefore, explicit legislation is typically required to allow local governments to restrict access to their geographic data sets. Thus, some state and local governments have altered their legislation accordingly and are now imposing intellectual property and ownership rights in the datasets created for public purposes and are attempting to generate revenue streams from secondary uses of the data (Onsrud, Johnson, and Winnecki 1996). However, to sell government data to a few private firms that can afford the data benefits primarily those privileged firms at the expense of the general public and the loss of widespread general benefits to the community. Those who seek to impose restrictions on citizen access should be required to overcome the underlying policy arguments on which such laws are based, foremost of which are that open access keeps government accountable and that open access to government information has far greater long-term economic benefits for a community than does pursuing revenue generation approaches.

(2) Database Legislation. Proposed Title V of the *Digital Millennium Copyright Act of 1998* (H.R. 2281 and S. 2037) if enacted into law would have severely constrained the 'fair use' provisions of copyright law relative to the use of commercial datasets and would have had a chilling effect on the academic and research communities. Of particular concern in the proposed legislation was the lack of definition of the term 'market harm', as it would be imposed against universities. The length of time for protection of databases would be perpetual since even minor updates in an electronic database would toll another 15 years of protection. Thus, if title V had been legislated in the 1998 session, data would never pass into the public domain. It is public domain datasets that have allowed U.S. businesses and scientists to make so much progress in advancing information system technologies and scientific discovery. There was broad consensus among the research and academic community that the bill as drafted would have had significant negative impacts on research and education in the nation. Due to heavy lobbying on the part of the academic, scientific, and library communities, the database provision was stripped from the *Digital Millennium Copyright Act of 1998* that was otherwise passed into law.

NOTE: Representative Howard Coble (R-NC) agreed to a commitment by Senator Orrin Hatch (R-UT) to revisit the database issue during the 1999 session of Congress. Senator Orrin Hatch introduced into the Congressional Record three major legislative models for discussion on 19 January 1999. At the time of this writing the House Committee on Commerce has introduced a slightly modified version of one of the models (see H.R. 1858, *The Consumer and Investor Access to Information Act of 1999*, 20 May 1999) while the House Committee on the Judiciary has passed a revised version of one of the competing models (see H.R. 354, *The Collections of Information Antipiracy Act*, 26 May 1999). Further action is expected in the 2000 session of Congress.

(3) Extension of Time for Copyright. The special genius of the United States copyright system has been its emphasis on an appropriate balance of public and private interests. U.S. dominance in international trade in current products of authorship has been made possible because of the rich and vibrant public domain passed down from earlier authors. Proposed legislation (H.R. 989) would extend the term of copyright protection for all copyrights, including copyrights on existing works, by 20 years. For individual authors, the copyright term would extend for 70 years after the death of the author, while corporate authors would have a term of protection of 95 years. Unpublished or anonymous works would be protected for a period of 120 years after their creation. The enactment of this legislation would impose substantial costs on the U.S. general public without supplying any public benefit. It would provide a windfall to the heirs and assignees of authors long since deceased, at the expense of the general public, and impair the ability of living authors to build on the cultural legacy of the past. The proposed extension would supply no additional incentive to the creation of new works – and it obviously supplies no incentive to the creation of works already in existence. The notion that copyright is supposed to be a welfare system to *two generations of descendants* has never been a part of American copyright philosophy. It is not *unfair* that a work enters the public domain 50 years after the death of its author. Rather, that is an integral part of the social bargain on which our highly successful system has always been based. After supplying a royalty stream for such a long time, these old works should be available as bases on which current authors can continue to create culturally and economically valuable new products (extracted and rearranged from Karjala (1995)).

NOTE: In spite of such arguments and widespread opposition by the academic and library communities, both the U.S. Senate and House passed on 7 October 1998 a 20-year extension of the then existing life-plus-50-year copyright term[4].

(4) Article 2B of the Uniform Commercial Code. Proposed Article 2B seeks to regulate almost all transactions in information. It could affect everything from whether book publishers start *shrink wrapping* books to restrict sharing to whether Internet robots can make legally binding contracts for computer users who unleash them. It is a so-called *model law* that each state legislature can accept or reject. States usually adopt model laws put forward in the Uniform Commercial Code so that business transactions across state lines remain consistent. Among other effects, Article 2B could chill the 'fair use' doctrine that the public and libraries depend on to share information. Like computer software, which is often packaged with the admonition that whoever breaks the seal is bound by the terms of an enclosed manufacturer's contract, so too books could be wrapped in cellophane and sold with all types of limitations. Article 2B would affect U.S. innovation – a vital engine driving business entrepreneurship and economic growth – by discouraging information sharing (extracted and rearranged from Samuelson 7/16/98).

---

[4] For background information, see http://www.public.asu.edu/~dkarjala/index.html and for information on a constitutional challenge to the new legislation, see http://cyber.law.harvard.edu/cc/press.html

NOTE: The proposed article to the Uniform Commercial Code has been withdrawn but now is being pursued by the Uniform Law Commissioners as an independent act under the title *Uniform Computer Information Transactions Act*[5]. Thus the proposal is still under active consideration even though widely opposed by very diverse groups ranging from film studios to consumer groups.

These are but a few illustrative examples of major attempts continually occurring in our legislative halls in attempts to restrict citizen access to public domain and government information. Thus, the assumptions of access to which we may have become accustomed when we operated in a paper world should not be taken for granted as data, information, and works of knowledge are transferred more and more by electronic means.

18.4 Expanding Citizen Rights in Information

In addition to fending off attempts to diminish citizen access to electronic information, there is a need to develop new approaches and models that might be used to expand citizen rights to information or alter the relationship between citizens and government in decision-making processes.

I have argued in past writings that we are witnessing tragedy of the information commons dynamics similar to tragedy of the commons dynamics witnessed in the environmental field (Onsrud 1998). Extending from this analogy I have argued that we may draw from the methods and techniques developed by environmentalists in combating the destruction of the environmental commons and apply them to combat the diminishment of public rights in data, information, and knowledge works. For instance, one of the favored and most effective techniques of environmentalists in protecting the environmental commons has been to expose a full-cost accounting of the effects of actions that diminish or despoil the commons. By example, pollution is often highly illogical for a community when the costs external to the decision-maker are added to the evaluation process. The same economic analysis techniques may be applied to information commons disputes. By further example, 'major Federal actions significantly affecting the quality of the environment' may only proceed after preparation of an environmental impact statement that must fully document the positive and negative impacts of the proposed action and possible alternatives to the proposed action. While preparation and thoughtful consideration of such statements was once quite controversial, this procedure is now an accepted practice with both agency personnel and the general public, and the approach has been copied by many other nations. Perhaps one way of revers-

[5] See http://www.2BGuide.com.

ing the trend of building walls around government information would be to require of government officials an 'information access impact statement' for any major state or local government action significantly affecting the quality of citizen access to government information. Many additional political and legislative lessons may be gleaned from the experiences of the environmental community in protecting the environmental commons.

Looking outside of the environmental-law realm, additional specific legal provisions or policies that might be advocated in expanding information access to scientific and technical data in the public interest might include the following:

(a) Legislate the *first sale* doctrine in networked electronic environments in instances where technology allows no more than one user of a purchased intellectual work at a time.

Discussion. Under copyright law as applied to traditional media, one is able to read or privately perform works without obtaining permission of the copyright owner. Some are suggesting that the same rule should be considered for application in the networked digital world (Samuelson 1996). One can envision a future situation in which a library might purchase five copies of a spatial data set from a publisher and use a technological solution whereby no more than five patrons could use the data set at any one time, whether physically at the library or from a distance. The control system might be similar to the current systems that are used on many university campuses to limit the number of simultaneous users of software on a network to the number of copies that have been purchased legally. Such systems are already being developed (Stefik 1997, Stefik and Lavendel 1997). A patron might check out a copy for two days to extract data and after that time period the data set automatically would become disabled while the copy at the library simultaneously would become enabled again. Through this technological arrangement it may be argued that the *first sale* doctrine and the lending arrangements that the doctrine enables could be applied by right in networked electronic environments to largely support the same principles that the doctrine supports in the current library environment of intellectual works contained on physical media.

(b) Legislate a depository library concept in which publishers to gain certain benefits must provide a digital copy of intellectual works and datasets to a national online collection that would then be accessible from public libraries across the nation (Nunberg 1998). By ensuring that all works to gain protection are publicly deposited, libraries would conclusively know the date when the copyright or other legislative protection of a digital product expires since they would have possessed a copy for the statutory period.

(c) Alternatively, if a licensing paradigm continues for access to online scientific and technical data, a portion of fees collected might be set aside (taxed) to subsidize access for schools and libraries in rural or under-served communities (Nunberg 1998).

(d) Development of standard licensing provisions and policies by libraries.

Discussion. The library community has been constructing its own set of standard licensing provisions with the implication that many librarians and the economic bloc they represent will no longer contract or license with electronic publishers that do not adhere to the library community's recommended licensing provisions[6]. Is this approach realistic? How may the library community position be strengthened without running afoul of antitrust laws and without devoting larger proportions of the library community's resources over time to licensing negotiations and the tracking of adherence to licensing provisions? Although the library as an institution may try to act in the best interests of its user community, in an economic marketplace negotiation environment, when push comes to shove, will there be a strong tendency for the library to act in its own best economic interests, and to sacrifice some social welfare interests of its user community?

(e) Development of university policies or funding agency policies that mandate that professors and researchers must maintain full non-exclusive rights in any works or datasets developed in their capacity as university professors or researchers (Guernsey 1998).

Discussion. Under this approach, virtually all copyright or other legal rights in authored works or data sets might be transferred by professors to publishers but, with the exception of the right of first publication, transfer of exclusive rights would not be permitted. Retention of full non-exclusive rights by universities would be for the purpose of letting the works or data enter the public domain at the option of the university, particularly under marketplace-failure circumstances. Because transfer of exclusive rights to private parties works against development of a common public domain in scientific works and technical data sets and thus works against scientific advancement, those professors and researchers transferring exclusive rights to publishers or others would have their works devalued by the university recognition and reward system processes. What would such a model policy for universities or funding agencies look like?

---

[6] See for example, http://www.library.yale.edu/consortia/icolcpr.htm.

I view all of the approaches mentioned so far as pragmatic proposals that may be used internal to the current legal system to make our society more responsive to protecting public access and the public commons in information. Within the environmental realm, another approach that has been far less successful to date has been to argue that there is something inherently wrong and unjust about the whole concept of real property ownership and to move to different models and concepts of ownership and rights in land. To deal with the environmental problems of the nation and the world, the argument is made that the current legal system can't support an appropriate solution and therefore we need to step outside of the constraints of the current legal and political system to arrive at systems that would be more responsive to the needs of the environment.

Vandana Shiva suggests that there is something rotten at the core of ownership claims in information and treatment of information as a commodity to be sought and sold. Rather than hone the existing legal and social models that assume that the current inequities in society are a given, she argues that there is a need to explore whole new models and theories of rights to access and use of data, information, and knowledge works. She offers a non-western, global, and community control perspective in which neither the state nor the market provide the organizing principles of how people live and how nature's wealth is owned and used (Shiva 1994, 1997).

Practical incremental approaches in expanding rights to information within the existing legal framework and wholesale reevaluation approaches are not mutually exclusive. Even though treatment of information as a commodity may be a social construct, information is, in fact, being treated as a commodity in our local communities, at the national level and throughout the globe on a day-to-day commercial basis as well as through the imposition of intellectual property rights laws. One may try to limit and adapt intellectual property rights laws to help ensure continued access to information and the continued development of public domain data or one may suggest complete new models or views on how control over information should be handled. The academic community should pursue both of these approaches.

As we already know, access, whether in the form of technically meaningful access or legal access does not guarantee power. Nor do consensus building or other participatory processes equalize political power. Measuring or modeling access, as well, does very little to affect power relationships. However, all of these are useful tools for aiding in struggles to gain political power. Paul Schroeder has noted that changing the conditions of power in society would have a substantial influence on changing the conditions of access to and handling of information (Schroeder 1998). Therefore, the implication is that to change the conditions of access one should focus primarily on altering power bases in society. Yet, the alternative approach of directly changing specific rights in information also has an influence over power and wealth in society. If this were not the case we would not see such intense lobbying in Congress over rights in information at the current time. Near the core of this power struggle lies the debate over development of a concept of human rights that could counteract or limit existing corporate and government agency powers in U.S. society.

18.5 Changing the Power Structure

In assessing the *justice* of a particular outcome of a participatory decision making process, Nyerges and Jankowski (1998) citing Lober (1995) refers to three different interpretations of *fairness*. In determining where to locate a hazardous waste facility the example is cited in which the use of three different definitions of fairness would have resulted in three different optimal locations for siting the waste facility. An egalitarian interpretation of justice benefits the most disadvantaged in society and, thus, may be characterized as an approach that minimizes pain. A libertarian interpretation of justice provides for unrestrained interactions among individuals and, therefore, may be characterized as maximizing liberty. A utilitarian interpretation of justice provides the greatest happiness for the greatest number and therefore may be characterized as maximizing happiness. I was struck by the fact that these three views of justice are mirrored very closely in the U.S. Declaration of Independence in the phrase 'life, liberty, and the pursuit of happiness.'

Charles Black Jr., former Dean of the Yale Law School, recently authored a book in which he states and supports 'my own life's conclusions' (Black 1997). After fifty years of professional thought and work surrounding Constitutional law issues, a major conclusion in his life has been that the 'foundations of American human-rights law are in bad shape.' He tempers this bold statement by going on to insist that a sound and well-reasoned basis for a national American law of human rights already exists under U.S. law in 'three imperishable commitments – the opening phrases of the Declaration of Independence, the Ninth Amendment, and the 'citizenship' and the 'privileges and immunities' clauses of the Fourteenth Amendment. He argues forcefully and convincingly that a nation that holds itself out as a power dedicated to securing human rights must provide a basis for sound human rights for its own citizens. If U.S. constitutional law was reinterpreted to support strong human rights for U.S. citizens based on the foundations advocated by Charles Black, the resultant legal framework would protect all three forms of *justice* described in the preceding paragraph and would balance these rights against each other.

Neither protecting access to information nor taking part in consensus building processes actually change power structures. However, a shift in constitutional law, such as suggested by Charles Black, towards strong protection of human rights would have a substantial and long-term effect on empowering groups and individuals and would greatly limit the ability of government and other powerful parties to marginalize other groups and individuals in society.

18.6 Conclusion

In the past, U.S. law has supported the proposition that citizens should have broad and open access to government information at local, state, and national government levels. In addition, U.S. laws have granted greater leeway than the laws of other nations to use the work products of others without permission in order that access and new innovations should be promoted and take precedence over wealth generated from old innovations. Both of these general principles are being severely challenged as publishers and government agencies use the threat of digital technology as an opportunity to limit the rights of citizens to access information.

Assuming that legal access to government information and other forms of information and intellectual works may be maintained, most of us want increased access to information as more and more of our daily activities are accomplished and relationships are established within digital environments. We want technical access to data, information or knowledge that is efficient, effective, and responsive to our specific needs. We want procedural capabilities and methods that will allow groups affected by decisions to be engaged with each other in constructive dialogue. We want access that is timely and understandable so that interested groups may constructively participate with government in more democratic decision making.

Although rights of access to information are insufficient conditions in themselves to achieve these goals, they are necessary and critical conditions. Providing and protecting legal access to information and knowledge works is at least as important as expanding effective and efficient technical access or developing means for measuring access. In addition, while each of these may be necessary and highly constructive societal activities, substantial gains in power for citizens and citizens groups will require new approaches to ownership (e.g., as suggested by Shiva) or major realignments in our existing constitutional framework (e.g., as suggested by Black).

Acknowledgments

Material for this chapter was developed in preparation for two specialist meetings funded by the Varenius project of NCGIA; Empowerment, Marginalization, and Public Participation GIS, NCGIA Varenius Specialist Meeting, Santa Barbara CA, Oct. 1998 and Measuring and Representing Accessibility in the Information Age, NCGIA Varenius Specialist Meeting, Asilomar, Pacific Grove CA, Nov. 1998.

References

Black, C.L., Jr. 1997. *A New Birth of Freedom: Human Rights, Named and Unnamed.* New York: Grosset/Putnam.

Guernsey, L. 1998. A provost challenges his faculty to retain copyright in articles, *Chronicle of Higher Education*, 18 Sept.

Karjala, D. 1995. WRITTEN TESTIMONY of DENNIS S. KARJALA, Professor of Law, Arizona State University representing UNITED STATES COPYRIGHT AND INTELLECTUAL PROPERTY LAW PROFESSORS before HOUSE OF REPRESENTATIVES COMMITTEE ON THE JUDICIARY SUBCOMMITTEE ON COURTS AND INTELLECTUAL PROPERTY on H.R. 989. A Bill to Amend Title 17, United States Code With Respect to the Duration of Copyright, and for Other Purposes. Rayburn House Office Building, Washington DC, 13 July.

Lober, D.J. 1995. Resolving the siting impasse: modeling social and environmental locational criteria with a Geographic Iinformation System, *American Planning Association (APA) Journal*, Autumn, 482-95.

Nunberg, G. 1998. Will libraries survive? *The American Prospect* Nov/Dec: 16-23.

Nyerges, T., and Jankowski. P. 1998. Empirical research strategies for investigating the use of public participation GIS, international workshop on groupware for urban planning, Universite C. Bernard Lyon I. Lyon, France: 3-4 Feb.

Onsrud, H.J. 1998. Tragedy of the Information Commons. In Taylor, D.R.F. (ed.) *Policy Issues in Cartography*, Pergamon: Elsevier Science Ltd., 141-58.

Onsrud, H.J., Johnson, J.P., and Winnecki, J. 1996. GIS dissemination policy: two surveys and a suggested approach, *Journal of Urban and Regional Information Systems* 8:8-23.

Samuelson, P. 1996. The copyright grab. *Wired Archive* 4.01 (Jan). http://www.wired.com/wired/archive/4.01/white.paper_pr.htm

Samuelson, P. 1998, E-mail to Cyberia-L, 16 July.

Schroeder, P. 1998. Statement made during the NCGIA Varenius Specialist Meeting on Empowerment, Marginalization, and Public Participation GIS, Santa Barbara,CA: Oct.

Shiva, V. 1994.The recovery of the commons, Alternative Radio, Colorado College, Colorado Springs, PO Box 551, Boulder, CO 80306, 24 Sept.

Shiva, V. 1997. *Biopiracy: The Plunder of Nature and Knowledge.* Boston: South End Press.

Stefik, M. 1997. Trusted systems, *Scientific American* 276(3):78-81.

Stefik, M., and Lavendel, G. 1997. Libraries and Digital Property Rights. In C. Peters and C. Thanos (eds.) Research and Advanced Technology for Digital Libraries, Proceedings of the First European Conference, ECDL '97, Pisa, Italy, 1-3 September 1997. Berlin: Springer.

19 Qualitative GIS: To Mediate, Not Dominate

Robert Mugerauer

Department of Geography, University of Texas at Austin, Austin TX 78712-1098. USA.
Email: drbob@mail.utexas.edu

19.1 Our Realm of Discourse

As Michael Goodchild reminds us[1], the Seventeenth-Century geographer, Bernard Varenius, produced a treatise focused on two views of geography. One, clearly related to the work of Newton, covered general geography (dealing with a general set of principles) and the other dealt with ideographic geography (having to do with the special character of places). Varenius' (1650) two-fold approach affirms what our society has forgotten, but what is in agreement with Newton himself: we need to conceive of – there is – both a*bsolute* and *relative* space. The former is assumed by physicists in the course of their abstractions and the latter is experienced by ordinary people in the course of making their way in the world. However, today, the powerful realm of Geographic Information Systems (GIS), for all its potential for human understanding and good, does substantial violence by requiring that all our transactions and uses translate (radically convert) our experiential realms into the coded terms of GIS as based on data provided and available only in Euclidean geometrical terms for Newtonian space.

This chapter does not in the least disparage the power of absolute space, Euclidean geometry, nor general geography; but it does argue that we must reaffirm what Varenius and Newton also contended: the specific characteristics of different places and our everyday life experiences relative to ordinary objects must be accepted as complementary to the dominant conceptions. For GIS, this means that we need to develop a *Qualitative GIS* system that allows us to access successfully one another's lifeworlds rather than build enclaves through information technology.

The much heralded *Digital Divide* between those who have access to information technology and those who do not is even deeper in the case of GIS because the cultural capital of marginalized groups is itself denied or cast aside when the foreign conceptualizations of GIS are used to access the systems according to the required technological formats. In contrast, Qualitative GIS could operate in two ways, though which way depends on major practical and theoretical outcomes.

---

[1] Welcoming remarks to the NCGIA Varenius Conference on Measuring and Representing Accessibility in the Information Age, Pacific Grove CA, November 1998.

Barbara Parmenter and I are conducting a series of projects to clarify logical and pragmatic alternatives. We begin with two assumptions: (1) GIS is structured on formal Euclidean geometry for spatial representation and on alpha-numeric database principles for informational content; and (2) current data bases represent Newtonian-Cartesian spatial conceptions and practices. What follows is our critical question: Given these two descriptively defining characteristics, is it the case, either theoretically and/or practically, *that GIS must operate on these Euclidean-Newtonian-Cartesian principles only*? If GIS is not so limited, then Qualitative GIS could be constructed on non-Newtonian, non-Cartesian, perhaps non-Euclidean databases – which can be found in or derived from the already existing, extensive ethnographic research literature and other existing data sources. On the other hand, if GIS is strictly contained within Euclidean-Newtonian principles of organization, then Qualitative GIS, strictly speaking, is impossible. The best that could be accomplished would be a translation of qualitative properties into Euclidean and alphanumeric representations. Even here, however, we have the possibility of two kinds of qualitative GIS. One such qualitative GIS would complement current GIS by inserting or encoding various kinds of hyper-media into standard GIS bases, actually superimposing qualitatively distinct information upon that standard base. The end result would be a kind of updated medieval, multi-perceptual mapping. Recall how medieval mappings regularly presented navigation information, along with glosses and drawings that surrounded or overwrote the basic cartography with story-telling, imaginative, theological, and other modes of information. A contemporary version of this would electronically insert personal, local, and imaginative narrations, images, and other perceptual-qualitative information over or through the standard GIS spatial layout. Alternately, it is possible to model mathematically various spatial configurations, for example, to represent qualitatively differentiated spatializations (raising the issue of whether such a format would be a mapping or a modeling, a question that does not need to be settled here). In either of the last two cases, though we would not have 'Qualitative' GIS strictly speaking, we nonetheless would have something close enough to it that, for non-specialized purposes, we would not have to apologize for and could drop the quotation marks, setting it off more rigorously according to its epistemological grounding.[2]

With either of these qualitative modes, the result would be a GIS that presents a set of alternative geographies and alternative ways of visualizing those spaces and

[2] I want to thank Professor Parmenter for her valuable contributions to this project. Not only did she keep me on the straight and narrow by providing normative control for correct use of concepts and technical terms, and provide helpful critique on the early drafts of this paper, but she continues to show a wonderful openness to theoretical and practical exploration of the topic. As I say to our students, we make a good team, since she knows what GIS actually is, while I, relatively unencumbered by facts, then can safely propose wild-eyed ideas. See at http://mather.ar.utexas.edu/students/cadlab/spicewood/ for more information on our current attempt at doing Qualitative GIS for a grass-roots, neighborhood natural and settlement environments project. If you have questions about the project, contact Barbara [parmentr@uts.cc.utexas.edu] or me [drbob@mail.utexas.edu]. I discuss the project briefly at the end of this paper.

places inhabited and experienced by diverse groups – in Varenius' terms, a new *geographis specificus*. This would enfranchise groups otherwise marginalized because it would allow them and the rest of us to begin to understand their worlds as articulated in their own terms and as embodying their own value systems. GIS then could manifest and affirm a multiplicity of worldviews and multiple geographies, rather than contribute to the reductive homogenization currently taking place. Our policy in regard to access would change: our professional and technical missions would be to help others say what they want to say in their own terms, so that GIS specialists could help others to delineate their own worlds. Together, we all could become conscious of our own lifeworlds, in their similarities and differences; consequently, we might learn to be more responsible toward all such lifeworlds, which in their intersections and tensions constitute the earth.

19.2 Problems

On whose behalf do GIS technical specialists gather and speak? It would seem presumptuous to say, since those who may be interested in more access to GIS would need to speak for themselves. But, they may not come forward unless invited, unless encouraged. So, we need to formulate an invitation, that is, to begin to create an opening in which they would be welcome and in which there would be a point to their coming. Of course, we are responding to the well-documented need that exists because of a gulf between those who have and use electronic telecommunication-information technology and those who do not.[3]

Researchers, practitioners, and policy makers recognize increasingly that the emerging division between 'haves' and 'have-nots' is no longer between geographically distinct first and third or fourth worlds divided according to degree of modernization-industrialization. Instead, it is between groups with and without access to the new information technologies and the power these bring. These new 'dual societies' often are found side by side, in the same cities and regions, in Washington, DC and Mexico City, in rural California and France (Castells, Mollenkkopf, and Robson 1998, Sanyal 1996).

Because highly developed information technologies, such as GIS, are both a product of and a means to develop our scientific and capital-intensive culture, we rightly assume that there are problems when these technologies are not as widespread as they might be. However, the motives and reasons of various constituencies using, not using, and promoting the spread of information technologies are varied.

---

[3] The evidence is presented in all formats and for many audiences: in books by and for academics and researchers (*Investors Business Daily* 1998, Loader 1998, NTIA 1997, Schiller 1996). There is a rapidly growing literature on the subject. Feenberg and Hannay (1995) is theoretically noteworthy.

Some have faith in the progress of civilization through technology – a guiding idea born in the scientific achievements and theories of the Renaissance, developed in the 19th Century's Darwinianism and Hegelian-Marxist ideas of historical-cultural-material change, and matured in the 20th Century's hope in technology and expertise, exemplified in progressivism. Should not everyone benefit from technology, which is making the entire world better?

Others, emphasizing the post-enlightenment development of democracy and its spread across the planet, believe that universal education and access to information are essential foundations for informed decisions, that is, for self-determination. Thus, healthy social and economic decisions and interactions would be possible if all share the same, maximum information.

Still others, in the expansion and intensification of international capital, realize that the future of capitalism depends on developing and increasing new markets. Even with the high demand for the latest hard- and software from the core elite groups, the large 'middle-class' population's enthusiastic participation is essential for mass consumption. And, of course, the largest portion of potential consumers, those at the bottom of the economic and class scales in the United States and around the world, constitute the real potential market for high-technology, just as for consumable goods. Since this last group, by definition, does not constitute a consumer group because it does not have money with which to purchase, it can become a market only when the other more economically powerful groups purchase on its behalf, authorizing expenditures by governments, non-profit agencies and foundations, charitable organizations, and so on. Naturally, this third mechanism works by appeal to the first two: in the name of social-material-economic progress through technology or because of hopes for the spread of democracy, the capital system of development and purchase may be mobilized by those with power on behalf of those without. As should be clear, whether well intended or motivated only by self-interest, our desires and practices have many layers of cultural, historical, economic, and individual assumptions and values.

19.3 Epistemology and Ontology

Information, telecommunication, and geographical systems all operate within the same family of electronic technology. In terms of the theories of knowledge and operational procedures, they actually constitute branches of one system, the electronic processing of information symbols based on the assumptions of classical and contemporary physical sciences and mathematics. Without rehearsing that background here, several key sets of assumptions can be noted. To that end, here is my brief description of GIS:

> 'Upon' or 'within' a topographically correct electronic mapping of a spatial area, other digitized data sets can be 'inscribed' or 'inserted,' so that we can examine the correla-

> tion of not only spatial elements, but economic, cultural, and any other kind of data we wish. Given the ability to see the way different dimensions of the world do or do not correlate, we can proceed with planning ways to change conditions to more closely realize the model we ultimately desire.

Among the assumptions, several concern the character of space. Though Newtonian physics has been supplanted by relativity, Newton's own theory of abstract and relative spaces remains pragmatically adequate to explain and operate within our earthly geo-political realms. Thus, we still use variations of Euclidean geometry and the concepts of absolute, abstract space in GIS because they are operationally correct and adequate. The definition of a 'good' map is one that corresponds correctly, point-by-point, with the features of the earth that objectively exist. This uses the Newtonian idea that in order for there to be a material world at all and for it to operate with law-like movements and forces, there first must be a containing envelope of space. This absolutely existing space (independent of and prior to the material bodies that come to occupy part of it) is not directly experienced, but abstractly understood through the mathematical sciences. Because this space is independent of material bodies, and a condition for their appearance, it itself is homogeneous--the same throughout. Differences within space are accounted for in terms of bodies and forces among bodies. Congruently, this homogeneous space is isotropic; that is, no 'direction' is inherently different than any other, much less privileged. Directional differences are purely a matter of our humanly oriented experiential-relational space.

Correlated with these spatial conceptions, developed over hundreds of years and displacing earlier Greek-based theories of relative, heterogeneous, and anisotropic space, there are epistemological assumptions of positive science. In brief, these include the following: (a) It is held that the world consists of at least space, physical materials, and forces of relation and change among the elements of matter. (b) The human mind (and parallel linguistic and symbolic systems) has the capacity to re-present objective states of affairs in our thought processes and symbolic representations. Thus, (c) what is true is what is a correct representation. Correct conceptual representations are held to work best (or perhaps only) in tight logically univocal concepts and in mathematics. In our area of concern – the visualization of data sets – the good/true map is one with features that correctly (completely and consistently) correspond to and re-present the topographical state-of-affairs. Similarly, data sets (e.g., the location of power and utility lines, land valuation and tax figures, zoning information, etc.) are true when they correspond correctly to the physical or social phenomena they represent and also fit correctly to the map itself. GIS then consists of a 'nested' series of representations that have their value in being correct and manipulatable re-presentations of objective states of affairs.

But, reality is not so simple. Certainly this is not the place to 'refute' or 'amend' the above assumptions. Here I can only assert that they are 'correct,' but incomplete and historically-politically constituted; that is, not straightforwardly anything like the whole and entire truth. This alternative position seems well established by current theoretical debates in the history of science, hermeneutics, and critical theory, to which we can refer should we wish (see Lefebvre 1991, Mugerauer

1991, D'Amico 1989, Gadamer 1989, McIntyre 1988, Heelan 1983, and Heidegger 1977). Suffice it to say that 'facts' or 'data' are not self-selective or self-validating; what becomes a fact or data-point or counts as 'information' does so only within the context of a conceptual-practical system, which itself has a historical, cultural context of limitations and aspirations, insights and blindness, fears and hopes. In short, *what counts as information or even as a geographic feature is a conceptual-pragmatic representation that results from discernment, selection, and suppression among alternatives within a historical, cultural world system.*

In addition, the lifeworld experience of places is primary and the conceptual constitution and grasp of abstract space a secondary and derivative development. As case studies in phenomenology, ethnology, and psychology demonstrate, in our experience, places appear as heterogeneous (not homogeneous), as a function of relationships to other people, places, and things (that is, relative, not absolute), and with directional differences of up and down, back and front, right and left, all of which are physiologically, psychologically, and symbolically charged (not isotropic).[4] Thus, our lived geographical experiences display features exactly the opposite of those attributed to space by the reigning conceptions of GIS. Since the similarities and differences of places experienced among individuals, groups, and entire cultures are among the chief sources of social cooperation and conflict, and of the opportunities and obstacles that we seek to consider, it is not politically sufficient or proper to operate from the limited conceptions of dominant positive science.

19.4 Alternative Geographies

That there are alternative geographies and alternative ways of visualizing spaces and places is patently obvious. One wonderful advantage of GIS is that it presents its diverse data *visually*. This is positive because, at some levels at least, those who are not fluent with concepts or numbers can interact with visual information – though social conventions are an enormous factor in sharing or mediating between 'creators' and 'users.' Further, cultural history shows that while some instruments are highly directive or limiting to those who use and interpret them, there are minimal limitations with simple drawing instruments (sticks scratching maps in the sand or dirt; drawing on hide, bark, paper, and stone with pencils or powdered-colored pigments; weaving various materials, or in the oral versions of mapping that specify places and routes with song and story. As to the latter, Bruce Chatwin (1987) nicely presents the Australian aboriginal tradition in which a physical-spiritual world is mapped by stories and songs; Inuit and other native American

[4] Within the large body of work in phenomenology and Gestalt psychology, of special note is the work of Rudolf Arnheim (1986), Thomas Thiis-Evensen (1984), and Maurice Merleau-Ponty (1979).

traditions of oral and visual mapping are covered in Bravo (1996), Rundstrom (1995), Moodie (1994), Brody (1988, 1989), Turnbull (1989), Aberley and Lewis (1998), Woodward and Lewis (1998), Walhus (1977), and Hisatake (1986).

Even a brief look at a variety of mappings makes clear that a range of compelling lifeworld geographies, rich in understanding, interpretations, and information, is brought forth by the designs and sayings/namings of many peoples where the visual and verbal systems are articulated in local or dialectical 'mother tongues' (which certainly are not the same as the systems of the univocal concepts in Western sciences, philosophy, and other discursive formations). The variation in mapping becomes obvious even in the simple set of figures provided in Figures 19.1–19.5.

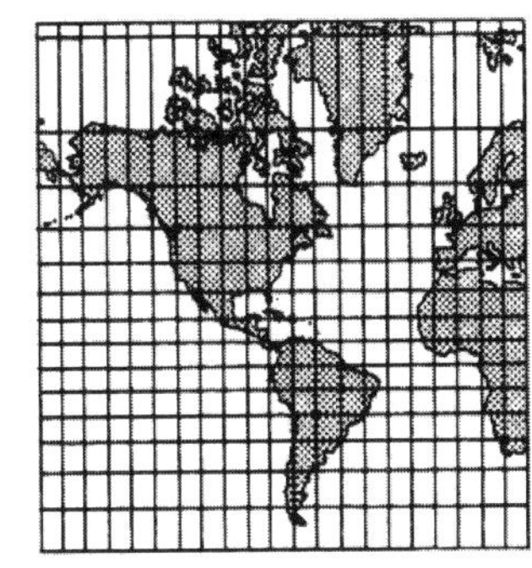

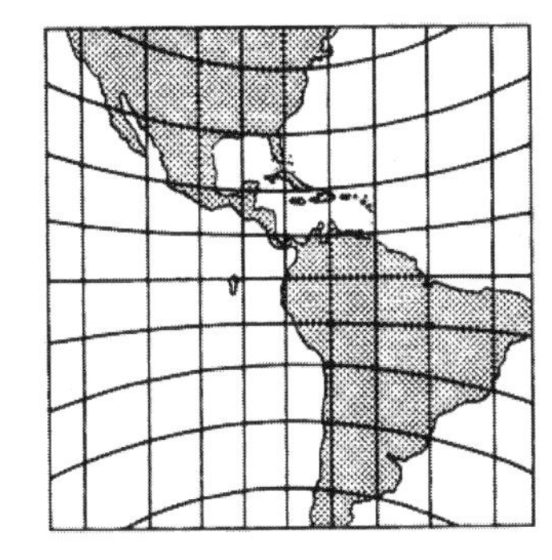

Figure 19.1. Standard Western cartographic representations: Mercator and Gnomic projections

Given the assumptions noted above that ground and drive GIS, it is clear how the now-standardized forms of scientific cartography provide the exemplars: they are taken to be the correct representations of the objective state of affairs. From this point of view, the other modes of mapping are interesting, perhaps, but 'incorrect,' or 'deviant,' or representative of some other dimension (such as the makers' dreams, feelings, impressions, limited perceptions, etc.), but not of the objective state of the world.

The assumptions discussed here and the attitude toward the 'incorrect' is nicely put by Peter Gould and Rodney White (1980).[5] Though they certainly are decent and well-intended persons, as are the rest of my positivistic colleagues, they ultimately display, use, and promote the austere judgments of positive science. When they examine 'the correlation between preferences and accuracy of location' they

[5] These authors, pioneers in perception and mental mapping research, are sensitive that class and economic resources make differences and they do describe the sociological understanding that comes from taking groups' perceptions as they are. Of course, their book is innocent of the theoretical sophistications developed here.

straightforwardly assume that the objective character of the terrain and correctness of representation are what matter.

> For our first, and to geographers, discouraging, plunge into spatial ignorance, we shall examine the situation in North Dakota. University students were asked to record the names of states on an outline map, and by recording the proportion of errors we can draw contour lines enclosing areas of equal misidentification. (Gould and White 1980, 82-83)

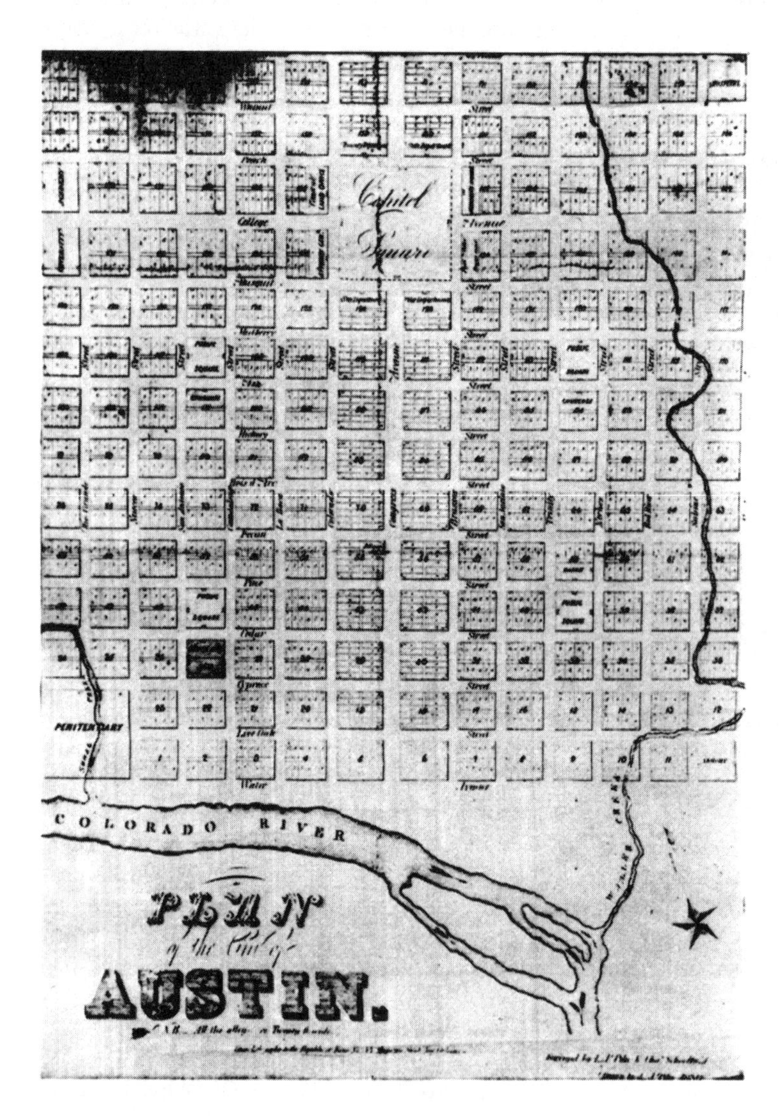

Figure 19.2. A standard city plan: Austin, Texas

Since they are scientific, they attempt to explain the cause of distortions and 'barriers to information flow' in terms commonly used in GIS and other information-communication studies and policies. They work to show that the number of transmissions (directly related to the number of people) and the degree of familiarity (directly related to proximity to geographical features described) explain the degree of accuracy or distortion of representations.

> After we have taken the logarithms of information, population, and distance from [our research area], we can write: log information = -1.38 + 0.87 log population - 0.40 log distance. . . . There is a very strong and significant relationship of information to both these predictive variables. (1980, 93)

The quiet force behind what Gould and White say lies in its comprehensive grasp of our increasingly global economic, legal, political, military, intellectual, and other institutions. Being able to present one's case in logical, linear terms, with quantitative evidence, is essential if one is to obtain a grant, be hired or promoted in the realms of research and technology, or convince a jury, city council, or government agency to grant one's request. This is why, no matter what one's epistemological or political position, *it is critical for those inside and outside the dominant realm to learn standard GIS.* Since the standard view is what exercises power in the world today, and increasingly so, one has to be able to understand it and participate in it or become excluded from power of all sorts. To argue against the importance of the reigning view concerning objectively arranged space and its technologies, including GIS, would be pointless. Thus, a first conclusion: those who have no access to GIS need to find a way to learn it, to acquire access to it, to use it. But, this is a minimal consequence of our reflections, for it is pragmatically harmful if one's lack of technology leads to being eliminated from the world, either effectively or actually.

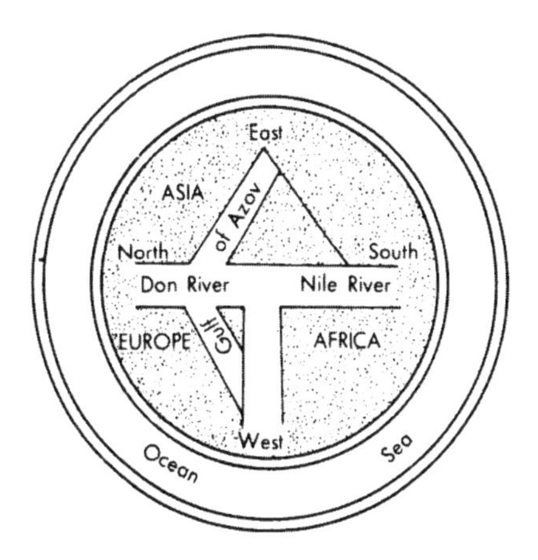

Figure 19.3. A medieval European T-O map

Figure 19.4. A Blackfoot tipi cover, painted with war episodes

Figure 19.5. A map from Paulo Freire's pedagogical exercises. Reprinted by permission of Continuum International Publishing Group from *Education for Critical Consciousness*. Copyright 1973 by Paulo Freire.

19.5 Obvious Issues

The problems with current GIS systems and their social uses form a relatively simple cluster, no matter what one's personal political or intellectual position. Since the dominant technological systems are grounded upon the post-renaissance, post-enlightenment system of rational-mathematical science, understanding the world depends, as Galileo already noted, on ability to do mathematics. Today, alternate symbolic systems are becoming available in forms we call 'user friendly' but which actually amount to translation of mathematical-logical codes into other representational forms, typically iconographic. Thus, while the 'driver' does not need to write code or understand the workings 'under the hood' she does need to have 'dashboard knowledge.' That is, the user must be literate in and dexterous at symbol distinctions, sequencing and other analytic-logical relations and operations, as well as in certain kinds of behavior routines.

Even setting aside the enormous pedagogical and political problems of how to help others become computer and GIS literate, there remain several bitter realities facing policy decisions. Given that there are many people who do not have access to the dominant GIS technologies and worldviews, there is not agreement on what to make of this fact. Currently, those with the information technologies live in a world where those without it largely are *ignored.* Apparently, many GIS specialists are or should be concerned with finding solutions to these problems. Most of us apparently believe in the value of inclusion of disenfranchised groups and in cooperation with other world systems. But, we need to be critically aware of our diverse motives and assumptions, lest we ourselves act imperialistically. Not surprisingly, even the well-meaning formal directives behind the NCGIA Varenius Project seem to consider the needs and possible remedies in terms of 'concepts' that 'reconceptualize, measure, represent, monitor, and plan for the new emergent geographies'[6], thus almost inevitably casting the project in the very terms of the dominant 'imperialistic' educational process. This is perverse since *it is precisely by their differences from the standard and dominant categories that the already marginalized groups constitute their identity.* In addition – though unavoidably – these 'have-nots' (the learners) are required to consciously or unconsciously conform by internalizing and using the very 'normative' concepts, maps, and images of the dominating groups (the teachers, fund-providers, and ultimately the 'host' social-conceptual-technological systems or cultures), of which more shortly. In its current form, it would appear that well-intended projects such as Varenius are reconceptualizing the issue in the same rationalistic terms that will perpetuate the inequality of accessibility opportunities, insofar as the latter have any substantial economic or political force, or further obliterate local, differentiated groups' identities.

[6] Cited from the NCGIA Varenius Project's 'Call For Participation' in 1998.

Further, we would need to be self-critical about our participation in the thoughtless confidence we likely have in the powers of communication and education. Communicating information, in itself, may release pent-up psychological or social pressures, but does not constitute or substitute for rationalized collective action (Blackburn 1989, Mazzioti 1984). Then, there is the undeniable fact that since knowledge and technology are forms of power, many factions in the world would prefer to *exclude* groups (so they remain powerless and unthreatening, or so that they constitute a larger unskilled group for the mining of relevant ores and elements or for the cleaning of toxic by-products of the industrial processes of high-technology).

In addition, there are two further dangers so serious, I believe, as to merit special attention. One is a version of the just-discussed *exclusion*. We have to deal with the fact that a great deal of the current interest in spreading access to information technologies stems from desires to exploit those without (or those who fund the fundless 'consumers-to-be'). The reasons are many: selling hardware and software and services to under-participating groups results in enormous profits and expanding professional job opportunities, banking of political good will, or power via image enhancement. Without denying the good that has come about within or from traditionally disenfranchised groups, we can not ignore the evidence that too much is solicited and sold largely for the sake of profit; too much is 'done unto others' by technical experts (even if well meaning or 'harmless and politically neutral'). What is the actual, positive accomplishment, in terms that matter to them, of GIS becoming available to the disadvantaged poor, homeless, veterans, migrant farm workers, and others? Does it allow them to do something they genuinely need or want, to become personally transformed to embody their own potential rather than the 'plans' of someone else? There is evidence that the Emperor of GIS often has no clothes (Forsher 1998).

We need more research and better policies concerning those on whose behalf we speak – a problem in itself--insofar as they are deemed important in our culture only, or largely, because they constitute the next market group to be exploited as consumers, whether they benefit from the newly installed equipment or not. *We need a fuller understanding and appropriate measures of what would matter* ***in their own terms and value systems*** *to those without technology, to those with different worldviews and geographies.*

Second, no matter that some of those in power seek cooperation and inclusion while others exploit and exclude, we cannot assume that the 'others' are indifferent or passive in these charged global issues. On the contrary, in addition to being pressed by those who 'want in', our dominant political-intellectual-cultural worldview or system already is under attack by groups who not only do not share our worldview, but who do not want to. They who actively want to defeat its spread across the world or even roll back its current influence. There is every reason to believe that the attacks will continue.

At the same time, it is reasonable to believe that confrontation is not necessary in every case. Sometimes conflict is preventable among people of good will, and we can better learn to interact positively with others. *It is an essential part of special-*

ist groups' deliberations and consequent actions to figure out how to cooperate with those who would affirm their own distinctive worldviews and geographical information systems; that is, with those who do not want to lose their own identities and ways of life just because they might have the opportunity to obtain access to ours.

19.6 Toward a Solution: Pluralistic-Democratic GIS for Mediating Individuals and Groups

The proposed partial solution to many of the problems outlined above aims at an affirmation of the identities and differences among individuals and groups within the context of a shared set of worlds. This would be the contemporary, information-age version of *e pluribus Unum*. Without the 'one,' we have chaos – anarchy, if not war; without the 'many,' we have totalitarianism.

The outcome envisioned here is intended to be simple and realistic. It is simple in that, opposite to a monoculture, which seems to spell doom to human social groups just as surely as to soil and crops, it envisions living in a multiply-cultured, non-isolated, set of worlds while maintaining several, possibly changing, identities. The vision is realistic in that it has operated across time and space for thousands of years. Very few people actually have been or have remained members of absolutely undifferentiated monocultures. Even within small primal groups there are multiple sub-cultures: men's and women's groups, earth and sky groups, monkey and snake people, children and post-initiates, gatherers, warriors, and shamans in dynamic relationships. Even among the earliest and most closed groups there are those who operate at the borders, learning and using the languages and material items of neighbors. The ancient trade of colored stones and weapons worked because groups with strong focal identities nonetheless had ways to interact with others who, in effect, lived in different worlds. The same phenomenon continues with the millions of migrants in today's world. Think of the worldwide phenomenon of children of immigrants mediating between the 'old world' culture of the transplanted grandparents and the host culture of the streets.

Transculturation does work. How? Note, here I am not talking about replacing one culture with another; whether freely chosen or forced. That phenomenon has to do with the operations of monoculturation. I mean the process whereby one maintains one's own initial cultural world and comes to participate in another, or several others, which also become one's own, while remaining able to pass back to inhabit, even to deepen, one's original 'home.' Again, most of us do this regularly, as would be apparent if we discussed our own lives as sons and daughters and parents, as Irish-American researchers studying Chilean economics, as academics who also repair and race motorcycles, and so on.

The process of which we are speaking is one of mediation, where some people open to each other, help each other to cross over and back, between cultures. This mediation is a *trans-lation*, which literally means a going over, across; a bridging. The persons with technical GIS expertise obviously have to be translators: they have to hear and understand what is being said to them about a world they do not genuinely inhabit and then try to help translate that into some kind of GIS presentation. But, first and foremost, those who articulate their own world do the translating, because they have to bring out of themselves the 'design' of their world and then name and reflect on the subtle, normally assumed and unspoken relationships among its elements. They have to go over across to the foreign, unavoidably falsifying, and dangerous formats of GIS, and then try to come back again, to their own worlds.

In addition to the question of whether and how GIS systems might be – or become – adequate to such a bridging, there is the more fundamental question of whether and how the different sets of people involved would be able to undertake and succeed at such a task. There is some reason to be hopeful if we consider the already developed and partially implemented theory and practices in planning and communication that are known as 'pluralism and advocacy' and 'critical theory,' which may be updated and newly implemented via GIS. As Davidoff and Reiner successfully argued, those with expertise must help those in need to articulate their own goals and visions, to translate and evaluate these into their own terms and into the coin of the current regime, and then to make their own decisions and, with the help of the expert, present their cases in terms of the group in power (Davidoff, 1965; Davidoff and Reiner, 1962). This does, I believe, need to be made less politically optimistic (or naive) by moving it in the direction of critical theory. As Habermas (1984, 1987), Forester (1980, 1982), Albrecht and Lim (1986), and others point out, the legitimacy of institutions as well as educational and political projects depend on the satisfaction of more complex criteria. There has been considerable work to show that there are at least four necessary and sufficient conditions to be met before an action can be considered legitimate: clarity, veracity, trust, and consent or validation by the groups affected.

How can we somehow deal nonarbitrarily with others in a way that results in genuine common understanding and a shared world, and that does not destroy actual and fruitful differences in the name of the unbearable sameness of forced monoculture? How can we be self-disciplined so as to respect others and thus ultimately enjoy their differences in our lives? How can we learn the non-intrusiveness and non-imposition that are crucial to understanding and practice?

Boundaries need to be acknowledged and respected. By letting the boundaries be, we mark differences, but are not separated by them. In pursuing personally important issues, we become able to pass over to other's concerns. We also necessarily pass back again, because (despite the claims of objectivist methodologies) we can not 'become' the other. By passing back again, we affirm our own and the other's identity.

Brazilian Educator Paulo Freire, through his life's work and his famous books, such as *Education for Critical Consciousness* (1968) and *Pedagogy of the Oppressed* (1973), argues that all of us, including educators, face the constant and grave danger of being tyrannical and imperial (cf., Putnam 1978, Collins 1977,

Lankshear, Paters, and Knobel 1996). To teach the corpus of knowledge and procedure that is the heart of any tradition, we need to teach students by way of standard, proven concepts, and methods. The very power and applicability of these concepts and methods ensures that they can be understood by everyone and passed on. Thus, in the rational, scientific world, Newton's concept of mass, Marx's concept of contradiction, or Rawles' concept of justice are univocal and precise. But, as noted above, this means that to educate our students we impose these concepts and practices upon them, consequently also forcing their experiences and actions, that is their worlds, into preexisting, standard concepts, which, after all, are not politically innocent.

The same is true of all learners' problems. The learner wants to know how to use GIS. To proceed with our expertise, which presumably is why we are valuable and have been brought onto the scene, we translate the learner's vague needs and general wants into precise terms. We supply or develop, and then apply, instruments that will give exact and irrefutable results and indications for practical procedures. We develop lesson plans that will be maximally functional, that fit with the correlate needs of the group and within the prescribed social-economic, aesthetic norms.

In these cases we exercise our power and accomplish things in the world precisely insofar as we get the learners to participate in and, thus, continue the preexisting and dominant system. To some extent this is good and unavoidable: learners want and need to learn GIS to become part of the powerful, dominating world. But, at the same time the result is oppressive to them and ensnares them in a cycle that continues the processes of oppression (with them now appropriated to continue what they have internalized). We know that we *also* need ways to *respond* to the worlds of individuals and groups so that what we come to understand and do together is generated out of the existential reality of these life-worlds. We are responsible for not stamping out their specific ways of being in the name of profitable and expedient homogeneity. We are responsible for developing ways to see, attune ourselves to, and nurture the life-worlds of others, including those who place themselves or are placed in the trajectory of our influence.

Freire agues that the primary way to do this is by disciplining ourselves so that we can listen to what others have to say and by *changing our professional mission to helping others to say what they want to say in their own terms*. In Freire's view, this means starting with the admission that we do not know what the other person's world is like, nor what their real problems and needs are, much less what acceptable solutions would be. Nor, likely, does the other person. If they did, they would not need or consult us. The vibrant relation between learner and teacher is generated insofar as teachers can help learners to name and become conscious of their worlds, their needs and possibilities. The process of helping learners to articulate their worlds in their own terms is a process of liberation and empowerment, for them and for the experts too.

Freire's approach integrates educational, political, and social theory with personal experience. He contends that the freeing transformation of *praxis* is achieved through dialogue in a process (in his words, *conscientizacao* – conscientization) that allows us to critically assess and understand society and our situation in it. The process begins with investigations that uncover what Freire calls *generative themes*, that is, the controlling postulates that are existentially and emotionally powerful to a group. These themes are then presented back to the group through a series of often-pictorial *codifications* in which the teacher elicits distinctions such as those between cultural and natural dimensions or relationships among inside and outside groups contending for power.

In this format, where problems are raised for people to discuss in their own terms, contradictions naturally are discovered; in turn, these can be codified and presented for further reflection. Thus, the educator can pose a problem to the group; through dialogue the group begins to surmount the initial limitations of the situation. Obviously, the only way for the project to work successfully is for the participants to engage in genuine dialogue together, for intensive and long periods of time. Together, and scrupulously avoiding thoughtlessly accepted concepts, what matters has to be allowed to be named and thought in its own terms, that is, in terms of the character of each thing and the webs of relationships among them. In the process, the learners can discover for themselves the contradictions among elements and systems of meaning, intent, and practice. They can begin to explore how the contradictions might be overcome in ways that allow the maximum nurture of their world as it discloses itself to them.

To have a more concrete sense of what this means, think of the fieldwork involved in understanding a given geographical realm. We know that it is easy to do research in the relevant literature, draw out the necessary concepts, devise a hypothesis, and formulate a questionnaire. After a pilot project or two, we are ready to go, to translate the not-yet-known into the known. But we also all know how we falsify the worlds we are studying when we do so--at least by leaving out so much, and I would agree with Freire, by violently translating everything into foreign, standard categories. To remedy this, it is increasingly common to try to go open-mindedly and see what is there. Then, from initial field observations and conversations, we devise open-ended interviews, and if that information is not precise enough, formulate questionnaires. But, these procedures have to do with *us coming to know their world. Freire's point is that the opposite needs to happen: the others need to articulate-delineate their own world, in their own terms*. Thus, though we can assist in the process with our expertise and technologies, our first obligation is to facilitate visualization and dialogue among the participants, who thereby articulate their world for themselves and us, as they explain it to us.

Comparative theologian John S. Dunne develops a very useful strategy that may help us to 'pass over' from ourselves to others, and then to pass back (1967). Dunne begins his reflections with a personal search for what some take to be the issues that matter most. 'How can I deal with my fear of my death?' 'Is there a God?' 'Am I all alone in facing life's difficulties?' These timeless questions have been encountered by many over the past centuries, but still are mine right now, to be answered by me, unavoidably. Though each of us has to answer such questions for ourselves, since others have asked these questions before us (and perhaps even found 'answers' or at least comforting resting places along the way), Dunne explores our issue of personal and shared understanding.

He argues that in pursuing personally important issues, we become able to pass over to other people's concerns. We necessarily also pass back again, because we cannot ever become the other. By passing back to our lives again, we affirm both our own and the other's identity. By passing over and back, based on shared struggles with the same genuine questions and realities (such as death and loneliness), our personal questions

> . . . can be broadened and followed in a much wider context than they ordinarily would be. The passing over and back, then, tends to bridge the gap between private knowledge and public knowledge and to give the seeking and finding that occurs on a strictly individual level something of the communicability of public knowledge. [Comparing one's personal questions and findings with those of others allows us to be] . . . able to pass from the standpoint of our lives to those of others, to enter into a sympathetic understanding of them, to find resonances between their lives and our own, and to come back once again, enriched, to our own standpoint. (Dunne 1967, viii-ix)

That such a process is reasonable theoretically and practically could be further established if we had the opportunity by referring to the non-ideological work of other diverse figures, such as Gadamer (1989), McIntyre (1988), and Heidegger (1966). Gadamer, for example, demonstrates how fusion of differing cultural-temporal horizons may happen when we encounter a 'text' with a genuine question. Our pressing concern may evoke new meanings, perhaps unintended by the original author, from the work which we address seeking insight. Heidegger and McIntyre account for parallel phenomena of mediation as *trans-lation.*

19.7 GIS Applications for Empowerment

GIS admirably suits itself to such a process. It can provide the means to graphically present the mapping of one's own world in most whatever way one wishes. (Remember that built into the very code systems and protocols there are deeper, fixed limitations that ultimately need to be overcome or removed.) What matters in a mapping, what is included and excluded (such as the relations among elements, the means and forms of graphic presentations) would be worked out in each original application of a system to a newly delineated and articulated world.

Importantly, the decisions that stem openly and responsibly from implicit and explicit value systems can be respected and built-in from the start. What was not self-consciously used can come to group consciousness so that its future importance may be decided. And, since learners would have to visualize-articulate their own world in their own way and then format that into GIS, they would start with their own world, pass over into the dominant one, and then back into their own (now bi-cultural realm). The teacher would begin in the dominant technological world of GIS (at least for purposes of the technical facilitation, but not necessarily), pass over, at least a bit, in dialogue and work to the world of the learners, and then pass back into her own.

Theoretically and practically, we are justified in holding that 'there is no absolute standpoint, since no standpoint would exhaust the truth of human culture and built reality, though there is the possibility of our passing over from one contextual horizon to another' (Dunne 1967, 5). Parallel with this, there is no purely relative standpoint, since though humans operate within specific traditions, disciplines, and cultural contexts, one's deep questions, patterns of thought and action, and way of life do connect with those in other traditions, disciplines, cultures, and times.

Boundaries need to be acknowledged and respected. By letting boundaries be, we mark or even celebrate the differences, but are not isolated by them. Crossing boundaries, then, is *not* a matter of scientific method achieving objectified knowledge; *nor* is it idiosyncratic voyeurism. Crossing over and back is possible because we face not the problem of the unintelligibility of the other, but the *inexhaustible intelligibility* of other people, practices, processes, GIS, and other information technology projects yet to come.

In contrast to the positivistic mental mapping procedures of the dominant GIS paradigm (recall the quotation above from Gould and White), examples of self-articulation exist that can be amplified. On the one hand, there are the many grass roots electronic communities that could implement GIS in the same spirit in which they now do operate electronically. We all have our favorite community Web sites.[7] Groups of specialists and ordinary users alike need to collate and share sources so that we all can learn from the entire set whose productions we value.

In addition, to focus on the basic operation of mapping, it would be interesting and fruitful to transfer to GIS the grass-roots mapping processes underway around the world, such as documented in *Boundaries of Home: Mapping for Local Empowerment* by Doug Aberley *et al.* (1993). In contrast to the criteria of good = true = correct in positive science, Aberley contends that

> It is important to repeat over and over that there is no 'good' mapping or 'bad' mapping. Leave the need for perfection to the scientists; what you are being encouraged to do is honestly describe what you already know about where you live in a manner that adds momentum to positive forces of change. . . . every region has the potential to be represented by as many unique interpretations as it has citizens. Reinhabitants will not only learn to put maps on paper, maps will also be sung, chanted, stitched and woven, told in stories, and danced across fire-lit skies. (1993, 5)

A moderate and seemingly unproblematic application would involve using existing, standard, and GIS base mappings to which personalized, or local, or biore-

---

[7] Among my favourites are Austin Free-Net <http:// www.austinfree.net> and the Community and Civic Network Discussion list <COMMUNET@LIST.UVM.EDU> archived at <http://list.uvm.edu/arch/archives/ communet.html>.

gional information would be added.[8] This is related to my own work with Barbara Parmenter and an interdisciplinary team of graduate students to generate a Qualitative GIS for a neighborhood planning project outside Austin, Texas. The residents in the Spicewood Corridor, off the Old Spicewoods Springs Road west of the city, are seeking a way to explore their own identity and that of their local place in order to begin to imagine ways to develop and keep safe the qualitatively distinctive environment in which they have chosen to live. This is a still-emerging version of a conservative Qualitative GIS, in which we are encoding information about the experiences of the natural environment and personalized individual and group information onto the standardized GIS databases.[9]

A more difficult and yet promising project would be to use basic GIS formats to generate customized combinations of not-necessarily-representational 'designs' and 'words' to *originarily* let a worldview emerge and be named in its own terms. There is no reason at all why a combination of Freire's proven pedagogy that combines visual representation and naming-dialogue in local, dialectical words cannot be transformed into visualization that presents other quantitative information and qualitative interpretations in a democratic, pluralistic GIS system.

References

Aberley, D., et al. 1993. *Boundaries of Home: Mapping for Local Empowerment.* Gabriola Island BC. and Philadelphia PA: New Society Publishers.

Aberley, D. and Lewis, G.M. 1998. *Cartographic Encounters: Perspectives on Native American Mapmaking and Map Use*. Chicago: University of Chicago Press.

Albrecht, J. and Lim, G-C. 1986. A search for alternative planning theory: Use of critical theory. *Journal of Architecture-Planning Research* 3:117-31.

Arnheim,. 1986. *Dynamics of Architectural Form*. Cambridge: Harvard University Press.

Blackburn, R. 1989. Defending the myths: The ideology of bourgeois social science. In Cockburn, A. and Blackburn, R. (eds.) *A Brief Guide to Bourgeois Ideology in Student Power*. New Left Review, 154-69.

Bravo, M.T. 1996. The accuracy of ethnoscience: A study of Inuit cartography and cross-cultural commensurability. Manchester *Papers in Social Anthropology*, no. 2. Manchester: Department of Social Anthropology, University of Manchester, 1-36.

Brody, H. 1988. *Maps and Dreams*. Vancouver: Douglas and McIntyre.

Brody, H. 1989. Maps and journeys. In Macleod, F. (ed.) *Togail Tir Marking Time: The Map of the Western Isles*. Stornoway, Scotland: Aeair Ltd. and Lanntair Gallery, 133-36

Castells, M., Mollenkkopf, R., and Robson, M. 1998. The Rise of the Network Society. Paper presented at Telecommunications and the City Conference, University of Georgia (March).

Chatwin, B. 1987. *The Songlines*. New York: Penguin.

Collins, D. 1977. *Paulo Freire: His Life, Works, and Thought*. New York: Paulist Press.

[8] Doug Aberley, Beatrice Briggs, Kai Snyder, Jonathan Doig, and George Tukel supply examples of successful case studies and techniques.

[9] The Web site for this project is <http://mather.ar.utexas.edu/students/cadlab/spicewood/> (For information on the Spicewoods Springs Road Project: drbob@mail.utexas.edu).

D'Amico, R. 1989. *Historicism and Knowledge*. New York: Routledge.

Davidoff, P. and Reiner, T.A. 1962. A choice theory of planning. *Journal of the American Institute of Planners* 28(May):11-39.

Davidoff, P. 1965. Advocacy and pluralism in planning. *Journal of the American Institute of Planners* 31(November):277-96.

Dunne, J.S. 1967. *A Search for God in Time and Memory*. New York: Macmillan Company.

Feenberg, A. and Hannay, A. 1995. *Technology and the Politics of Knowledge*. Bloomington: Indiana University Press.

Forester, J. 1980. Critical theory and planning practice. *Journal of the American Planning Association* 46:(July) 91-112.

Forester, J. 1982. Planning in the face of power. *APA Journal* Winter: 67-80.

Forsher, J. 1998. Access and exclusion in the age of information and communication technology. Paper presented at Telecommunications and the City Conference, University of Georgia (March).

Freire, P. 1968. *Pedagogy of the Oppressed*. New York: Seabury Press.

Freire, P. 1973. *Education for Critical Consciousness*. New York: Seabury Press.

Gadamer, H-G. 1989. *Truth and Method*. New York: Continuum.

Gould, P. and White, R. 1980. *Mental Mapping*. London: Little, Unwin.

Habermas, J. 1984, 1987. *Theory of Communicative Action*. Boston: Beacon.

Heelan P. 1983. *Space Perception and the Philosophy of Science*. Berkeley: University of California Press.

Heidegger, M. 1966. *Discourse on Thinking*. New York: Harper & Row.

Heidegger, M.1977. *The Question Concerning Technology*. New York: Harper & Row

Hisatake, T. 1986. Indigenous maps, cosmology, and spatial recognition of the North American Indians: With special reference to the Ojibway around Lake Superior, In Nozawa, Hideki (ed.) *Cosmology, Epistemology, and the History of Geography: Japanese Contributions to the History of Geography*. Fukuoka: Institute of Geography, Kyushu University, 3:1-25.

Investor's Business Daily. 1998. Does U.S. face a 'Digital Divide'?, August 14:B1-2.

Lankshear, C., Paters, M., and Knobel, M. 1996. Critical pedagogy and cyberspace. In Giroux, H.A., *et al. Counternarratives: Cultural Studies and Critical Pedagogies in Postmodern Spaces*. New York: Routledge, 149-88.

Lefebvre, H. 1991. *The Production of Space*. New York: Blackwell.

Loader, B. (ed.) 1998. *Cyberspace Divide: Equality, Agency and Policy*. New York: Routledge.

MacIntyre, A. *Whose Justice? Which Rationality*? Notre Dame: University of Notre Dame Press.

Mazzioti, D.F. 1974. The underlying assumptions of advocacy planning: pluralism and reform. *Journal of the American Institute of Planners* 40(1):207-25.

Merleau-Ponty, M. 1979. *Phenomenology of Perception*. Atlantic Highlands, NJ: Humanities Press.

Moodie D.W. 1994. The role of the Indian in the European exploration and mapping of Canada. *Zeitschrift der Gesellschaft fur Kanada-Studien* 14(2):79-94.

Mugerauer, R. 1991. Post-structuralist planning theory. Austin: University of Texas Graduate Program in Community and Regional Planning *Working Papers*.

Mugerauer, R. 1995. *Environmental Interpretations: Tradition, Deconstruction, Hermeneutics*. Austin: University of Texas Press.

NTIA (National Telecommunications and Information Administration). 1997. *Falling Through the Net II: New Data on the Digital Divide, A Survey of Information 'Haves' and 'Have Nots' in 1997*. See <http://www.ntia.doc.gov/ntiahome/net2/falling.html>.

Putnam, C. 1978. *Adult Literacy Education: Paulo Freire's Ideas and a Program for the United States*. Unpublished manuscript, Ann Arbor: University of Michigan.

Rundstrom, R.A. 1995. *GIS,* Indigenous peoples, and epistemological diversity. *Cartography and Geographical Information Systems* 22(1):63-75.

Sanyal, R. 1996. *Report of a colloquium held at department of urban planning*, Cambridge: MIT.

Schiller, H.I. 1996. *Information Technology: The Deepening Social Crisis in America.*

Thiis-Evensen, T. 1984. *Archetypes in Architecture*. Oslo: Norwegian University Press.

Turnbull, D. 1989. *Maps Are Territories: Science is an Atlas*. Geelong, Victoria, Australia: Deakin University.

Varenius, B. 1650. *Geographia* Generalis. Amsterdam: Apud Ludovicum Elzervirium.

Warhus, M. 1977. *Another America: Native American Maps and the History of Our Land.* New York: St. Martin's Press.

Woodward, D. and Lewis, G.M. (eds.) 1998. *Cartography in the Traditional African, American, Arctic, Australian, and Pacific Societies*, vol. 2, bk. 3 of *The History of Cartography*. Chicago: University of Chicago Press.

Part IV

Conclusion

20 From Sustainable Transportation to Sustainable Accessibility: Can We Avoid a New *Tragedy of the Commons*?

Helen Couclelis

Department of Geography, University of California, Santa Barbara CA 93106-4060, USA.
Email: cook@geog.ucsb.edu

20.1 Introduction

Accessibility is the geographic definition of opportunity. The opportunity individuals have to participate in necessary or desired activities, or to explore new ones, is contingent upon their ability to reach the right places at the appropriate times and with reasonable expenditure of resources and effort. Up until recently the history of the increase in accessibility at local, regional, and global scales has largely been the history of improvements in transportation. With the advent, spread, and now merging of telecommunications and digital information technologies there exist for the first time viable and often preferable alternatives to physical movement for accessing and engaging in economic, social, or cultural activities. These developments combine with advances in the design and management of physical transportation to create substantially altered forms of accessibility landscapes reflecting profound changes in the meaning of that term itself and its implications for urban and regional structure and function.

In the geographic literature accessibility has been studied mostly as a property of individual locations or as a locational requirement of individuals or groups. A location (such as the traditional city center) is accessible if it is easily reachable from most other locations. Conversely, accessibility is a critical consideration for people who typically require access to jobs, services and other activities and opportunities. In both these cases there is a clear spatial (and eventually also temporal) distinction between an origin, where the need for a contact originates, and a destination, where that need is satisfied, the two being separated by a physical (spatial or spatio-temporal) distance. Accessibility then is a question of how easy (fast, cheap, comfortable…) it may be for that distance to be overcome, and the answer has traditionally been sought in the availability and quality of transportation.

The societal changes described by the term 'the information age' modify this picture in at least two respects. First is the widely advertised substitution of telecommunications contact for physical movement. This is indeed happening locally, regionally and globally at an accelerating pace and if things were to end there we would be enjoying by default the age of sustainable transportation. That this is not the case is due to the fact that profound reorganizations of activity patterns are taking place at all scales such that the net number of interactions that involve physical movement rather than electronic contact appears to be increasing rather than decreasing. As this paper argues, a large part of this phenomenon is due to the *fragmentation of activity* that is taking place such that activities that used to be associated with a single location (e.g., my workplace) are now increasingly scattered among geographically distant locations (e.g., my office, home, associate's home, hotel room, car, train, or plane). Thus the *contact set* of individuals, the number of places they interact with, explodes from one location per activity to a potentially indefinite number of locations. While many of the corresponding interactions are carried out in virtual space, many others are very much physical, placing old-fashioned demands on transportation systems to meet old-fashioned accessibility needs.

With more people traveling more often than ever before, sustainable transportation is emerging as a major theme in contemporary transportation research. The notion of sustainable transportation is closely associated with mobility, accessibility, urban form and function, environmental quality, and social and economic life. Of these, accessibility may be seen as the most central concept, linking the others together. Hodge (1995) argues that the recent shift in emphasis from mobility to accessibility in transportation research may not be quite enough to ensure sustainability. The reason is that in the information age the demand for contacts in general and especially of the physical kind may be increasing faster than is necessary for a well functioning economy, society, and culture: striving to meet without further ado the associated proliferating accessibility demands may not be in the best long-term interest of our cities, the environment, or society. Thus we may speak of 'sustainable accessibility' as the goal of meeting reasonable mobility and accessibility requirements of individuals while reducing the need for ever larger contact sets now and in the future.

This paper argues that sustainable accessibility is a precondition for sustainable transportation, and explores the meaning of that phrase within the context of rapidly spreading information and communication technologies (ICTs). It begins by highlighting the complexity of the notion of accessibility and its relation with different views of transportation technology. It then discusses how accessibility may be redefined in the information age, in pace with other momentous changes in how places, activities, and spatial relations are being conceptualized within a new hybrid space-time that is partly traditionally geographic, partly virtual. Central to these changes is the fragmentation of activity brought about by ICTs. That observation leads to an exploration of the notion of sustainable accessibility within the framework of the 'tragedy of the commons' dilemma. The paper closes with a brief discussion of some of the many research challenges, both theoretical and practical, lying ahead in this brave new domain.

20.2 Transportation, ICTs, and the Fragmentation of Activity

Accessibility and Transportation: Multiple Perspectives

Though central to transportation and spatial interaction research, accessibility is a notoriously difficult concept to pin down. Scott (See Chapter 3) provides a comprehensive review of how accessibility has been variously defined and treated in the literature; this discussion will not be repeated here. Instead, I propose in this section an informal typology of perspectives on the concept, useful for clarifying the connections with transportation (and beyond, with ICTs). What is highlighted here above all is the complexity of the notion of accessibility, the impossibility to pigeonhole it within some well-defined domain. In a similar spirit, Occelli (see Chapter 17, p. 280) speaks about the 'junction' function of accessibility:

> Accessibility is difficult to translate into a single meaningful notion… [it] shares features of the two fundamental components of any spatial system: the spatio-temporal pattern of activities, and the spatio-functional pattern of interdependencies. In bridging these two components, the concept of accessibility provides a 'junction' between them.

There is at least one other aspect to the 'junction' function of accessibility, beyond the spatio-temporal and functional. That is the urban economics perspective, with its distinctive treatment of accessibility as a good. In addition, there is a scale issue: one can talk about accessibility with respect to a place, or an individual, or an urban or regional system as a whole. These observations suggest the following scheme, based on these three distinct perspectives on accessibility (spatio-temporal, functional, and economic), and the three views of place, individual, and system (Table 20.1). There are arguably several more perspectives on accessibility (social, cultural, affective, cognitive, etc.) but these will not be considered here.

Table 20.1. Diverse views and perspectives on accessibility

	Spatial (spatiotemporal) perspective	Functional perspective	Economics perspective
Place	Locational attribute	Enabling attribute of locations	Market good
Individual (micro)	Behavioral explanatory principle	Determinant of space-time activity patterns	Resource for individuals and groups
System (macro)	Aggregate spatial property of urban areas	Structural efficiency measure of urban areas	Positive externality

Thus, from the spatio-temporal perspective, accessibility is a situational attribute of places, an explanatory principle of individual spatial behavior, or an aggregate property of spatially compact or central urban areas and regions. From a functional perspective, accessibility is what makes locations more suitable for some land uses rather than others; at the micro-scale it is a major determinant of the space-time activity patterns of individuals, and a measure of the structural efficiency of urban areas and regions at the macro-scale. Finally, from an economics perspective, accessibility governs land values in the urban land market: it is a market good that can be bought and sold (an *exchange value*). For individuals and groups by contrast it is a *use value*, a resource that makes life and work easier and more productive, while at the macro level of the system as a whole it is a positive externality, a public good arising from the myriad of micro-level arrangements and decisions making up an urban area (Papageorgiou 1987).

The close affinity between accessibility and transportation is reflected in Table 20.2, which exactly parallels Table 20.1 in structure. Table 20.2 summarizes the roles of transportation technologies from the spatio-temporal, functional, and economic perspectives, and from the distinct views of place, the individual or group, and the urban or regional system as a whole. It looks so similar to Table 20.1 as to appear almost redundant, and yet there are some critical differences. For example, at the system level, while accessibility is always a positive externality, transportation is fully capable of generating negative externalities in the form of congestion, pollution, accidents and the other well-known banes of an excessively mobility-oriented society. Similarly the land-use-allocating function of transportation often ends up producing reduced overall accessibility, as when home and work are pushed so far apart as to add millions of wasteful commuting miles to already overburdened networks. Thus, while the relationship between accessibility and transportation appears to be linear at the micro level, this is not necessarily the case at the system level, where accessibility and transportation can sometimes be at odds.

Table 20.2. Views and perspectives on transportation

	Spatio-temporal perspective	Functional perspective	Economics perspective
Place	Place-bridging	Interaction-generating	Land-value-enhancing
Individual (micro)	Mobility-generating, time-saving	Activity-enabling	Utility-producing
System (macro)	Space-time distorting	Land-use-structuring	Externality-generating

The Death of Distance, the Explosion of Place?

There is mounting evidence that aspects of these traditional relationships between accessibility and transportation are changing rapidly within the context of the profound socio-economic, institutional, technological and cultural changes variously designated as the information age, the global age, the post-modern, the post-industrial, or the network society (Castells 1996). The emerging new geographies that accompany these broader transformations range from the continental scale where marked metropolitan and regional restructuring is taking place (Office of Technology Assessment 1995, Graham and Marvin 1996), to the intimate scale of changing individual lifestyles and their spatial consequences. Undoubtedly the defining technologies of the age are the information and communication technologies. For those who are part of the information society, the technical ease of remote communication and information transfer and the virtual annihilation of the marginal contact cost has led to an explosion of the size of the contact sets of individuals and firms, and to the much ballyhooed 'death of distance' (Cairncross 1997). 'Darling, the telecoms revolution is finally happening ' says the woman on the telephone. And the banner at the top of the cartoon reads: 'Suddenly ... distance no longer mattered'. Every geographer should have that cover of the *Economist* (30 September 1995) posted on her wall as it summarizes the major theoretical challenge geography as a spatial discipline has to face at the turn of the millennium. Although news of the demise of distance may be premature, unquestionably its role in structuring spatial interaction is increasingly a different one from what traditional geographical thinking had assumed (Couclelis 1996a, Johnston 1997). Considering that all definitions and measures of accessibility involve some notion of distance, one wonders what happens to accessibility in a world where many critical distances appear to vanish.

It is indeed reasonable to expect that the spread of ICTs in urban regions will result in profoundly altered, hybrid physical-virtual accessibility landscapes requiring a new conceptual vocabulary for their description. Not only distance but also *place*, another fundamental concept in geography that is closely connected with traditional definitions of accessibility (see Table 20.1), appears to be undergoing a momentous shift (Curry and Eagles 1999). Adams (1998) reviews the changing understanding of the concept from the standpoint of social theory and structuration theory, noting that we are witnessing 'a serious ontological challenge to geography' (p. 89). Thus place has been variously described as a 'dynamic system of connections', as 'process', as 'defined by the communication acts that structure both personal activities and collective social processes', as an 'interweaving of communication and action', (Adams 1998, 94), and as 'network' (Bolton 1997). Underlying all of these definitions is a negation of the traditional cartographic representation of place as a point or contiguous zone in Euclidean 2-d space. What this may mean for the notion of distance between places, and hence for accessibility, is anybody's guess.

It is in the context of these conceptual shifts, grounded in the wide-ranging empirical transformations taking place around us, that we need to consider one of the major paradoxes of the information age: *that there is at the same time an increasing substitution of telecontact for travel, and an increasing demand for travel.* While the former is a natural consequence of the spread of ICTs, the increased travel demand observable at all scales is counter-intuitive (Bureau of Transportation Statistics 1997). The phenomenon is undoubtedly due to multiple and complex reasons, including the well-established increases in individual leisure time and incomes in industrialized regions of the world and the greater efficiency and availability of transportation, both public and private. Of particular concern to this paper however is that part of the increased travel demand is due directly or indirectly to the ICTs themselves and their synergistic relations with transportation. Distance is clearly not dead for those growing numbers of travelers stuck in city traffic or enduring endless cramped hours on crowded airplanes. However, each one of these conventional trips is increasingly likely to be a component in a broader nexus of communications and interactions – some physical, some virtual – that together define a single basic activity. I am flying to a conference and look forward to meeting colleagues I have been working with over the Internet. I am stuck in traffic and discussing a problem with an associate on my cell phone. Once back home I turn on my computer and respond to e-mail from my students. I am participating in a teleconference that is saving me a trip for now, but which concludes with planning one for next month. I am writing a paper on my laptop while in a foreign country, and share an electronic draft with my co-author in California. I am working: but when am I 'at work', and where is my workplace? (Mitchell 1995, Couclelis 1998).

The Disintegration of Activity

Underlying virtually all conventional theories and models of urban land-use structure and planning is the assumption that an activity is associated with one place at one time: work with a workplace, shopping with a commercial location, schooling with an educational facility, and so on. Tell me where you are, I can tell you what you are doing. What is more, tell me where you work and I may be able to tell you where you are likely to live, because I know that you are trying to optimize accessibility to your most important destinations. In the information age by contrast, to the extent that an activity involves exchange of information between or among people, or between people and machines, that activity may be fragmented into tasks that are widely distributed over space and across time. Whatever activity they may be engaged in: work, recreation, shopping, education, people are increasingly likely to rely on a variety of physical and digital means to gain access to the people, tools and information they need. Some of these contacts can only be realized through physical movement, some can only be realized electronically, and some others may be realized in either way. Each of these contacts involves a geographic location where the contacted person or device is, so that several locations may have to be accessed by the appropriate combination of travel and telecoms for a single activity to be carried out. It is this interweaving and mutual dependence of physical mobility and electronic communication, not merely the spreading use of ICTs, that defines the major information-age challenge for accessibility research.

Because of these changes the familiar one-to one mapping: *activity* => *place* becomes a one-to-many mapping: *activity* => *places*, or even a many-to-many mapping: *activities* => *places.* The same is true of time, as the proverbial nine-to-five weekday job gradually gets fragmented into chunks spread out over arbitrary hours of the day (and many of the night), interspersed with tasks from other activities occurring at equally odd – by traditional standards – times as well as places. The colonization of time, as this phenomenon has been called, does not concern only the business executive who is talking with London or Tokyo in the middle of the night just as the financial markets start trading in distant time zones. The secretaries and the janitors, the computer support personnel, the food service and transportation workers, the gym and the grocery store and the plumbers and baby sitters are all part of the same nexus of interdependent activities that drive one another into unprecedented kinds of spatio-temporal configurations (Couclelis 1996). Table 20.3 summarizes the possible spatial structures of activities in the information age and highlights the complete symmetry with their possible temporal distributions. Converting Table 20.3 into a matrix with 25 different cells (Table 20.4) conveys the richness of possible space-time structures of activities enabled by spatial technologies. Take, for example, work. At least since the industrial revolution the 'all at one location, all at one time' cell has corresponded to the overwhelmingly preponderant type of urban job, with the odd traveling salesman or itinerant piano tuner providing a few interesting exceptions. While the majority of today's jobs may still fall within that one cell, the colonization of the remainder of the matrix is proceeding rapidly in a diagonal wave of less and less space-time-bound economic activities, reaching towards the 'anywhere, any time' outer limit that some think is just around the corner from today (Knoke 1996). The influence on the economy and society of the highly mobile and multiply connected 'information workers' falling in these newly occupied cells is already acknowledged to be disproportionate to their numbers and growing. For many areas of socioeconomic research, filling out that kind of matrix with appropriate data is bound to be a critically important task for the years to come.

Table 20.3. A taxonomy of possible space and time distributions of activities in the information age

Location of activity	Time of activity
all at one location	all at one time
at alternate locations	at alternate times
distributed along a route	in time sequence
distributed across space	at several different times
ubiquitous (anywhere)	any time
nowhere	never

Table 20.4. A matrix of space-time distributions of activities in the information age

Time Location	*one-time*	*alternating*	*sequential*	*fragmented*	*anytime*
localized					
alternate					
linear sequence					
distributed					
ubiquitous					

Hence the central thesis of this paper: *It is not distance that is dead; it is activity that is disintegrating.* Activity becomes a distributed space-time process, a network of material movements and digital contacts, an interweaving of (electronic) communication and (physical) action. We thus rejoin the earlier mentioned discussions of place as process, communication, or network. Understood from the functional, socioeconomic, cultural, rather than the strictly spatial point of view (see Table 20.1), these outlandish-sounding notions of place amplify the idea of the fragmentation of activity, and raise some difficult issues regarding the meaning of territory, community, and identity in the information age (Curry and Eagles, 1999). The purview of this paper is however much more modest, as it only focuses on the spatio-temporal and functional dimensions of the restructuring of place and activity, and on the implications for accessibility of that fluid but still somewhat tangible context. From this restricted perspective I propose the following as a working definition of accessibility in the information age: *the ability to access, either physically or electronically, and at the appropriate time(s), all the locations that are necessary or desirable for participating in a given activity.* This implies that there are three kinds of knowledge necessary for the study of accessibility: (1) the distribution in physical space-time, relative to actual or potential activity participants, of the loci of the component tasks constituting specific activities; (2) the structure and characteristics of the access-enabling technologies, both physical and electronic; and (3) the relations of substitution, complementarity, and synergism among available physical and electronic options (Abler and Falk 1981). The terms *space-adjusting technologies* (Abler 1975, Janelle 1991) or *spatial technologies* (Couclelis 1994), designating the complex of information, communication and transportation technologies that together work to modify spatial relations, serve to underline the need for a common approach to two materially very different but functionally very similar approaches to overcoming spatial separation.

20.3 Sustainable Accessibility and the Tragedy of the Commons

Sustainable Accessibility: Two Hypotheses

As we think ahead towards a research agenda for making sense of accessibility in the information age it seems appropriate to collect the speculative arguments developed so far into a couple of tentative but eventually testable hypotheses. These hypotheses are linked and together they help define a novel notion, sustainable accessibility, which I will discuss in the following sections.

Hypothesis 1: The fragmentation of activity enabled by the spatial technologies of the information age is one of the reasons for the widely observed increases in travel demand in the industrialized world. There are three components to this hypothesis: (a), the per capita demand for contacts in general (by all means, physical and digital, and for all reasons) is growing; (b), the physical-contact (travel-dependent) part of the demand is also growing; and (c), the fragmentation of activities enabled by spreading ICTs is partly responsible for that increase in travel-

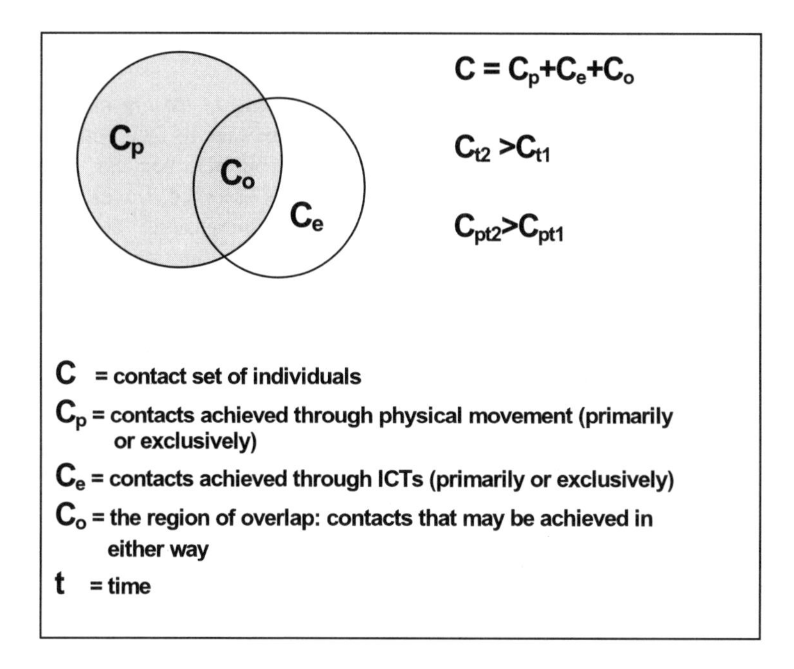

Figure 20.1. Physical and electronic contacts and the possibility for substitution

dependent contact demand. The first two components of the hypothesis, which are already well supported by data, can be expressed in simple notation and illustrated with a Venn diagram in Figure 20.1.

Can, and should transportation keep up with these ever-increasing demands? In recent years the traditional goal of transportation planning has shifted from an almost exclusive emphasis on facilitating mobility to a greater concern for accessibility, in recognition of the fact that travel is (for the most part) a means towards the end of accessing a desired or needed destination. Mobility demands are thus better seen in the context of more fundamental accessibility requirements, which in turn are closely linked with spatio-temporal patterns of activities at both the individual and the systemic scales. While the goal of facilitating accessibility may seem beyond reproach, some researchers doubt that striving to meet without further ado the ever-mounting accessibility demands of the post-industrial or information society is a viable strategy for transportation planning (Hodge 1995). One may argue instead that the goal of sustainable transportation should be to meet *reasonable* accessibility demands, that is, to facilitate mobility where it is really necessary. Mobility is not *really* necessary where there are good alternatives to it, as they are in the case where appropriate ICTs can substitute for physical movement. Referring to the diagram above, there are two ways to slow down the growth of the demand for physical contacts, (C_p): increase the possibility for travel substitution, or increase the attractiveness (relative utility) of substitution. Both aim at enlarging the overlap area (C_o) on the diagram: The first will ensure that people have a choice; the second will increase the likelihood of people making the correct choice from the viewpoint of sustainable transportation (Janelle 1995). This leads to the second hypothesis:

Hypothesis 2: The utility of travel remains too high relative to its alternatives. People like to travel not just because they still tend to value face-to-face contacts and real-world destinations over their virtual counterparts but also because, contrary to what traditional transportation research has always assumed, travel itself has considerable intrinsic utility. Modern automobiles are increasingly comfortable, pleasant and safe moving 'places' where people enjoy spending part of their day. Inside our cars we entertain and educate ourselves, work and daydream, smoke and snack, watch our kids or socialize with friends, nurture our self-esteem as smart and skilled drivers. Style, pleasure, comfort and luxury are increasingly becoming selling points in public transportation as well, at least for those who can afford the first-class ticket. ITCs save time and money but have yet to meet the psychic utility of driving a late-model luxury car or even of sharing a bus seat with a favorite fellow commuter. Undoubtedly the modal choice problem within the growing set of spatial technologies is a much more complex one than that of choosing among physical transportation alternatives. Fischer, Ricco, and Rammer (1992) and Fischer and Rammer (1992) have investigated these kinds of choices involving both physical and electronic options within contact networks of academics. It is clear that the utilities leading to such choices are multidimensional and are influenced by a variety of characteristics pertaining to the initiator of the contact, the contacted destination, the costs and benefits of the available options, and the familiarity of the contact initiator with these options and their availability. This is not much different from what is known from numerous studies of discrete choice behavior in general and traditional modal choice research in transportation in particular. These studies demonstrate that well-established methods of analysis can in principle be extended to the study of information-age behaviors within the ever-expanding set of spatial technologies. The highly interdependent choices of contact mode made in weaving together component tasks of fragmented activities will of course pose novel, additional challenges. These are likely to be practical much more than conceptual, as these micro-level approaches appear well suited in principle for studying how individuals cope with these more complex choice situations.

Individual Choices, Collective Outcomes

Complementing micro-level analyses, other approaches bridge the gap between individual choices and collective outcomes and lead to intriguing insights regarding emerging phenomena and externalities in a society. There is in particular a vast class of choice situations where individual rational action backfires on the rational actors themselves because it brings about a collective outcome that is suboptimal for everybody. An extensive literature exists in economics, policy science, and game theory on this kind of paradox, of which the *prisoner's dilemma* (PD), first studied in the 1950s, is the best-known formulation. The prisoner's dilemma describes a binary choice situation involving two players having to decide independently from one another whether to cooperate or defect. The structure of payoffs is such that while the collective optimum is reached when both players opt to cooperate, individual rationality in ignorance of the other's choice dictates defection. More specifically, a prisoner's dilemma situation occurs whenever the structure of payoffs obeys the following ordering, regardless of the actual numerical values of the terms (Hardin 1982):

$$T > R > P > S$$

where

R = reward for cooperating (collective optimum)

T = temptation to defect for the individual decision maker

P = punishment for not cooperating (collective cost)

S = 'sucker's payoff' for being the only nice person

The prisoner's dilemma and its various extensions, especially the iterated and *n*-person versions, have been used widely in social choice research to investigate individual cooperation and defection and the ensuing generation of negative externalities in a collectivity. Real-world examples abound: a water or power system collapses during a period of peak use, despite dire warnings, because people are not prepared to inconvenience themselves by cutting down on their own consumption if they believe that most others will not. In a similar vein, in the absence of external coercion, few car buyers will be persuaded to buy the less glamorous but less polluting model, few fishermen to limit their harvest so as not to deplete the fisheries, few developers to spare the area they live in from over-development, if they know or believe that most others will not. These are all cases where individuals have a choice to cooperate, at some personal cost, towards some common goal, or to do what is best for them individually – best, that is, as long as all or most of the others do not decide to do the same thing. The PD-like structure of payoffs is such that even well meaning, not necessarily selfish individuals are deterred from attempting to cooperate by the belief that their small sacrifice will be in vain. These kinds of phenomena often arise in a spatial context, as several geographers and planners have found out (see, for example, Herniter and Wolpert 1967, Odland 1985, Couclelis 1989).

The case of spatial technology choices discussed in this paper is analogous to a particularly common version of the *n*-person PD known as the *tragedy of the commons* (Hardin 1968). This describes the depletion or degradation of a common property resource or public good to which people have free and unmanaged access. Here the common resource threatened by degradation is the transportation system in its environmental and socioeconomic context (its 'sustainability'), and the binary choice for individuals is whether to achieve a desired contact through a physical trip (*defect*) or ICTs (*cooperate*), assuming that the option is available. The other terms of the payoff structure are interpreted as follows:

T = temptation to defect, i.e., make a trip where telecontact is possible

R = societal reward for cooperating: sustainable transportation!

P = societal punishment for not cooperating: increasing social costs of transportation

S = sucker's payoff: you stay glued at your computer while your associates are enjoying a pleasant trip and face-to-face contacts

The well-known work of Axelrod (1984, 1997) and others on the iterated PD tempers the cynicism of the original PD by showing that cooperation makes good sense for individuals involved in repeated transactions with others. Also, Schelling's (1978) classic analysis of the *n*-person version demonstrates that a collective optimum is often possible even when (or provided that) some fraction of the participants opt to defect. Schelling's work has especially interesting implications for policy making because it shows that in the *n*-person case several different outcomes are likely depending on the relative numerical values of the payoffs, rather than just their rank orderings. In some cases manipulating slightly the values of individual rewards and punishments (in this case: the utility of telecontacts versus the utility of physical travel) can help steer the system towards its collectively optimal state despite a number of 'free riders' who will take advantage of others' restrain. Indeed, the optimal transportation conditions are not achieved when no one is traveling but when just enough are traveling at any one time so as to keep both the traffic flowing and the gas stations in business. The value of this kind of analysis lies in the possibility to quantify the obvious qualitative fact that deliberate policy measures that make travel less attractive or that increase the utility of telecontacts are likely to encourage ICT substitution for travel. On the other hand, the analysis also suggests that changing relative utilities and feedbacks within a dynamic system are likely to bring about adjustments towards the collective optimum even in the absence of any external intervention. In a reasonably rational world, as more and more people are deterred by the individual costs and negative externalities of increasing congestion (decreasing T, increasing P), more and more should be seeking out alternatives. The self-correcting property of high congestion has often been argued but less often observed on our transportation networks. It is conceivable that in the not too distant future ICTs will offer the higher-utility alternatives that other conventional measures of trip reduction have not been able to provide.

The implications of the notion of sustainable accessibility range well beyond transportation systems to urban structure, the environment, the economy, and society. Conventional planning wisdom holds that compact rather than diffuse urban forms, mixed rather than segregated land uses, reliance on public rather than private transport –for example, European-type rather than American-type cities – help achieve higher levels of overall accessibility with lower negative socioeconomic and environmental impacts. This may no longer hold true in the information age. Indeed, it is likely that in settings with high overall physical accessibility there is less incentive for individuals to meet the proliferating demands for contacts through virtual rather than physical interaction, leading to increasing pressures on the transportation infrastructure with all the associated societal costs. We have seen something like this happen time and time again in the best laid out modern cities where neither accessibility nor transportation proved sustainable in the long run. What is new is that planners now have even less control over the location of today's fragmented activities (and the ensuing growing transportation demands) than they did when living, working, shopping, learning, or recreating were each to be found in their own pre-assigned places. Is it a feasible goal for transportation to make everywhere easily accessible from everywhere? Are accessibility and sustainability public goods vulnerable to the 'tragedy of the commons'? If so, what societal and planning responses are appropriate? The research challenges are formidable, and we have just begun to scratch the surface.

20.4. In Lieu of Conclusions: Some Broad Research Questions

This paper has made several claims that demand to be tested through both empirical research and more rigorous theoretical development. These are as follows: First, that a useful conceptualization of accessibility in the information age may be based not on the 'death of distance' metaphor but on a new functional conception of activity (and perhaps also place) as spatially fragmented. Second, that the fragmentation of activity directly contributes to the somewhat paradoxical rapid growth in travel demand especially noticeable in countries with the highest rates of ICT development. Third, that because of the growing spatio-temporal plasticity and fragmentation of activities, planners have less control than ever before on what activities take place where (and when). Fourth, that planners should perhaps not strive to keep up uncritically with proliferating mobility demands but espouse instead the notion of sustainable accessibility. Fifth, that physical travel remains too attractive relative to viable ICT-based alternatives, but that the 'tragedy of the commons' some fear will result may turn out to be to some degree self-correcting.

To what extent these sweeping statements may be true or helpful for understanding accessibility in the information age, or for informing planning and policy making in related domains, cannot be said for sure at this point. As the chapters in this book have shown there are too many defining aspects of the issue where either the data or the appropriate conceptual frameworks are simply lacking. Concerning the notion of sustainable accessibility highlighted in this essay, perhaps the most critical empirical question is finding out how the relative growth of physical and digital contacts in people's contact sets may be affected by a range of factors that may differ widely from place to place. This chapter is based on the U.S. experience and may not reflect how the issues discussed may be perceived from the viewpoints of other nations. Comparative studies are very appropriate here, examining, for example, regions of the world that are similar in general socio-economic terms but different in terms of some obviously important variables such as spatial structure, ICT presence, or degree of planning control. Such research could look at the generally different structures of – say – European and American cities (e.g., high-density vs. low-density, public-transit oriented versus automobile oriented); the different rates of Internet connectivity, access and use in Europe and the United

States; the different degrees and effectiveness of planning intervention in Europe and the United States; or the different degrees of socio-spatial segregation observable in European and American cities. Other major research challenges derive from the choices people will make among expanding sets of functionally similar but materially extremely dissimilar contact alternatives, and the implications of these choices at all geographic scales for the economy, the environment, further technological development, social welfare and social equity, our cities and regions, politics and culture, and practically every aspect of human life (Castells 1989, 1996, 1998). As material, psychological, socio-economic and cultural inertia meet the still unexplored world of cyberspace, we find ourselves increasingly challenged by new mixtures of the familiar and relatively well understood with the outlandish and incomprehensible. In trying to figure out what is going on, extrapolating from past experience will need to go hand in hand with inventing new conceptual frameworks for phenomena that we currently have no name for, and that only recently were in the realm of science fiction.

Acknowledgements

This paper was first presented at the conference on Social Change and Sustainable Transport (SCAST), co-sponsored by the U.S. National Science Foundation (NSF) and the European Science Foundation (ESF). My thanks to Professor Bill Black and the other SCAST conference organizers for their generous support, and to Professor Hugo Priemus and his group at the TU Delft, the Netherlands, for thoughtful comments and another opportunity to air these ideas.

References

Abler, R. 1975. Effects of space adjusting technologies on the human geography of the future. In Abler R., Janelle, D., Philbrick, A. and Sommer, J. (eds.) *Human Geography in a Shrinking World.* North Scituate MA: Duxbury Press, 35-56.

Abler, R.F., and Falk, T. 1981. Public information services and the changing role of distance in human affairs. *Economic Geography,* 57: 10-22.

Adams, P. 1998. Network topologies and virtual place. *Annals of the Association of American Geographers,* 88: 88-106.

Axelrod, R. 1984. *The Evolution of Eooperation.* New York: Basic Books.

Axelrod, R. 1997. *The Complexity of Cooperation: Agent-based Models of Competition and Collaboration.* Princeton Studies in Complexity. Princeton NJ: Princeton University Press.

Bolton, R. 1997. 'Place' as 'network': Applications of network theory to local communities. Volume in honor of T.R. Lakshmanan, Department of Economics, Williams College, Williamstown MA 01267, December.

Bureau of Transportation Statistics (1997). *American Travel Survey 1995*, BTS/ATS95-US. US Department of Transportation, Washington DC.

Cairncross, F. 1997. *The Death of Distance: How the Communications Revolution will Change Our Lives*. Boston MA: Harvard Business School Press.

Castells, M. 1989. *The Informational City: Information Technology, Economic Restructuring, and the Urban-regional Process*. Oxford: Blackwell.

Castells, M. 1996. *The Rise of the Network Society.* The Information Age: Economy, Society, and Culture, 1. Oxford: Blackwell.

Castells, M. 1998. *End of Millennium.* The Information Age: Economy, Society and Culture, 3. Oxford: Blackwell.

Couclelis, H. 1989. Macrostructure and microbehavior in a metropolitan area. *Environment and Planning B* 16:141-54.

Couclelis, H. 1994. Spatial technologies. *Environment and Planning B: Planning and Design* 21:142-43.

Couclelis, H. 1996a. The death of distance. *Environment and Planning B: Planning and Design* 23:387-89.

Couclelis, H. 1996b (ed.). *Spatial Technologies, Geographic Information, and the City.*, TR 96-7. National Center for Geographic Information and Analysis, Santa Barbara CA.

Couclelis, H. 1998. The new field workers. *Environment and Planning B: Planning and Design* 25:321-23.

Curry, M., and Eagles, M. 1999 (eds). *Place and Identity in an Age of Technologically Regulated Movement,* National Center for Geographic Information and Analysis, Santa Barbara CA.

Fischer, M.F., Ricco, M. and Rammer, C. 1992. Telecommunication media choice behavior in academia: an Austrian-Swiss comparison. *Geographical Analysis* 24:1-15.

Fischer, M.F. and Rammer, C. 1992. Kommunikationsnetze von Wissenschaftlern: Ergebnisse einer Fallstudie and Wiener Universitaeten. *Mitteilungen Der Oesterreichischen Geographischen Gesellschaft* 134:159-76.

Graham, S. and Marvin, S. 1996. *Telecommunications and the City: Electronic Spaces, Urban Places.* London: Routledge.

Hardin, G. 1968. The tragedy of the commons. *Science* 162:1243-48.

Hardin, R. 1982. Collective action as an agreeable *n*-prisoner's dilemma. In Hardin, R., and Barry, B. (eds.) *Rational Man, Irrational Society.* Beverly Hills: Sage, 123-35.

Herniter, J. and Wolpert J. 1967. Coalition structures in the three-person non-zero-sum game. *Peace Research Society: Papers* 8:97-108.

Hodge, D.C. 1995. Intelligent transportation systems, land use, and sustainable transportation. Paper presented at the ITS America Alternative Futures Symposium on Transportation, Technology, and Society, Washington DC.

Janelle, D.G. 1991. Global interdependence and its consequences. In Brunn, S.D. and Leinbach, T.R. (eds.) *Collapsing Space and Time: Geographic Aspects of Communications and Information.* London: Harper Collins Academic, 49-81.

Janelle, D.G. 1995. Metropolitan expansion, telecommuting, and transportation. In Hanson, S. (ed.) *The Geography of Urban Transportation.* 2d ed. New York: Guilford Press, 407-35.

Johnston, R.J. 1997. The end of distance but a continuing bounded journey through space? *Environment and Planning B: Planning and Design,* 24:319-22.

Knoke, W. 1996. *Bold New World: The Essential Road Map to the Twenty-first Century.* New York: Kodansha International.

Mitchell, W.J. 1995. *City of Bits: Space, Place, and the Infobahn.* Cambridge MA: The MIT Press.

Odland, J. 1985. Interdependence and deterioration in the housing stock of an American city. In Haynes, K. E., Kuklinski, A., and Kultalahti, O. (eds.) *Pathologies of Urban Processes.* Tampore: Finn Publishers, 323-39.

Office of Technology Assessment (1995). *The Technological Restructuring of Metropolitan America*. US Congress, OTA-ATI-643. Washington DC: US Government Printing Office.

Papageorgiou, G. 1987. Spatial public goods. 1: theory. *Environment and Planning A*, 19:471-92.

Schelling, T. 1978. *Micromotives and Macrobehavior*. New York: Norton.

Figures

Tables

Author Index

Subject Index

Contributors

PAUL ADAMS is Assistant Professor of Geography at Texas A&M University and holds a Ph.D. in Geography from the University of Wisconsin, Madison. Research has focused on communications theory and the mapping of media, the extensibility of everyday life in cities, the politics of telecommunications, and the emergent role and meaning of virtual places.

MICHAEL BATTY is Professor of Spatial Analysis and Planning, and Director of the Centre for Advanced Spatial Analysis at University College London. Previous positions include Director of the National Center for Geographic Information and Analysis (NCGIA) at the State University of New York at Buffalo, and Head of the Department of City and Regional Planning and Dean of the School of Environmental Design at the University of Wales at Cardiff. Dr. Batty is editor of the journal *Environment and Planning B: Planning and Design.* Research interests involve the development of computer models and computer graphics in land use and transport planning, the spatial analysis of urban form, GIS technology, the impact of information technology on cities, and formal methods of decision making in policy analysis.

HELEN COUCLELIS is Professor of Geography at the University of California, Santa Barbara. She holds a Doctorate from the University of Cambridge, a Diploma in Urban and Regional Planning from the Technical University of Munich, and an MA.-equivalent in Architecture from the Technical University of Athens. She spent several years as a professional planner and policy advisor in Greece. Research interests are in the areas of urban and planning theory, behavioral geography and spatial cognition, and geographic information theory. Recent research includes work on cellular automata models of spatial dynamics, representations of space in both human cognition and computers, and the development of GIS-based approaches to resolve locational conflicts in planning. She is Co-editor of the journal *Environment and Planning B: Planning and Design.*

MARTIN DODGE is a researcher in the Centre for Advanced Spatial Analysis, University College London, with a background in social geography and GIS. A major research interest is the geography of the Internet. He maintains the Cyber-Geography Research web site at http://www.cybergeography.org/ and he has written the book *Mapping Cyberspace* (Routledge, 2000) with Rob Kitchin.

PHILIP CHARLES (PIP) FORER is Professor of Geography and Geographic Information Studies at the University of Auckland. His Ph.D. in Geography is from the University of Bristol. He is Co-editor of the *New Zealand Geographer and* edits the journal *Transactions of GIS.* Research specialties include modeling individual human activity patterns and space-time trade offs; applications of GIS for spatial modeling, visualization and education; applications of GIS in tourist-flow modeling; and transport and economic geography.

ARTHUR GETIS is Professor and Birch Foundation Endowed Chair of Geographical Studies at San Diego State University. His Ph.D. is from the University of Washington. He served previously as Head of the Department of Geography and Director of the School of Social Sciences at the University of Illinois. He is Co-editor of the *Journal of Geographical Systems* and has published extensively on spatial statistical modeling, urban growth and change, and general geography. Currently, his interests are in GIS, spatial statistics, urban transportation modeling, disease transmission modeling, and geographic education.

MICHAEL F. GOODCHILD is Professor and Chair of Geography at the University of California, Santa Barbara, Chair of the Executive Committee of the National Center for Geographic Information and Analysis, Associate Director of the Alexandria Digital Library Project, and Director of the Center for Spatially Integrated Social Science. He is former Chair of the Department of Geography at the University of Western Ontario, former Editor of *Geographical Analysis*, Editor for Methods, Models, and Geographical Information Sciences for the *Annals of the Association of American Geographers*. His research addresses the accuracy of spatial databases, modeling within GIS, the development and application of location-allocation models, and the theory and methodology of spatial analysis.

SUSAN HANSON is Professor of Geography and former Director of the Graduate School of Geography at Clark University. Her Ph.D. is from Northwestern University. She is currently Co-editor of *Economic Geography,* is a former editor of *The Professional Geographer* and the *Annals of the Association of American Geographers*, and a past President of the Association of American Geographers. Her research interests are in urban, social, and economic geography. She has written extensively on transportation and on gender and labor markets. Recent studies concern gender and entrepreneurship.

ANDREW S. HARVEY is Professor of Economics and Director of the Time Use Research Program at Saint Mary's University in Canada. His Ph.D. is in Economics from Clark University. His research has explored time-use and human activity patterns cross-nationally, focusing mainly on Canada, Japan and The Netherlands. Specific issues include activity-based approaches to travel demand modeling and episodal and contextual analysis of time-use data.

ERIC J. HEIKKILA is Associate Professor in the School of Urban Planning and Development at the University of Southern California. His Ph.D., in Economics; is from the University of British Columbia. His research is in the areas of regional science and urban economics, with a focus on computer applications in planning and the development issues that face East Asian cities and cultures.

DAVID C. HODGE is Dean of the College of Arts and Sciences at the University of Washington. His Ph.D. is in Geography from Pennsylvania State University. He served previously as Chair of the University of Washington's Geography Department, as Program Director for the National Science Foundation's Geography and Regional Science Program, and as Editor of *The Professional Geographer*. Research has contributed to understanding equity issues in urban social geography and transportation, and to the impact of information and intelligent transport technologies on the spatial form of metropolitan regions.

OTTO HUISMAN is a Doctoral Candidate and Tutor at the University of Auckland and holds a M.Sc. (Honors) from Canterbury University. His present research interests include the application of time geographic concepts to urban spatial processes, with a particular focus on issues of accessibility and spatial interaction. His Ph.D. research employs GIS and other spatial modeling tools to operationalize these concepts and adopts time as a co-equal dimension in modeling human activities.

DONALD G. JANELLE is Research Professor and Program Director for the Center for Spatially Integrated Social Science at the University of California, Santa Barbara. He served previously as Assistant Vice Provost of the University of Western Ontario and as Chair of its Department of Geography. His Ph.D., in Geography, is from Michigan State University. He is a former editor of *The Canadian Geographer*. Research specializations focus on temporal patterns of human spatial behavior in cities, and on social issues associated with transportation and communication technologies.

MEI-PO KWAN holds a Ph.D. in Geography from the University of California, Santa Barbara, and is Associate Professor of Geography at The Ohio State University. Her research is directed to application of space-time measures of individual accessibility, the use of GIS to study travel behavior and gender differences in accessibility to urban opportunities, and 3-D geovisualization of activity-travel patterns.

PAUL MACNAB is a Lecturer in Geographic Information Science in the Department of Geography at Saint Mary's University in Canada. His MA is in Geography from the University of Waterloo. Research interests are in issues relating to fisheries and marine conservation.

HARVEY J. MILLER has a Ph.D. in geography from The Ohio State University and is currently Associate Professor of Geography at the University of Utah. Dr. Miller is also North American Editor of the *International Journal of Geographical Information Science*. His research and teaching interests focus on Geographic Information Systems for Transportation (GIS-T) and geocomputational methods for spatial analysis. He is well known for his work on modeling accessibility using space-time prism concepts.

MITCHELL L. MOSS is Director of the Taub Urban Research Center and is the Henry Hart Rice Professor of Urban Policy and Planning in the Robert F. Wagner Graduate School of Public Service at New York University. His Ph.D. is in Urban Studies from the University of Southern California. He has written extensively about the diffusion of telecommunications technologies, the role of telecommunications in altering urban landscapes, and on policies regarding economic development, information cities, and the global economy.

ROBERT MUGERAUER is a specialist on the impacts of electronic technology on communities, especially regarding transformations in the development of urban and suburban areas and in changes in the sense of place, quality of life, location decisions, and modes of work. He holds a Ph.D. in Philosophy from the University of Texas and is presently the Meadows Foundation Sid Richardson Centennial Professor at the University of Texas at Austin, with affiliations in Architecture, Geography, Philosophy, and Urban Planning.

SHANE MURNION is Head of Geographic Information Systems in the Department of Geography at the University of Portsmouth and holds a Ph.D. in Geography from Queen's University of Belfast. His research has considered applications of neural and genetic algorithms in GIS and the modeling of distance-decay effects in web-server information flows.

SYLVIE OCCELLI is a Senior Researcher with the Socio-Economic Research Institute of Piedmont and holds a 'Laurea' in Architecture from the Polytechnic of Turin. Her research interest is in regional analysis, with special concern for housing, transportation and mobility, urban modeling, and spatial analysis. The current emphasis of her work is in 'modeling activity' as a way to promote modernization in planning practices at metropolitan and regional levels.

HARLAN J. ONSRUD is Professor of Spatial Information Science and Engineering at the University of Maine and a research scientist with the National Center for Geographic Information and Analysis. He is currently a member of the Mapping Science Committee of the U.S. National Research Council. He holds degrees in Civil Engineering from the University of Wisconsin and Juris Doctorate from the University of Wisconsin Law School. His research focuses on the analysis of legal, ethical, and institutional issues affecting the creation and use of digital spatial databases and the assessment of the social impacts of spatial technologies.

LAUREN MARGARET SCOTT is a product specialist at Environmental Systems Research Institute (ESRI). She recently completed her Ph.D. in the Joint Doctoral Program at San Diego State University and the University of California, Santa Barbara. Her research has addressed social fragmentation issues in urban areas, the use of Exploratory Data Analysis for evaluating spatial data accuracy within GIS, and most recently, changes in accessibility patterns in the spatial structure of employment within the metropolitan region of Los Angeles.

QING SHEN is the Mitsui Associate Professor of Urban Planning at the Massachusetts Institute of Technology. His Ph.D. is from the University of California, Berkeley. His research has concerned the modeling of relationships of urban spatial structure, transportation, and telecommunications. In addition he has investigated the office growth of metropolitan America, and in the impact of metropolitan restructuring on employment accessibility and central cities.

ERIC SHEPPARD is Professor of Geography at the University of Minnesota and holds a Ph.D. from the University of Toronto. His research adopts a political-economy perspective on understanding the capitalist space-economy, patterns of corporate investment, and economic development. He Chairs the Varenius Panel on Geographies of the Information Society, which sponsored the conference that formed the basis for this book.

DANIEL Z. SUI is Associate Professor of Geography at Texas A&M University and holds a Ph.D. from the University of Georgia. His research explores the theoretical foundations of GIS, the integration of spatial analysis and modeling with GIS, and GIS applications in urban and regional studies. Recent work includes the integration of artificial intelligence (particularly expert systems and neural networks) with GIS and remote sensing, the simulation modeling of urban development, telecommunications and virtual cities, and the geography of the emerging information society.

ANTHONY M. TOWNSEND is a Ph.D. candidate in Urban Planning at the Massachusetts Institute of Technology. His research investigates methods for tracking Internet traffic among U.S. cities, and the significance of the information economy in altering the role and status of metropolitan centers.

MARK I. WILSON is Associate Professor at Michigan State University in the Department of Geography and Urban and Regional Planning and in the Institute for Public Policy and Social Research. His academic training is in Economics and Regional Science, and his Ph.D. is from the University of Pennsylvania. His research is in the areas of economic geography, political economy of cyberspace, urban and regional planning, and the non-profit sector. He is co-convener with Kenneth Corey of E*SPACE: The Electronic Space Project, which investigates the spatial dimensions of cyberspace (http://www.ssc.su.edu/~Dean/espace.htm). The E*SPACE initiative serves as a network for scholars interested in the spatial impact of information technologies.

Papers in Regional Science
Journal of Geographical Systems
The Annals of Regional Science
An International Journal of Urban, Regional and Environmental Research and Policy
Volume 34 Number 1 2000
Springer

Springer

Druck: Strauss Offsetdruck, Mörlenbach
Verarbeitung: Osswald, Neustadt